T0178575

Physical Properties of Materials

CRC Press

Companion Website

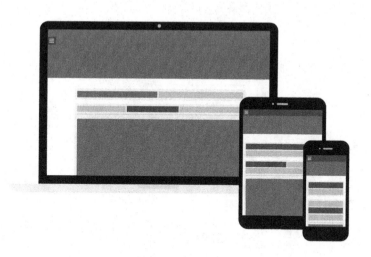

Enhancing online learning and teaching.

Physical Properties of Materials

Third Edition

Mary Anne White

CRC Press
Taylor & Francis Group
Boca Raton London New York

CRC Press is an imprint of the
Taylor & Francis Group, an **informa** business

CRC Press
Taylor & Francis Group
6000 Broken Sound Parkway N W, Suite 300
Boca Raton, FL 33487-2742

Printed in the United States of America on acid-free paper

International Standard Book Number-13: 978-1-138-56917-1 (Hardback)
International Standard Book Number-13: 978-1-138-60510-7 (Paperback)

Library of Congress Cataloging-in-Publication Data

Names: White, Mary Anne, 1953- author.
Title: Physical properties of materials / Mary Anne White.
Description: Third edition. | Boca Raton : Taylor & Francis, CRC Press, 2019.
Includes bibliographical references and index.
Identifiers: LCCN 2018023452 | ISBN 9781138605107 (pbk. : alk. paper) | ISBN 9781138569171 (hardback : alk. paper) | ISBN 9780429468261 (ebook)
Subjects: LCSH: Materials.
Classification: LCC TA404.8 .W457 2019 | DDC 620.1/12—dc23
LC record available at https://lccn.loc.gov/2018023452

Visit the Taylor & Francis Web site at
http://www.taylorandfrancis.com

and the CRC Press Web site at
http://www.crcpress.com

and the Book Web site at
www.physicalpropertiesofmaterials.com

Dedicated to my first teachers, my parents

Contents

Preface to the Third Edition xiii
Preface to the Second Edition.. xv
Preface to the First Edition.. xvii
Acknowledgments xix
About the Author xxi

Part I   Introduction

1. Introduction to Materials Science 3
   1.1   History 3
   1.2   More Recent Trends 4
   1.3   Impact on Daily Living 6
   1.4   Future Materials and Sustainability Issues.................................. 6
   1.5   Structures of Materials 9
   1.6   Learning Goals 12
   1.7   Problems.. 12
   Further Reading 13

Part II   Color and Other Optical Properties of Materials

2. Atomic and Molecular Origins of Color... 19
   2.1   Introduction 19
   2.2   Atomic Transitions... 22
   2.3   Black-Body Radiation 24
   2.4   Vibrational Transitions as a Source of Color.............................. 26
   2.5   Crystal Field Colors 27
   2.6   Color Centers (F-Centers).. 29
   2.7   Charge Delocalization, Especially Molecular Orbitals 32
   2.8   Light of Our Lives: A Tutorial 36
   2.9   Learning Goals 38
   2.10 Problems.. 38
   Further Reading 44

3. Color in Metals and Semiconductors 47
   3.1   Introduction 47
   3.2   Metallic Luster.. 47
   3.3   Colors of Pure Semiconductors.. 52

3.4   Colors of Doped Semiconductors 54
3.5   The Photocopying Process: A Tutorial..59
3.6   Photographic Processes: A Tutorial...61
3.7   Learning Goals 63
3.8   Problems..64
Further Reading 67

4. Color from Interactions of Light Waves with Bulk Matter..................71
4.1   Introduction 71
4.2   Refraction 71
4.3   Interference 79
4.4   Scattering of Light..85
4.5   Diffraction Grating 87
4.6   An Example of Diffraction Grating Colors: Liquid Crystals 88
4.7   Fiber Optics: A Tutorial..93
4.8   Learning Goals 94
4.9   Problems..95
Further Reading 100

5. Other Optical Effects 105
5.1   Introduction 105
5.2   Optical Activity and Related Effects 105
5.3   Birefringence..109
5.4   Circular Dichroism and Optical Rotatory Dispersion.................110
5.5   Nonlinear Optical Effects 112
5.6   Transparency: A Tutorial 118
5.7   Learning Goals...118
5.8   Problems..119
Further Reading 122

Part III   Thermal Properties of Materials

6. Heat Capacity, Heat Content, and Energy Storage................................127
6.1   Introduction 127
6.2   Equipartition of Energy 127
        6.2.1   Heat Capacity of a Monatomic Gas....................................129
        6.2.2   Heat Capacity of a Nonlinear Triatomic Gas....................129
6.3   Real Heat Capacities and Heat Content of Real Gases 130
        6.3.1   Joule's Experiment 134
        6.3.2   Joule–Thomson Experiment..135
6.4   Heat Capacities of Solids...138
        6.4.1   Dulong–Petit Law 138
        6.4.2   Einstein Model 139
        6.4.3   Debye Model..140

	6.4.4	Heat Capacities of Metals	143
6.5		Heat Capacities of Liquids	144
6.6		Heat Capacities of Glasses	145
6.7		Phase Stability and Phase Transitions, Including Their Order	149
6.8		$(C_p - C_v)$: An Exercise in Thermodynamic Manipulations	155
6.9		Thermal Energy Storage Materials: A Tutorial	158
6.10		Thermal Analysis: A Tutorial	160
6.11		Learning Goals	162
6.12		Problems	163
		Further Reading	169

7. Thermal Expansion 173
7.1	Introduction	173
7.2	Compressibility and Thermal Expansion of Gases	173
7.3	Thermal Expansion of Solids	177
7.4	Examples of Thermal Expansion: A Tutorial	186
7.5	Learning Goals	187
7.6	Problems	188
	Further Reading	192

8. Thermal Conductivity 195
8.1	Introduction	195
8.2	Thermal Conductivity of Gases	195
8.3	Thermal Conductivities of Insulating Solids	200
8.4	Thermal Conductivities of Metals	204
8.5	Thermal Conductivities of Materials: A Tutorial	207
8.6	Learning Goals	208
8.7	Problems	208
	Further Reading	212

9. Thermodynamic Aspects of Phase Stability 215
9.1	Introduction	215
9.2	Pure Gases	215
9.3	Phase Equilibria in Pure Materials: The Clapeyron Equation	216
9.4	Phase Diagrams of Pure Materials	219
9.5	The Phase Rule	230
9.6	Liquid–Liquid Binary Phase Diagrams	234
9.7	Liquid–Vapor Binary Phase Diagrams	236
9.8	Relative Proportions of Phases: The Lever Principle	239
9.9	Liquid–Solid Binary Phase Diagrams	241
9.10	Compound Formation	248
9.11	Three-Component (Ternary) Phase Diagrams	252
9.12	A Tongue-in-Cheek Phase Diagram: A Tutorial	256
9.13	Applications of Supercritical Fluids: A Tutorial	257
9.14	Learning Goals	258

9.15 Problems 259
Further Reading 270

10. Surface and Interfacial Phenomena 275
10.1 Introduction 275
10.2 Surface Energetics 277
10.3 Surface Investigations 278
10.4 Surface Tension and Capillarity 279
10.5 Liquid Films on Surfaces 285
10.6 Nanomaterials: A Tutorial 289
10.7 Learning Goals 290
10.8 Problems 291
Further Reading 294

11. Other Phases of Matter 299
11.1 Introduction 299
11.2 Colloids 299
11.3 Micelles 301
11.4 Surfactants 305
11.5 Inclusion Compounds 306
11.6 Hair Care Products: A Tutorial 310
11.7 Applications of Inclusion Compounds: A Tutorial 311
11.8 Learning Goals 312
11.9 Problems 312
Further Reading 314

Part IV   Electrical and Magnetic Properties of Materials

12. Electrical Properties 321
12.1 Introduction 321
12.2 Metals, Insulators, and Semiconductors: Band Theory 321
          12.2.1 Metals 324
          12.2.2 Semiconductors 325
          12.2.3 Insulators 328
12.3 Temperature Dependence of Electrical Conductivity 328
          12.3.1 Metals 329
          12.3.2 Intrinsic Semiconductors 330
12.4 Properties of Extrinsic (Doped) Semiconductors 335
12.5 Electrical Devices using Extrinsic (Doped) Semiconductors 336
          12.5.1 p,n-Junction 336
          12.5.2 Transistors 342
12.6 Dielectrics 344
12.7 Superconductivity 347
12.8 Thermometry: A Tutorial 352

12.9 Learning Goals 355
12.10 Problems 355
Further Reading 362

13. Magnetic Properties 371
13.1 Introduction 371
13.2 Origins of Magnetic Behavior 371
13.3 Magnetic Induction as a Function of Field Strength 378
13.4 Temperature Dependence of Magnetization 382
13.5 Magnetic Devices: A Tutorial 387
13.6 Learning Goals 389
13.7 Problems 389
Further Reading 392

Part V Mechanical Properties of Materials

14. Mechanical Properties 397
14.1 Introduction 397
14.2 Elasticity and Related Properties 402
14.3 Beyond the Elastic Limit 407
14.4 Microstructure 411
14.5 Defects and Dislocations 411
14.6 Crack Propagation 415
14.7 Adhesion 420
14.8 Electromechanical Properties: The Piezoelectric Effect 422
14.9 Shape-Memory Alloys: A Tutorial 425
14.10 Cymbals: A Tutorial 427
14.11 Learning Goals 431
14.12 Problems 431
Further Reading 440

Appendix 1: Fundamental Physical Constants 447

Appendix 2: Energy Unit Conversions 449

Appendix 3: The Greek Alphabet 451

Appendix 4: Sources of Lecture Demonstration Materials 453

Glossary 455

Index 479

Preface to the Third Edition

Materials science continues to fascinate with new materials discovered weekly, leading to new technologies that advance humanity. At present, we hold up much hope for future students to discover new materials to address such important problems as storage of renewable energy and purification of water. My wish is that this book will lay the foundations to be part of your journey!

It is a privilege to be asked to write a third edition of a book, and such an invitation speaks to the growing importance of the subject, and the welcome receipt of the second edition among its readers. Continuing interest in materials science has made writing this third edition a pleasurable part of my growth as a scientist. Having taught "Materials Science" at the undergraduate and graduate levels for many years, I have recently retired from teaching but continue toward my goal of life-long learning, and the book in your hands is a large part of that experience.

In my most recent years teaching, it was my special privilege to work with Dr. Gianna Aleman Milán to develop a laboratory program to accompany the undergraduate Materials Science course for junior-year Chemistry and Physics students at Dalhousie University, for which this book was originally written. The combination of the fascinating physical properties of materials from the lecture course was fully complemented by hands-on labs, and the students regularly commented that it was their favorite course in their program. Please contact us if you have questions or comments about this book or the laboratory program.

This third edition has been updated from the second edition by inclusion of new materials and processes, such as topological insulators, 3-D printing, and more information on nanomaterials. Another change is that all pressures are now given in SI units; just remember that 101 kPa is 1 atmosphere.

The biggest changes are the addition of Learning Goals at the end of each chapter and a Glossary with more than 500 entries. It is my hope that these features will make it easier for students to keep track of new concepts. In addition, the Glossary should make this book an enduring resource.

With the assistance of CRC Press, we have put together an excellent website, PhysicalPropertiesOfMaterials.com to complement this book. Many of the features (all those under Student Resources) are freely available to all. In particular, we made approximately 30 videos specifically to complement the contents of the book. These videos are highlighted at the appropriate points in the text. The book website also has many links to relevant websites around the world, sorted by chapter, to be used by students and instructors.

Enjoy!

Mary Anne White
Halifax, Nova Scotia, Canada

Preface to the Second Edition

Materials science offers a wonderful opportunity to introduce students to basic principles of matter through forefront research topics of immediate or near-future direct application to their lives.

Materials research is advancing rapidly. New topics in this second edition include materials and sustainability, carbon nanotubes, other nanomaterials such as nanocomposites, quantum dots, spintronics, and magnetoresistance. In addition, in response to comments from other professors who taught from the first edition, this new edition includes an introductory section on structures of materials and more discussion concerning polymers.

All aspects of the text have been edited and revised to correct some minor shortcomings in the first edition and to clarify points where readers kindly indicated a need. The format is similar to the first edition, in that the text is brought alive through Comments and Tutorials that illustrate the role of materials in our lives. In this second edition, several new Comments and a new capstone Tutorial on the materials science of cymbals have been added. References at the ends of chapters, to be used by the reader for further depth, have also been updated. In addition, more than 60 new end-of-chapter problems have been added, bringing the total number of problems to 300. To guide students through the myriad equations, a margin marker for the most important equations has been introduced in this second edition.

Since 1991, I have been teaching a materials science course at the junior (third-year) undergraduate level, cross-listed as a chemistry/physics course. Initially, I taught with the notes that became the first edition and then I taught from the first edition, most recently supplanted with additional information that is now in the second edition. This is a 13-week course, with three hours of lectures per week. Within that envelope, I also include Tutorials.* For the most part, I cover all the topics, in the order of the textbook. Sometimes in my lectures, I have to omit or abbreviate topics, and, if so, what is covered in Chapters 10 and/or 11 is reduced. Mostly, I include the introductory parts of Chapter 14 after presenting Chapter 5, so the students have a foreshadowing of the importance of mechanical properties. This course has now been enthusiastically taken by about 800 students, mostly in chemistry and physics programs but also in engineering, earth science, and biochemistry, and they frequently comment on how much they appreciate the links to their everyday lives. I have managed to find lecture demonstrations for each and every lecture, many from everyday life. And I enjoy learning more examples from the students!

---

* For details concerning presentation, see M. A. White, 1997. Tutorials as a teaching method for materials science. *Journal of Materials Education*, 19, 23–26.

The background colors on the front cover were produced by viewing a CD case through crossed polarizers, with backlighting. See Chapter 14 for discussion of color from stress-induced polarization.

We will encounter many equations in this book. The most important are designated with ▌ to the left of the equation.

Finally, to emphasize that this book concerns the physical properties of materials, the word "physical" has been added to the first-edition title of *Properties of Materials*.

A website for *Physical Properties of Materials* is maintained by the publisher at PhysicalPropertiesofMaterials.com. Further updates and contact information for reader suggestions are available at www.physicalproperties-ofmaterials.com.

Enjoy learning about physical properties of materials!

Mary Anne White
Halifax, Nova Scotia, Canada

Preface to the First Edition

The idea for this book germinated at a public lecture in about 1989. The lecturer had mentioned in passing, "and you all know how a photocopier works." Most of the remainder of the lecture was lost on me, in wondering what fraction of that educated audience knew how a photocopier worked. Then I began to realize that there was no place in our curriculum where such an important concept was taught to our students. After that lecture, I decided to take advantage of revision of the physical chemistry curriculum going on in my department at that time, to prepare a curriculum for a class based on properties of materials. The class was launched in 1991. Finding no appropriate textbook, I wrote out my lecture notes for the students to use. With subsequent revisions and additions, these have become the book before you.

The purpose of this book is to introduce the principles of materials science through an atomic and molecular approach. In particular, the aim is to help the reader to learn to think about properties of materials in order to understand the principles behind new (or old) materials. A "perfect goal" would be for the reader to be able to learn about a new material in the business pages of the daily newspaper, giving minimal scientific information, and from that decipher the scientific principles on which the use of the material is based.

Properties of materials have interested many people. When speaking of his youth, Linus Pauling[*] said: "I mulled over the properties of materials: why are some substances colored and others not, why are some minerals or inorganic compounds hard and others soft."[†] It is noteworthy that such deep considerations of the world around us eventually led Pauling to make seminal contributions in many areas. It is hoped that encouragement of wonder in the variety and nature of properties of materials will be of benefit to those who read these pages.

This textbook should be viewed as an introductory survey of principles in materials science. While this book assumes basic knowledge of physical sciences, many of the concepts are presented with introductory mathematical and theoretical rigor. This approach is largely to refrain from use of tensors (avoided in all but a few places) in this presentation; the interested reader will find more detailed presentations and derivations in the bibliography.

This book is divided into sections based on various properties of materials. After a general introduction to materials science issues, origins of colors and other optical properties of materials are considered. The next part concerns thermal properties of materials, including thermal stability and phase

---

[*] Linus Pauling (1901–1994) was an American chemist and winner of the 1954 Nobel Prize in Chemistry for research into the nature of the chemical bond and its applications to elucidation of structure of complex substances. Pauling also won the 1962 Nobel Peace Prize.
[†] John Horgan, 1993. Profile: Linus Pauling. *Scientific American*, March 1993, 36.

diagrams. A further part is about electrical and magnetic properties of materials, followed by a final part on mechanical stability.

This book has been used for the basis of a one-semester class in materials science, offered in a chemistry department, and taken by students in chemistry, biochemistry, physics, engineering, and earth sciences. The prerequisite for the class is prior introduction to the laws of thermodynamics.

A feature of this book is the introduction of tutorials, in which the students, working in small groups, can apply the principles exposed in the text to work out for themselves the physical principles behind applications of materials science.

An Instructors' Supplement, containing complete discussion of all the points raised in the tutorials, and complete worked solutions to all the end-of-chapter problems, is available through the publisher to instructors who have chosen the book for class use.

This book has aimed to present the principles behind various properties of materials and, since it is a survey of a large subject, additional reading suggestions are given at the end of each chapter as sources of more detailed information. References to recent developments also are given, in order to expose the readers to the excitement of current materials science research. Updates to these references will be made available through the publisher's web site, www.physicalpropertiesofmaterials.com.

The presentation for this book is by property—optical, thermal, electrical, magnetic, and mechanical. Types of materials—metals, semiconductors (intrinsic and extrinsic), insulators, glasses, orientationally disordered crystals, defective solids, liquid crystals, Fullerenes, Langmuir–Blodgett films, colloids, inclusion compounds, and more—are introduced through their various properties. As new types of materials are made or discovered, it is hoped that the approach of exposing principles that determine physical properties will have a lasting influence on future materials scientists and many others.

Acknowledgments

This book has been helped through the comments of many people, especially those who have taught from the first and second edition and those who have used it as students at Dalhousie University and elsewhere all around the world. I hope that readers will help me to continue to refine this text by bringing suggestions for improvement and inclusion to my attention.

I particularly want to thank N. Aucoin, P. Bessonette, R. J. C. Brown, R. J. Boyd, W. Callister, P. Canfield, D. B. Clarke, A. Cox, J. Dahn, S. Dimitrijevic, R. Dumont, A. Ellis, J. E. Fischer, H. Fortier, J. M. Honig, M. Jakubinek, N. Jackson, B. Kahr, C. Kittel, B. London, M. Marder, B. Marinkovic, T. Matsuo, M. Moskovits, K. Nassau, G. Nolas, M. Obrovac, N. Pelot, R. Perry, J. Pöhls, A. W. Richardson, L. Schramm, J.-M. Sichel, E. Slamovich, T. Stevens, T. Swager, I. Tamblyn, M. Tan, and J. Zwanziger for constructive comments. Special thanks to A. Inaba for the detailed comments as a result of his translation of the first edition to Japanese and to chemistry/music student P. MacMillan, who introduced me to the materials science of cymbals. Thanks also to M. LeBlanc for preparation of most of the diagrams for the first edition and to J. E. Burke, D. Eigler, F. Fyfe, Ch. Gerber, M. Gharghouri, M. Jericho, M. Jakubinek, C. Kingston, A. Koch, M. Marder, K. Miller, J.-M. Phillipe, R. L. White, and K. Worsnop for photographs or other assistance with graphical contributions. I thank the Dalhousie University Faculty of Science for a teaching award that made it possible to hire an assistant to help prepare the first edition. In addition, I am grateful to S. Gillen, J. S. Grossert, K. J. Lushington, M. Meyyappan, A. J. Paine, D. G. Rancourt, and R. L. White for providing useful information, and to members of my research group, past and present, A. Bourque, C. Bryan, A. Cerqueira, J. Conrad, L. Desgrosseilliers, S. Ellis, S. Kahwaji, K. Miller, M. Johnson, J. Noël, J. Niven, C. O'Neill, J. Pöhls, A. Ritchie, C. Romao, and C. Weaver, for assistance. Thanks also to D. White, H. Stubeda, and J. MacDonald for assistance in making the videos for the website. Special thanks to G. Aleman Milán who developed labs to accompany the Materials Science course for which this book was written. Thanks also to J. Jurgensen, J. Lynch, A. Shatkin, and the other staff at CRC Press for their support.

Writing a book takes time. My now-grown children, David and Alice, have never really known a time when I was not working on this book! For their forbearance while I have been preoccupied, I sincerely thank them and also other members of my family, especially my husband Rob, as this book would not have reached fruition without his loving support.

About the Author

Mary Anne White is a materials research chemist and a highly recognized educator and communicator of science. Dr. White holds the distinguished title of Harry Shirreff Professor of Chemical Research (Emerita) at Dalhousie University, Halifax, Nova Scotia, Canada, where she has been since 1983, following a BSc in honors chemistry from the University of Western Ontario, a PhD in chemistry from McMaster University, and a postdoctoral fellowship at University of Oxford. From 2002 to 2008, she was the founding director of the Institute for Research in Materials at Dalhousie University, and from 2010 to 2016, she was director of the multidisciplinary graduate program, Dalhousie Research in Energy, Advanced Materials and Sustainability (DREAMS). She has been Professor (Emerita) since 2017.

Dr. White's research interests are in energetics and thermal properties of materials. She has made significant contributions to understanding how heat is stored and conducted through materials. Her work has led to new materials that can convert waste heat to energy, materials that trap solar energy, and materials that reversibly change color on heating. Her research contributions have been recognized by national and international awards, and she is an author of more than 200 research papers and several book chapters. She has trained more than 50 graduate students and postdoctoral fellows and more than 80 undergraduate research students.

Dr. White enjoys sharing her knowledge with students and with the general public. She is especially well known for presenting clear explanations of difficult concepts. Mary Anne's outstanding abilities as an educator have been recognized by the Union Carbide Award for Chemical Education from the Chemical Institute of Canada. Mary Anne has given more than 170 invited presentations at conferences, universities, government laboratories, and industries around the world.

Dr. White has been active throughout her career in bringing science to the general public. This includes helping establish a hands-on science center; many presentations for schools, the general public, and others (including a lecture for members of Canada's parliament and senate); booklets on science activities for children (published by the Canadian Society for Chemistry); serving as national organizer of National Chemistry Week; more than 150 articles for educators or the general public; and appearances on television and especially on CBC Radio. For her contributions to public awareness of science, Mary Anne was awarded the 2007 McNeil Medal of the Royal Society of Canada.

Dr. White holds honorary doctorates from McMaster University, the University of Western Ontario, and the University of Ottawa. She is an Officer in the Order of Canada.

> Knowledge comes, but wisdom lingers.
>
> *Alfred Lord Tennyson*

Part I

Introduction

Good material often stands idle for want of an artist.

Lucius Annæus Seneca

1

Introduction to Materials Science

1.1 History

In some sense, materials science began about two million years ago when people began to make tools from stone at the start of the Stone Age. At that time, emphasis was on applications of materials, with no understanding of the microscopic origins of material properties. Nevertheless, the possession of a stone axe or other implement certainly was an advantage to an individual.

The Stone Age ended about 5000 years ago with the introduction of the Bronze Age in the Far East. Bronze is an *alloy* (a metal made up of more than one element), mostly composed of copper with up to 25% tin, and possibly other elements such as lead (which makes the alloy easier to cut), zinc, and phosphorus (which strengthens and hardens the alloy). Bronze is a much more workable material than stone, especially since it can be hammered, beaten, or cast into a wide variety of shapes. After a surface oxide film forms, bronze objects corrode only slowly.

Although bronze is still used today, about 3000–3500 years ago, iron working began in Asia Minor. The Iron Age continues to the present time. The main advantage of iron over bronze is its lower cost, bringing metallic implements into the budget of the ordinary person. The Iron Age ushered in the common use of coins, which greatly improved trade, travel, and communications. Even today these three activities are strongly tied to materials usage.

Throughout the Iron Age many new types of materials have been introduced. Today we may take for granted the properties of glass, ceramics, semiconductors, polymers, composites, etc.

One major change over the millennia of the "Materials Age" has been our understanding of the properties of materials and our consequent ability to develop and prepare materials for particular applications. Materials research has been defined as the relationships between and among the structure, properties, processing, and performance of materials (see Figure 1.1). Understanding the principles that give rise to various properties of materials is the aim of our journey.

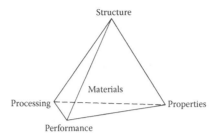

FIGURE 1.1
Materials research is the investigation of the relationships between and among structure, properties, processing, and performance of materials.

1.2 More Recent Trends

A report written in 1953* shows how far we have come in materials research in the last decades. It listed objectives for future materials research: materials to behave well at extremely high and extremely low temperatures; more basic knowledge about the behavior of metals (e.g., to predict strength, fatigue); production of metals of higher purity; preparation of new alloys to replace stainless steel (due to the then-shortage of nickel); preparation of better conductors and heat-resistant insulators for miniature electronic circuits; improvements in welding and soldering; and development of adhesion methods to allow applications of fluorocarbons and other new alloys and plastics. Most of these goals have been achieved, but the list, which now is quite dated, shows the emphasis on metals and metallurgy that was prevalent at that time.

About 20 years later, another report† again listed current topics of interest to materials researchers. This time, the issues were quite different: strategic materials, fuel availability, biodegradable materials, and scrap recovery. This list shows the close connection between materials science and economic and political issues.

In the 1980s, it was said that the "field of materials science and engineering is entering a period of unprecedented intellectual challenge and productivity."‡ Although metals will always be important, emphasis in the subject has moved from metallurgy to ceramics, composites, polymers, and other molecular materials. Important considerations include semiconductors, magnetic properties, photonic properties, superconductivity, and biomimetic properties. Furthermore, perceived barriers between types of materials are now falling away (e.g., metals can be glasses and molecular materials can be

---

* Science and the Citizen. *Scientific American*, August 1953.

† *Materials and Man's Needs.* National Research Council's Committee on the Survey of Materials Science and Technology, 1975.

‡ *Materials Science and Engineering for the 1990s.* National Academy Press, Washington, DC, 1989.

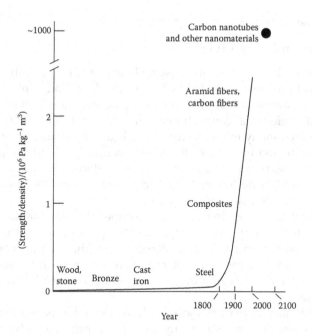

FIGURE 1.2
The dramatic progress in strength-to-density ratio (i.e., specific strength) of materials has allowed a wide variety of new products, from dental materials to tennis racquets.

conductors or magnets). Following the clues that nature provides, we now understand that different end products depend on assembly and addition of other phases. This is an example of complexity depending on the different length scales—from nanostructures to macrostructures.

A 2017 report from MIT* argues that advances in materials science will determine the future of civilization, solving such major problems as the energy crisis.

The purpose of this book is to expose the underlying principles, to allow understanding of new (and old) materials and their properties. With structure–composition–properties relationships now more fully understood, some materials can be tailor-made to have specific properties. Nevertheless, new materials with novel properties will continue to surprise us, and the guiding principles will allow us to understand them also.

Modern materials research is both important to our economies and intrinsically fascinating. The subject cuts across many traditional disciplines, including chemistry, physics, engineering, and earth sciences. For example, the dramatic recent improvement in strength-to-density ratio in materials (also known as specific strength; see Figure 1.2) has required input from a wide variety of subjects and has led to considerable improvements in materials available to consumers.

---

* *How Materials Science Will Determine the Future of Human Civilization.* www.technologyreview.com/s/608662/how-materials-science-will-determine-the-future-of-human-civilization/.

1.3 Impact on Daily Living

Modern materials have had immeasurable impact on our daily lives. New materials and processes have led to diverse products and applications: fiber optics; better computers; durable outdoor plastics; cheaper metal alloys to replace gold in electrical connections; more efficient LED, liquid crystal and plasma displays; many microelectronic applications. Materials development is important to many industrial sectors: aerospace, automotive, biological, chemical, electronic, energy, metals, and telecommunications. Indeed, economic growth is no longer linked to production of basic materials, but rather to their use in goods and services.

The understanding of behavior and properties of materials underlies every major technology. For example, in the space industry, the following material properties must be considered: strength, thermal conductivity, outgassing, flammability, effect of radiation, and stability under thermal cycling. Every manufacturing industry depends on materials research and development.

While fields such as plastics started to have a major impact on consumers a few decades ago, there are still many areas for improvement. New materials are being developed for applications such as bulletproof vests, and pop bottles that can be recycled as polyester clothing, and food containers that can be composted. Smart materials have led to products such as airplane wings that deice themselves and buildings that stabilize themselves in earthquakes. The drive to miniaturization, especially of electronic components, is opening new fields of science. For example, the development of blue light-emitting diodes has allowed more compact storage of digital information. New materials are seeing applications as diverse as artificial body parts and new applications in semiconductor technology.

1.4 Future Materials and Sustainability Issues

Future developments in materials science could include the predictable (cotton shirts that never require ironing!) and the unimaginable. We will learn some lessons from nature: spider silk mimics; composites that are based on the structure of rhinoceros horn, which is a natural composite similar to that used on the Stealth aircraft, and shells, which are as strong as the most advanced laboratory-produced ceramics; and biocompatible adhesives fashioned after mollusk excretions. We might also learn from nature how molecules assemble themselves into complex three-dimensional arrays. Electronic components will certainly be miniaturized further, exploiting new properties when the length scale is comparable to atomic sizes and no longer divisible. Quantum mechanics will govern the processes, and studies of these nanomaterials are already leading to new materials and devices.

COMMENT: MATERIALS SUSTAINABILITY

It is humbling to realize that everything around us is made of fewer than a hundred types of atoms, giving rise to enormous varieties of materials and objects. As materials scientists, we should embrace the concept that virtually all of our building materials come from the earth and atmosphere and will return there after we are done with them.

However, precious materials can become "lost" through use (i.e., so widely distributed that they are not worth recovering), a matter to consider when we take into account that the earth's supply of materials is not limitless (see Table 1.1). So many of our natural resources have been mined and then implemented into "disposable" products that landfill sites now are one of the richer sources of elements such as copper. At the present rate of consumption, it is estimated that the world supply of indium could be exhausted—that is, dispersed widely in items such as indium tin oxide coatings and indium gallium arsenide semiconductors—within a few decades.

With these factors in mind, forward-thinking architect William McDonough, along with chemist Michael Braungart, have introduced the "Cradle to Cradle" concept of ecologically intelligent design. In their vision, the "waste" of one process becomes the input material of another. For example, they have been working with the chemical company BASF to retrieve and reuse Nylon 6 in closed-loop cycles. The "used" Nylon 6, (e.g., from old carpets) is depolymerized and then "upcycled" into a product of higher quality, rather than being downcycled into a material with less value. There is considerable promise for other manufacturing and industrial processes making use of similar concepts.

Many new materials are being developed to capture and store renewable energy, and this is an area in which materials science will have considerable impact in decades to come. However, it also takes energy to make such materials, and a useful concept to consider is a material's embodied energy. This is the energy required to make this material. Some values of embodied energies for common materials are presented in Table 1.2. Clearly, if a material is to be useful to capture renewable energy, it should recover more than its embodied energy during its lifetime.

TABLE 1.1

Natural Abundance of Elements in the Earth's Crust

Element	Natural Abundance in the Earth's Crust (%)
O	46.1
Si	28.2
Al	8.2
Fe	5.6
Ca	4.2
Na	2.4
Mg	2.3
K	2.1
Ti	0.57
H	0.14
Cu	0.005
B	0.001
In	2×10^{-5}
Ag	1×10^{-5}
Bi	2×10^{-6}
Ru	1×10^{-6}
He	5×10^{-7}
Au, Pt, Ir	1×10^{-7}
Te	8×10^{-8}
Rh	4×10^{-8}

Source: Data from Encyclopedia Britannica, Inc. Elements with lower abundance in the Earth's crust are not listed.

Note: As originally deduced by the geochemist, Frank W. Clarke (1847–1931), and extended by others.

Some questions are of immediate importance. For example, we need a fundamental understanding of high-temperature superconductors. With this, room-temperature superconductivity might be achieved. We also need to understand complex structures, such as quasicrystals, composites, and nanomaterial composites, as they often exhibit properties that are outside of the bounds of the properties of the materials from which they are composed. In the interests of our planet, we need to find materials that can be produced with less waste, using abundant elements that can be recycled with greater efficiency. In the development of new chemical sensors, issues are sensitivity, longevity, and selectivity. In the automotive area, high power density, low fuel consumption, low emission of greenhouse gases, and lightweight aerodynamic bodies with a high degree of recyclability are some of the goals. Other questions are too preposterous to even pose at present!

With a firm understanding of the principles on which materials research is based, we can proceed confidently into the future with new materials and processes.

TABLE 1.2

Embodied Energies of Common Materials

Material	Embodied Energy/(MJ kg^{-1})
Straw bale	0.2
Concrete	1.1
Brick	2.8
Ceramic tile	3.0
Recycled steel	7.3
Plywood	10
Glass	16
Steel (virgin)	27
Lead	27
Paper	45
Brass	54
Copper	60
PVC	68
Aluminum (43% recycled)	131
Aluminum (virgin)	210
Transistor	3000
Single crystal silicon for electronics	6000

Source: Data from *Materials and the Environment*, 2nd Edition, by M. F. Ashby, Elsevier (2013).

1.5 Structures of Materials

One of the most basic properties of materials, underlying almost all material properties, is *structure*. This can range from the random "organization" of an amorphous material to the beautiful morphology associated with a crystal. As a basis for the properties of materials presented in the coming chapters, here we present some of the basics of crystal structures. Noncrystalline structures, such as quasicrystals and glasses, are introduced in later chapters.

It has been known since ancient times that the smallest building blocks in nature, when packed together in regular structures, often result in beautiful crystals. We now know, especially based on x-ray diffraction studies, that the number of ways in which atoms can pack in a regular fashion in a crystal is not unlimited, and can be categorized according to the shape of the smallest building block that will fill space by repetitive translation, known as the *unit cell*. The lengths of the unit cell are designated a, b, and c, and the corresponding angles are α, β, and γ, as defined in Figure 1.3.

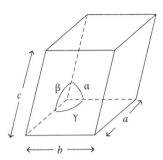

FIGURE 1.3
Dimensions and angles for a generalized unit cell.

COMMENT: UNITS AND UNIT PRESENTATION

Units in this book are given according to Système International (SI) conventions. Fundamental physical constants are given in Appendix 1, and energy unit conversions are presented in Appendix 2.

The presentation of units in tables and graphs in this book is according to quantity calculus (i.e., the algebraic manipulation method), as recommended by IUPAC. Briefly, a quantity is treated as a product of its value and its units:

$$\text{quantity} = \text{value} \times \text{units} \tag{1.1}$$

so, for example, the value of the strength/density ratio (i.e., specific strength) for composites (Figure 1.2) is 1×10^6 Pa kg^{-1} m^3:

$$\text{strength/density} = 1 \times 10^6 \text{ Pa kg}^{-1} \text{ m}^3 \tag{1.2}$$

so the value (= 1) plotted on the y-axis in Figure 1.2 is given by:

$$1 = (\text{strength/density}) / (10^6 \text{ Pa kg}^{-1} \text{ m}^3). \tag{1.3}$$

Labels on graphs and tables are treated similarly throughout this book.[*]

---

[*] See M.A. White, 1998. Quantity calculus: Unambiguous presentation of data in tables and graphs. *Journal of Chemical Education* 75, 607.

Of all the possible unit cell shapes, only certain categories pack together tightly to fill space. These are known as the seven *crystal classes,* as described by Table 1.3.

TABLE 1.3

Crystal Classes

System	Unit Cell
Cubic	$a = b = c$
	$\alpha = \beta = \gamma = 90°$
Tetragonal	$a = b \neq c$
	$\alpha = \beta = \gamma = 90°$
Orthorhombic	$a \neq b \neq c$
	$\alpha = \beta = \gamma = 90°$
Rhombohedral (also called trigonal)	$a = b = c$
	$\alpha = \beta = \gamma \neq 90°, <120°$
Hexagonal	$a = b \neq c$
	$\alpha = \beta = 90°, \gamma = 120°$
Monoclinic	$a \neq b \neq c$
	$\alpha = \gamma = 90°; \beta \neq 90°$
Triclinic	$a \neq b \neq c$
	$\alpha \neq \beta \neq \gamma \neq 90°$

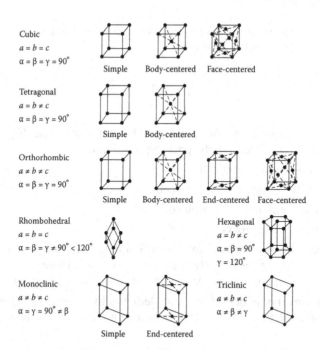

Cubic
$a = b = c$
$\alpha = \beta = \gamma = 90°$
Simple   Body-centered   Face-centered

Tetragonal
$a = b \neq c$
$\alpha = \beta = \gamma = 90°$
Simple   Body-centered

Orthorhombic
$a \neq b \neq c$
$\alpha = \beta = \gamma = 90°$
Simple   Body-centered   End-centered   Face-centered

Rhombohedral
$a = b = c$
$\alpha = \beta = \gamma \neq 90° < 120°$

Hexagonal
$a = b \neq c$
$\alpha = \beta = 90°$
$\gamma = 120°$

Monoclinic
$a \neq b \neq c$
$\alpha = \gamma = 90° \neq \beta$
Simple   End-centered

Triclinic
$a \neq b \neq c$
$\alpha \neq \beta \neq \gamma$

FIGURE 1.4
The 14 Bravais lattice types.

For several of the members of the crystal classes, there are different types of unit cells, giving a total of 14 possible Bravais lattices, as shown in Figure 1.4.

By combining all possible symmetry elements with the 14 Bravais lattices, we get 230 space groups that can be used to classify crystal structures.

1.6 Learning Goals

- History of materials development and driving forces for materials research
- Sustainability issues, including element availability
- Concept of embodied energy
- Crystal structure types for crystalline materials (Bravais lattices)
- Expression of numbers with units in an unambiguous way, using quantity calculus

1.7 Problems

1.1 The packing fraction, F, of a crystal lattice is defined as the fraction of space that is filled, that is,

$$F = \frac{V_{filled}}{V_{total}} \qquad (1.4)$$

where V_{filled} is the volume occupied by the atoms, and V_{total} is the total volume. Calculate the packing fraction for:

a. a simple cubic unit cell;

b. a body-centered cubic unit cell;

c. a face-centered cubic unit cell; and

d. a hexagonal unit cell.

1.2 In 2002, the price of indium was ~US\$100 kg^{-1} and the price reached over US\$700 kg^{-1} within the following few years. However, the price dropped in subsequent years. Suggest reasons for these trends.

1.3 Polonium (Po) is one of the few elements to exhibit a simple cubic crystal structure. Why is this structure not more prevalent among the elements?

1.4 The kiln at a brick production plant uses 300 GJ of energy per day of production. On a production day, the plant turns out 1.6×10^5 kg of bricks, which corresponds to 60,000 bricks, filling 16 train cars. From this information, calculate the embodied energy for a brick (in MJ kg^{-1}) and compare your answer with the value presented in Table 1.2. Discuss the difference and its origin.

1.5 The Herfindahl–Hirschman index (HHI) is a way to quantify the market availability of the elements as it takes into account geological data for the known elemental reserves and geopolitical data. The HHI values for all the elements have been tabulated (M. Gaultois et al., Chemistry of Materials 25, 2911–2920 (2013)). A low HHI value means that the element is readily available. Consider the HHI values for rare earth elements neodymium, lanthanum and terbium, that are required in large quantities for wind turbines and discuss the importance of recycling.

Further Reading

See also the lists at the end of individual chapters.

General References

General sources include: *Advanced Materials, Annual Review of Materials Science, Chemical and Engineering News, Chemistry World, Chemistry of Materials, Journal of Chemical Education, Journal of Materials Education, Materials Research Society (MRS) Bulletin, Materials Today, Physics Today, Scientific American.*

Dictionary of Materials Science, 2003. McGraw-Hill, New York.

Scientific American, September 1967. Special Issue: Materials.

Scientific American, October 1986. Special Issue: Advanced Materials and the Economy.

H. R. Allcock, 2008. *Introduction to Materials Chemistry.* John Wiley & Sons, Hoboken, NJ.

M. Alpert, S. Ashley, G. P. Collins, C. Ezzell, M. Fischetti, W. W. Gibbs, M. May, P. E. Ross, and I. Amato, 2006. The gorgeous inside stories of metal. *Chemical and Engineering News*, May 15, 14.

M. Ashby and K. Johnson, 2002. *Materials and Design.* Elsevier, Oxford.

P. Ball, 1997. *Made to Measure.* Princeton University Press, Princeton, NJ.

A. Barton, 1997. *States of Matter: States of Mind.* Institute of Physics Publishing, Bristol.

R. W. Cahn, P. Haasen, and E. J. Kramer, Eds., 2006. *Materials Science and Technology: A Comprehensive Treatment.* John Wiley & Sons, Hoboken, NJ.

W. D. Callister, Jr. and D. G. Rethwisch, 2013. *Materials Science and Engineering: An Introduction*, 9th ed. John Wiley & Sons, Hoboken, NJ.

R. Cotterill, 2008. *The Material World.* Cambridge University Press, Cambridge.

M. de Podesta, 2002. *Understanding the Properties of Matter*, 2nd ed. Taylor & Francis, Washington, DC.

J. P. Droske and C. E. Carraher, Jr., 2006. Polymers: Cornerstones of construction. *Journal of Chemical Education*, 83, 1428.

A. B. Ellis, M. J. Geselbracht, B. J. Johnson, G. C. Lisensky, and W. R. Robinson, 1993. *Teaching General Chemistry: A Materials Science Companion.* American Chemical Society, Washington, DC.

A. C. Fischer-Cripps, 2008. *The Materials Physics Companion.* Taylor & Francis, Washington, DC.

J. I. Gersten and F. W. Smith, 2007. *The Physics and Chemistry of Materials.* John Wiley & Sons, Hoboken, NJ.

D. L. Goodstein, 1985. *States of Matter.* Dover, New York.

M. Gross, 2005. Where biology meets chemistry. *Chemistry World*, November, 46.

L. V. Interrante, L. A. Casper, and A. B. Ellis, Eds., 1995. *Materials Chemistry.* American Chemical Society, Washington, DC.

L. V. Interrante and M. J. Hampden-Smith, Eds., 1998. *Chemistry of Advanced Materials: An Overview.* Wiley-VCH, Hoboken, NJ.

I. P. Jones, 2001. *Materials Science for Electrical and Electronic Engineers.* Oxford University Press, New York.

J. S. Langer, 1992. Issues and opportunities in materials research. *Physics Today*, October, 24.

D. C. MacLaren and M. A. White, 2003. The chemistry of cement. *Journal of Chemical Education*, 80, 623.

J. P. Mercier, G. Zambelli, and W. Kurz, 2004. *Introduction to Materials Science.* Elsevier, Oxford.

W. J. Moore, 1967. *Seven Solid States.* W.A. Benjamin, New York.

R. J. Naumann, 2009. *Introduction to the Physics and Chemistry of Materials.* Taylor & Francis, Washington.

G. A. Ozin, A. C. Arsenault, and L. Cademartiri, 2009. *Nanochemistry.* Royal Society of Chemistry, Cambridge.

D. W. Pashley, Ed., 2001. *Imperial College Inaugural Lectures in Materials Science and Engineering.* Imperial College Press, London.

S. L. Sass, 1998. *The Substance of Civilization: Materials and Human History from the Stone Age to the Age of Silicon.* Arcade Publishing, New York.

L. E. Smart and E. A. Moore, 2012. *Solid State Chemistry*, 4th ed. CRC Press, Boca Raton, FL.

G. Stix, 2002. Chemical and materials. *Scientific American*, December, 48.

M. A. White, 2006. Concrete. *Journal of Chemical Education*, 83, 1425.

A. Wold and K. Dwight, 1993. *Solid State Chemistry: Synthesis, Structure, and Properties of Selected Oxides and Sulphides.* Chapman & Hall, New York.

Biomaterials

Special issue on biomaterials. *Chemistry in Britain*, March 1992.

Sustainable Materials Approaches

M. Ashby, 2013. *Materials and the Environment*, 2nd ed. Elsevier, Oxford.

D. Cohen, 2007. Earth's natural wealth: An audit. *New Scientist*, May 23, 34.

M. W. Gaultois, T. D. Sparks, C. K. H. Borg, R. Seshadri, W. D. Bonificio, and D. W. Clarke, 2013. Data-driven review of thermoelectric materials: Performance and resource considerations. *Chemistry of Materials*, 25, 2911.

T. E. Graedel, 2011. On the future availability of the energy metals. *Annual Review of Materials Research*, 41, 323.

T. E. Graedel, E. M. Harper, N. T. Nassar, P. Nauss and B. Reck, 2015. Criticality of metals and metalloids. *Proceedings of the National Academy of Sciences*, 112, 4257.

S. Hadlington, 2014. Rare element substitution a tricky proposition. *Chemistry World*, 6 January.

T. Letcher and J. L. Scott, Eds., 2012. *Materials for a Sustainable Future*. Royal Society of Chemistry, London, UK.

W. McDonough and M. Braungart, 2002. *Cradle to Cradle: Remaking the Way We Make Things*. North Point Press, New York.

Reference Textbooks

M. F. Ashby and D. H. R. Jones, 2006. *Engineering Materials 2: An Introduction to Microstructures, Processing and Design*, 3rd ed. Elsevier, Oxford.

D. J. Barber and R. Loudon, 1989. *An Introduction to the Properties of Condensed Matter*. Cambridge University Press, Cambridge.

R. S. Berry, S. A. Rice, and J. Ross, 2000. *Physical Chemistry*, 2nd ed. Oxford University Press, New York.

R. J. Borg and G. J. Dienes, 1992. *The Physical Chemistry of Solids*. Academic Press, Oxford.

D. W. Bruce and D. O'Hare, Eds., 1997. *Inorganic Materials*. John Wiley & Sons, Hoboken, NJ.

A. K. Cheetham and P. Day, Eds., 1992. *Solid State Chemistry: Compounds*. Clarendon Press, Oxford.

J. H. W. de Wit, A. Demaid, and M. Onillon, Eds., 1992. *Case Studies in Manufacturing with Advanced Materials*, Vols. 1 and 2. North-Holland, Amsterdam.

H. F. Franzen, 1994. *Physical Chemistry of Solids: Basic Principles of Symmetry and Stability of Crystalline Solids*. World Scientific, Singapore.

D. Gay and S. V. Hoa, 2007. *Composite Materials*. CRC Press, Boca Raton, FL.

J. I. Gersten, 2001. *The Physics and Chemistry of Materials*. John Wiley & Sons, Hoboken, NJ.

G. G. Hall, 1991. *Molecular Solid State Physics*. Springer-Verlag, New York.

R. A. Higgins, 1994. *Properties of Engineering Materials*, 2nd ed. Industrial Press, New York.

R. E. Hummel, 2004. *Understanding Materials Science*, 2nd ed. Springer, New York.

C. Kittel, 2004. *Introduction to Solid State Physics*, 8th ed. John Wiley & Sons, Hoboken, NJ.

J. P. Mercier, G. Zambelli, and W. Kurz, 2003. *Introduction to Materials Science*. Elsevier, Oxford.

R. J. Naumann, 2008. *Introduction to the Physics and Chemistry of Materials*. CRC Press, Boca Raton, FL.

D. V. Ragone, 1995. *Thermodynamics of Materials*, Vols. I and II. John Wiley & Sons, Hoboken, NJ.

C. N. R. Rao, Ed., 1993. *Chemistry of Advanced Materials*. Blackwell Scientific Publications, Oxford.

C. N. R. Rao and J. Gopalakrishnan, 1997. *New Directions in Solid State Chemistry*. Cambridge University Press, Cambridge.

R. M. A. Roque-Malherbe, 2009. *The Physical Chemistry of Materials*. CRC Press, Boca Raton, FL.

J. F. Shackelford, 2004. *Introduction to Materials Science for Engineers*, 6th ed. Prentice Hall, Upper Saddle River, NJ.

J. F. Shackelford and W. Alexander, 2000. *The CRC Materials Science and Engineering Handbook*, 3rd ed. CRC Press, Boca Raton, FL.

J. P. Sibilia, 1996. *A Guide to Materials Characterization and Chemical Analysis*, 2nd ed. Wiley-VCH, Hoboken, NJ.

L. E. Smart and E. A. Moore, 2012. *Solid State Chemistry*, 4th ed. CRC Press, Boca Raton, FL.

W. F. Smith, 2004. *Foundations of Materials Science and Engineering*, 3rd ed. McGraw-Hill, New York.

D. Tabor, 1991. *Gases, Liquids and Solids and Other States of Matter*, 3rd ed. Cambridge University Press, Cambridge.

R. J. D. Tilley, 2004. *Understanding Solids: The Science of Materials*. John Wiley & Sons, Hoboken, NJ.

L. H. Van Vlack, 1989. *Elements of Materials Science and Engineering*, 6th ed. Addison-Wesley, Reading, MA.

J. K. Wessel, Ed., 2004. *The Handbook of Advanced Materials: Enabling New Designs*. John Wiley & Sons, Hoboken, NJ.

Structure/Crystallography

S.-W. Chen and M. Kotlarchyk, 2007. *Interactions of Photons and Neutrons with Matter*. World Scientific, Singapore.

MRS Bulletin, 1999. Special Issue: Synchrotron radiation techniques as tools for in situ characterization in materials processing. January.

MRS Bulletin, 1999. Special Issue: Neutron scattering in materials research. December.

Units

E. Gibney, 2017. New definitions of scientific units are on the horizon. *Nature*, 550, 312.

D.B. Newell, 2014. A more fundamental International System of Units. *Physics Today*, July, 35.

Websites

For links to relevant websites, see PhysicalPropertiesOfMaterials.com

Part II

Color and Other Optical Properties of Materials

It is the pure white diamond Dante brought
To Beatrice; the sapphire Laura wore
When Petrarch cut it sparkling out of thought;
The ruby Shakespeare hewed from his heart's core;
The dark, deep emerald that Rosetti wrought
For his own soul, to wear for evermore.

Eugene Lee-Hamilton
"What The Sonnet Is"

2

Atomic and Molecular Origins of Color

2.1 Introduction

For many, the colors in nature and in the laboratory are one of the inspirations for studying science. Perhaps this is partly because of our perception of color. Including hue, saturation, brightness, and intensity, we can distinguish about 10 million colors with our eyes. Few laboratory-built detectors are as sensitive!

Color in materials is caused by interactions of light waves with atoms, and especially with their electrons. In fact, color is a manifestation of many subtle effects that are important in determining the structure of matter.

In practical terms, it is the relative contributions of light of various wavelengths that determine the color of a material. In order to pursue this further, it is useful to review the electromagnetic spectrum (Figure 2.1), keeping in mind that light is an electromagnetic wave. The energy of the radiation, E, is related to its wavelength, λ, and frequency, ν, by[*]

$$E = h\nu = \frac{hc}{\lambda} \tag{2.1}$$

where h is Planck's constant and c is the speed of light. (See Appendix 1 for values of physical constants.)

As shown by Equation 2.1, when the wavelength is longer, the energy per photon is lower. Very long wavelength radiation (e.g., a radio wave) passes through us without damage, whereas radiation of much shorter wavelength and higher energy can do biological damage. For example, high doses of x-rays can be very damaging because their high energy can initiate unfavorable chemical reactions in the body. It is easy to see from Equation 2.1 why it is safe to stand under an infrared lamp to dry oneself after a bath, but an ultraviolet (UV) light requires more safeguards.

---

[*] The most important equations in this chapter are designated with ▌ to the left of the equation.

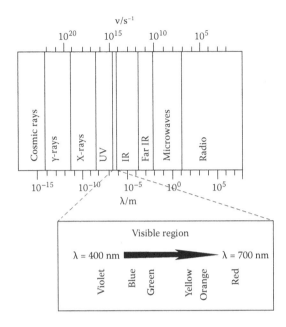

FIGURE 2.1
The electromagnetic spectrum. Note that the boundaries between regions are not sharp and the spacing of the visible colors is not even.

Color is often the result of transitions between electronic states; this can involve energy absorption or emission. Most sources of color can be categorized as originating from absorption or emission of light, and it is useful to keep in mind which of these dominates in the discussion in this chapter and in Chapters 3, 4 and 5. A useful general consideration is that emission can be seen in the dark, whereas absorption requires a source of light. (Color also can arise from either transmission or reflection of light [see Comment: Transmission or Reflection?]. Other sources of color are light scattering, dispersion, and interference; these are discussed in Chapter 4.) In considering these changes in states, it is useful to keep in mind that electronic states are much more widely separated than vibrational states, which are themselves more widely separated than rotational states. This is shown schematically in Figure 2.2. If an electronic transition involves core electrons or localized electrons in a closed-shell electron configuration, the energy involved is so high that it will be in the UV or even in the x-ray region of the electromagnetic spectrum. Many colors that we see are caused by electronic transitions involving valence electrons, the electrons that form chemical bonds, as these transitions can be at low enough energy for us to see their effects with our eyes.

In this chapter, we introduce atomic and molecular origins of color in insulating materials.

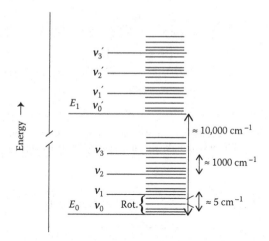

FIGURE 2.2

Schematic representations of the relative spacings of electronic energy levels (marked E_0 and E_1), vibrational energy levels (marked v_0, v_1, ... and v_0', v_1', ...), and rotational energy levels (marked Rot.). The approximate energies are given in units of cm^{-1}, called wavenumbers and abbreviated $\tilde{v}$, where $\tilde{v} = 1/\lambda$. 1 cm^{-1} = 1.239 × 10^{-4} eV = 1.985 × 10^{-23} J.

COMMENT: TRANSMISSION OR REFLECTION?

When an opaque material derives its color primarily from absorption of light, the color observed is due to the absorption of certain colors from the ambient white light. The light that reaches the eye and causes color perception is due to refection and scattering, and it is depleted of the absorbed color(s). The color observed is the *complementary color* (see Figure 2.3) to the color that is absorbed.

If a material is transparent, and we observe it with a light on the other side, we perceive its color due to the wavelengths of light that are transmitted. Again, relative to ambient light, the transmitted light will be depleted in the colors of the absorbed light. Looking through this material, it will appear as the color that complements the absorbed color or colors.

If the color of a transparent object is due only to absorption, the transmitted and reflected colors would be the same. However, surface reflectivity can have some dependence on wavelength, and this can lead to different colors in refection and transmission. Refection and transmission are illustrated in Figure 2.4.

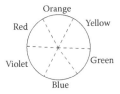

FIGURE 2.3
The "color wheel." Colors opposite to each other are complementary. If one color is absorbed by a material, the reflected light will make the material appear the complementary color to the color absorbed.

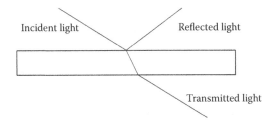

FIGURE 2.4
Some light is reflected from a material and other light is transmitted. Surface scattering and internal scattering are omitted here.

2.2 Atomic Transitions

Electronic transitions within a given atom can give rise to color through emission of light. A well-known example of atomic emission is the yellow color of sodium in the flame test for this element. The ground elec-

See the video "Complementary Colors" under Student Resources at PhysicalPropertiesOfMaterials.com

tronic configuration of Na is $1s^2 2s^2 2p^6 3s^1$, so the "outermost" electron is the $3s^1$ electron. The ground state of this valence electron has two closely spaced energy levels above it, one 2.105 eV above the $3s^1$ level, and one 2.103 eV above the $3s^1$ level. (The electron volt, or eV, is a convenient method for measuring energies of electrons. 1 eV = 1.602×10^{-19} J; see Appendix 2 for more energy conversions.) The corresponding wavelengths of light of these energies are 589.1 and 589.6 nm, where, of course, the longer wavelength corresponds to the lower energy emission shown in Figure 2.5.

These wavelengths of light emitted from excited sodium atoms, at about 590 nm, correspond to the yellow part of visible light. Therefore, when a sodium-containing material is heated in a flame such that $3s^1$ electrons are promoted to the 3p excited levels, each electron can then return to the ground state ($3s^1$ configuration), concurrent with emission of yellow

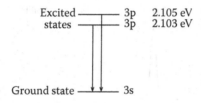

FIGURE 2.5

The electronic energy levels of Na. In the Na flame test, excitation is caused by heat, and emission of yellow light results. The separation between the excited states is exaggerated in this diagram.

light. This light is often referred to as the sodium D line, in reference to being the fourth prominent line (hence D line) recorded by Fraunhofer[*] as missing from the otherwise nearly continuous emission from sunlight. We now know that this wavelength of light is missing because it is absorbed by sodium atoms present in the sun's atmosphere. Close examination also shows that the D "line" is really two closely spaced absorptions (see Figure 2.5).

The light emitted in this way from an atom is characteristic of the particular element because each element has its own characteristic electronic energy levels. For example, the red glow of a neon sign is characteristic of the electronic energy levels of Ne. The blue color of a mercury lamp is caused by two emissions, one green and one violet (green and violet emissions together produce blue). Other examples of colors arising from atomic emission include the colors of lasers making use of monatomic gases (such as the Ar-ion laser). In addition, lightning and arcs (e.g., around spark welding or a skate-sharpening device) derive some of their color from electronic excitations of the atoms in the surrounding gases. Northern lights (*aurora borealis*) arise from atomic emissions caused by interaction of atmospheric atoms or molecules (mostly oxygen, nitrogen, and hydrogen) with solar wind, that is, charged species from the sun entering our atmosphere. Emissions from various elements are used in fireworks—for example, strontium for red and barium for green.

In general, gases have relatively sharp emission and absorption lines (see Figure 2.6). This is, at least in part, caused by the low density of gas molecules, because at higher pressures there would be more collisions that would increase the linewidth.

---

[*] Joseph Fraunhofer (1787–1826) was a German physicist who studied optical properties of matter in order to design and produce fine optical and mechanical instruments.

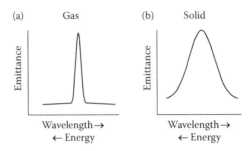

FIGURE 2.6
Schematic emissions from (a) a gas and (b) a solid from, for example, fluorescence. Both emissions are narrower than the black-body emissions shown in Figure 2.7. However, atomic emissions, such as shown in Figure 2.5, would be sharper (narrower in energy range) than those shown above.

2.3 Black-Body Radiation

As the name implies, black-body radiation also involves emission. Whereas gases have sharp emission lines, condensed matter (solids and liquids) usually gives a broad emission over a wide range of wavelengths. This situation is shown schematically in Figure 2.6.

A "black body" is an idealized material that absorbs light of all wavelengths (it does not reflect or transmit any light, hence the adjective "black") and it also is a perfect emitter of light of all wavelengths. The study of the intensity distribution of radiation from black bodies by Max Planck[*] in 1900 led to the development of the quantum theory. An ideal black body emits a "spectrum" of light, and this spectrum depends only on the temperature of the black body. Black-body radiation is often referred to as *incandescence*.

For example, at $T = 0$ K, all the atoms and subatomic particles in the black body are in their ground state, so there can be no light emitted at any wavelength.

At a temperature T, where $T > 0$ K, the occupancy of states from which the black body can emit light depends on the energy of the state. The body both absorbs and emits light, but when its temperature exceeds the temperature of the surroundings, emission will dominate. Therefore, there is a distribution of emission (i.e., a spectrum) as shown in Figure 2.7.

At higher temperatures, the distribution of emitted light moves, as shown in Figure 2.7, such that the maximum intensity is at shorter wavelength (higher energy) than at lower temperature. Of course, because of the higher

---

[*] Max Karl Ludwig Planck (1858–1947) was a German theoretical physicist who made major contributions to the fields of thermodynamics, electrodynamics, radiation theory, and relativity. He also contributed to the philosophy of science and was an accomplished musician. He was awarded the 1918 Nobel Prize in Physics for the discovery of energy quanta.

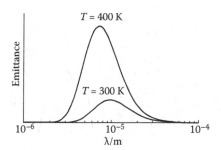

FIGURE 2.7

Typical ideal black-body radiation spectra at two temperatures. Note the increase in the emitted radiation (curve area) and the shift in the peak to lower wavelength (higher energy) when the temperature is increased.

occupancy of excited states at the higher temperature, the overall intensity of emitted light (i.e., the area under the entire curve) is greater at higher temperature than at lower temperature.

The exact spectrum is determined solely by the temperature of the ideal black-body emitter, independent of the type of material. This information can be of considerable importance.

At room temperature, black-body radiation is negligible to our eyes. At a temperature of about 700°C, the emission maximum is still in the infrared (IR) region, but some long wavelength (red) light is perceptible, and an object at this temperature glows red. We are familiar with this in glowing red embers in a fire. As the temperature of the black-body emitter increases, the "glow" moves from red to orange to yellow to white (the spectrum covers the whole visible range, and when all colors are emitted, it appears to be white*) to very pale blue. The progression of the emission over a range of temperatures is shown in Figure 2.8.

Examples of color caused by black-body radiation can be found in high-temperature bodies in our world. For example, carbon of a candle flame or log on a fire glows orange-yellow at a temperature of about 1700°C. The tungsten filament of an incandescent light bulb glows yellow-white at a temperature of about 2200°C. The filament of a camera flash provides virtually "white" light because it is at a higher temperature still, about 4000°C. The light appears white because its emission peak covers the whole visible range and almost exactly matches the sensitivity of the eye to various colors.

As mentioned earlier, the ideal black-body radiation spectrum is characteristic of the temperature of an object and does not depend on the type of material. This knowledge, combined with a chance observation of unexpectedly high levels of radiation at a particular wavelength, led researchers from Bell Labs to determine that the average temperature of our universe is 3 K, a

---

* When white light is broken into its parts (e.g., with a prism), it can be seen to be composed of red, orange, yellow, green, blue, and violet constituents. In a similar but reverse manner, when all of the colors of the rainbow are emitted, they combine to produce white light.

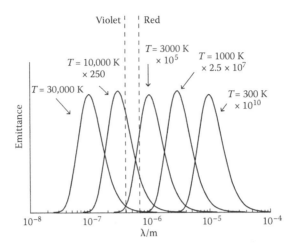

FIGURE 2.8
Black-body radiation as a function of temperature, highlighting how the color changes as the temperature changes from $T = 300$ K to $T = 30,000$ K. To have all the peaks appear to have approximately the same area, the emittance values at lower temperatures have been multiplied by the factors shown in the figure. Lower-temperature peaks are much smaller due to less overall emission. Note also the shift in the peak to longer wavelength (lower energy) with lower temperature.

souvenir of the Big Bang that is thought to have created the universe 12 billion years ago. The discovery of this black-body radiation has considerably advanced theories of the origins of our universe.[*]

2.4 Vibrational Transitions as a Source of Color

In general, vibrational transitions are relatively rare as a source of color to our eyes. The energies of most vibrational excitations fall in the infrared region, so this does not noticeably influence the color of white light that falls upon a sample. However, there are some exceptions, and one rather prominent one in our lives is H_2O, both in the liquid and solid state.

Excitation of a complex many-molecule motion in the hydrogen-bonding network in water or in ice involves absorption of a small amount of red-orange light. This is a rather high-energy excitation (higher energy than the usual IR range for molecular vibrations) because the excitation is to a highly excited vibrational overtone (i.e., leading to a much higher vibrational level). The absorption of red-orange light depletes white light of red-orange and

---

[*] For further details, see Jeremy Bernstein, 1984. *Three Degrees above Zero: Bell Labs in the Information Age.* Charles Scribner's Sons, New York.

leaves the complementary green-blue color. (The visible colors and their complements were shown in Figure 2.3.)

The absorbance, A, is related to the incident light intensity, I_0, and the exiting light intensity, I, through the Beer[*]–Lambert[†] law (also known as the Beer–Bouger[‡]–Lambert law):

$$\log_{10}\left(\frac{I_0}{I}\right) = A = \varepsilon c l \qquad (2.2)$$

where ε is the absorption coefficient, c is the concentration, and l is the length of the sample in the light path. Because ε is rather small for this excitation in H_2O, it takes a large path length, l, to produce a detectable color. We are familiar with this, as a glass of water and an ice cube both appear colorless, whereas the blue color of light observed through a large body of water or through ice in a glacier is readily apparent. For example, with a 3-m path length, 56% of incident red light is absorbed by water and the liquid appears blue-green in transmitted light. (The color of water seen in refection is caused by both absorption of red light and light scattering; see Chapter 4 for a discussion of the latter effect.)

Although vibrational excitation is not a common source of visible color, it gives water its familiar color.

2.5 Crystal Field Colors

Another case of absorption as a source of color involves ions present in solids. These can be ions in a pure solid or impurity ions.

When atoms are free, as in the gas phase, we have seen that their valence electrons can be thermally excited, giving rise to emission colors, as in the case of sodium in a flame. However, when valence electrons combine with one another to form chemical bonds, the ground state energy of the atom is lowered and much more energy is then required to promote the electrons to excited states. In many compounds the outermost electrons are so stabilized by chemical bonding that their excitation takes energy in the UV region, so no visible color is associated with these materials, although they do absorb UV light.

One exception is compounds containing transition metals with incomplete d-shells. The excitation of these valence electrons can fall in the visible region

---

[*] August Beer (1825–1863) was a German physicist and optical researcher who worked at Bonn University.
[†] Johann Heinrich Lambert (1728–1777) was a German mathematical physicist who discovered a method for measuring light intensities.
[‡] Pierre Bouger (1698–1758) was a French mathematician who invented the photometer.

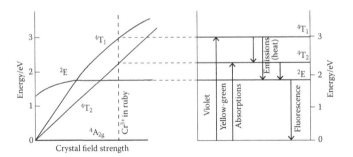

FIGURE 2.9
The effect of crystal field strength on the splitting of the energy levels of Cr^{3+} ions in a lattice. (Styled after K. Nassau, 1980. *Scientific American*, October 1980, 134.)

of the spectrum, and these compounds can have beautiful colors. Such colors also can arise from compounds containing transition metal impurity ions. For example, the well-known cobalt blue pigment derives its color from excitation of the cobalt ions.

All electronic energy levels are sensitive to their environment; electrostatic and other interactions can make the levels move relative to one another. An ion in the electric field in a crystal (i.e., in a crystal field, or, if the field arises from the presence of ligands, it is called a ligand field) can have different energy levels from the free ion. The interaction of an ion with an electric field can provide information concerning the electronic environment of the ion. In the cases considered here, the crystal field can give rise to observable colors.

The Cr^{3+} ion in ruby lies at the center of a distorted octahedron of O^{2-} ions. The Cr–O distance is about 1.9 Å, and the Cr–O bonds are quite ionic (estimated at more than 50% ionic character), so the Cr^{3+} ion exists in a strong electric potential (i.e., crystal field). The influence of the crystal field on the electronic energy levels of Cr^{3+} is shown in Figure 2.9.

At the crystal field strength of Cr^{3+} in ruby, the higher-energy absorption is in the violet region of the spectrum, and the lower-energy absorption is yellow-green. Therefore, white light is depleted of violet and yellow-green, and ruby appears red. The main color of ruby is due to absorption of violet and yellow-green light; it is only coincidental that the color of *fluorescence*[*] in ruby (Figure 2.9) also is red.

Interestingly, emeralds, which are green, also derive their color from Cr^{3+} ions, this time in a material with composition $Be_3Al_2SiO_6$ (known as beryl when pure). The Cr^{3+} in this structure is in a very similar environment to that in ruby, but it experiences a different crystal field strength. The difference in color subtly illustrates the importance of the magnitude of the crystal field strength. Determination of whether the crystal field strength is less than or greater than in ruby is left as an exercise for the reader.

---

[*] See Section 2.7 for a discussion of fluorescence.

Electronic transitions of other ions in solids can give rise to crystal field colors. Some further examples are alexandrite (Cr^{3+} in $BeAl_2O_4$), aquamarine (Fe^{3+} in $Be_3Al_2SiO_6$), jadeite form of jade (Fe^{3+} in $NaAl(SiO_3)_2$), citrine quartz (Fe^{3+} in SiO_2), blue and green azurite ($Cu_3(CO_3)_2(OH)_2$), malachite ($Cu_2CO_3(OH)_2$), and red garnets ($Fe_3Al_2(SiO_4)_3$). In the last three cases, the color comes from the transition metal ions (Cu^{2+} or Fe^{2+}) of the pure salt rather than impurity ions. When the color of a mineral is inherent (said to be *ideochromatic*; the color will remain even if finely divided), it is often caused by its pure composition, which is usually independent of the origin of the mineral. A mineral of variable color (called *allochromatic*) generally derives its color from impurities (also called *dopants*) or other factors that depend on location and geological history.

COMMENT: DEFECTS IN CRYSTALS

In contrast with the perfect crystalline structures shown in Figure 1.4, real crystals have imperfections. These can be *point defects* (i.e., occupational imperfections) or *line defects* (i.e., imperfections in translational symmetry). Some examples are shown in Figure 2.10. (See also Chapter 14.)

Defects play an important role in crystals and can greatly modify optical, thermal, electrical, magnetic, and mechanical properties. For example, photovoltaic solar cells rely on defects (see Chapter 3).

2.6 Color Centers (F-Centers)

Electronic excitation can give rise to visible colors, but the excited electrons do not need to be bound to ions; they also could be "free." For example, the "electric blue" color of Na dissolved in liquid ammonia comes from the

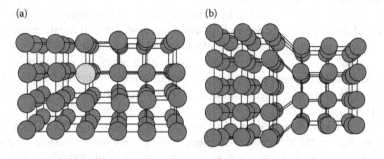

(a)   (b)

FIGURE 2.10
Defects in a solid: (a) point defect due to impurity (shown as a lighter-colored atom); (b) line defect due to a packing fault.

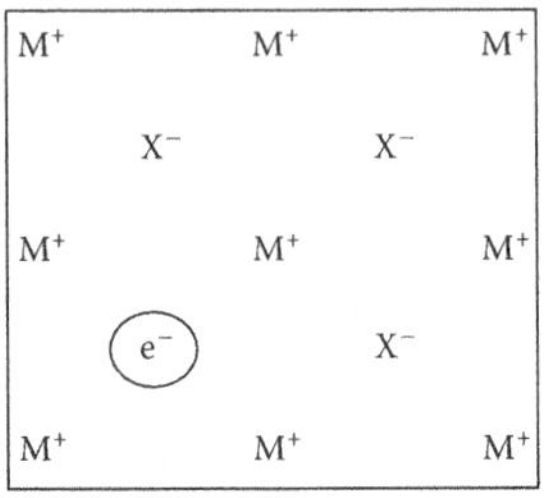

FIGURE 2.11
Presence of a trapped electron in an idealized two-dimensional lattice of salt MX.

FIGURE 2.12
Presence of an extra electron associated with a hypervalent impurity (i.e., an impurity with an extra charge, shown here as Mg^{2+}) in the idealized two-dimensional salt MX.

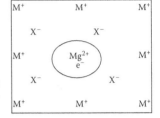

FIGURE 2.13
A hole, h^+, is present in a lattice of MX to compensate for the charge of the impurity Y^{2-}.

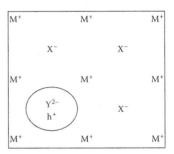

absorption of light in the yellow region by solvated electrons in solution (the solution contains NH_3, Na^+, and e^-).

Unattached electrons also can exist in solids. For example, if there is a structural defect, such as a missing anion, an electron can locate itself in the anion's usual position, as shown in Figure 2.11 for the salt M^+X^-.

An "extra" electron also could be associated with an impurity of mismatched valence. For example, if Mg^{2+} is an impurity in the M^+X^- lattice, there would be an extra electron located near Mg^{2+} to balance electrical charge, as shown in Figure 2.12.

Similarly, the absence of an electron, which is termed a *hole* (abbreviated as h^+; see Figure 2.13), locates itself near an impurity ion or structural defect. Holes are as important as electrons in discussion of electrical properties of solids because both electrons and holes are charge carriers. Whereas electrons carry negative charge, holes carry positive charge (see Comment: Holes).

COMMENT: HOLES

In some ways, it is difficult to imagine holes since they are really not matter but more like the absence of matter (absence of electrons, to be specific). However, there are instances where we are familiar with the motion of the absence of matter. One example is the bubbles that are seen to rise in a fish tank. Although most of the matter in view is water, the bubbles can be considered to be an "absence of water"; as a bubble rises from the bottom to the top of the tank, the net flow of water is in the opposite direction (i.e., toward the bottom). This situation is similar to the case of electrons and holes: While the negative charges (electrons) flow in one direction, the positive charges (holes) "flow" in the opposite direction.

Extra charge centers (electrons or holes) are called *color centers* or sometimes *F-centers* ("F" is an abbreviation of "Farbe," the German word for "color"). They give rise to color centers because the excitation of electrons (or holes) falls in the visible range of the spectrum.

As an example, consider the sometimes violet color of fluorite, CaF_2. This color arises when some F^- ions are missing from their usual lattice sites, giving rise to an extra electron in the lattice. This can happen in one of several ways: the crystal might have been grown in the presence of excess Ca; the crystal might have been exposed to high-energy radiation that displaced an ion from the usual lattice site; or the crystal might have been in an electric field that was strong enough to remove F^- electrochemically. In any case, electrons exist in the places where F^- would be in a perfect lattice. These electrons give rise to the color, as the electrons absorb yellow light, leaving behind a violet color.

Because a color center could arise from radiation, the intensity of the color of a material such as fluorite can be used to monitor radiation dosage. This coloration has laboratory applications (e.g., in radiation safety badges worn by people who work with x-rays), and it can be used in the field to determine the radiation dosage that a geological sample has received from its environment.

The color of amethyst also is caused by a color center, this time by the absorption of energy associated with the electronic energy levels of a hole. The hole exists because of the presence of Fe^{3+} in the SiO_2 lattice. Fe^{3+} substitutes for Si^{4+} in the lattice, requiring the presence of a hole for electrical neutrality. The energy levels of the hole allow absorption of yellow light, giving amethyst its familiar violet color. The intensity of the color is an indication of the concentration of iron impurity ions.

Color centers are usually irreversibly lost on heating because increased temperature can mobilize imperfections and cause them to go to more regular sites. For example, amethyst turns from violet to yellow or green on heating; the loss of the violet color is because of the loss of the color centers, and the remaining color, which was there all along but was not as intense as the violet, is because the Fe^{3+} transition metal ion absorbs light due to crystal field effects. If it is yellow, this form is citrine quartz, described previously under crystal field effects.

2.7 Charge Delocalization, Especially Molecular Orbitals

When electrons are paired in chemical bonds, their excited electronic states are usually so high that electronic excitations are in the UV range, and these materials are not colored (i.e., there is no absorbance in the visible range). This is especially true if the pair of electrons is localized to a particular chemical bond.

In molecular systems with extensive conjugation such as alternating single and double carbon–carbon bonds, the electrons are delocalized throughout a number of chemical bonds. The electronic excitation is from the highest occupied molecular orbital (HOMO) to the lowest unoccupied molecular orbital (LUMO). For a conjugated system, the HOMO–LUMO transition is of lower energy than when electrons are localized to one bond; if the transition in the conjugated system is low enough in energy, absorption of visible light can cause the promotion of an electron to this excited state. If visible light is absorbed, the material with conjugated bonds will be colored. (If only more energetic UV light is absorbed, the material might have promise as a component in a sunscreen.)

Many organic molecules derive their colors from the excitation from electronic states that involve large degrees of conjugation. For example, the more than 8000 dyes in use today derive their color from this mechanism. The part of the molecule that favors color production is called a *chromophore*, meaning "color bearing." The main distinction between a dye and a pigment is the state of matter: Dyes are used in liquid form, whereas pigments are used as solids, often suspended in solvent.

The color of organic materials can be changed substantially by the presence of electron-withdrawing and/or electron-donating functional groups, as this changes the electronic energy levels. These are called *auxochromes* ("color enhancers").

As an example of the dependence of the color of an organic compound on structure, we can consider the case of an organic acid-base indicator. These materials are often weak organic acids, which will be deprotonated in basic solution. The loss of the proton can be detected visually, as the color of the acid and its conjugate base can be substantially different, especially due to different degrees of delocalization. The pH-dependence of color is taken up further in Problem 2.9 at the end of this chapter.

Retinal is the pigment material of the photoreceptors in our eyes. Its color changes in the presence of light, as retinal undergoes a photo-induced transformation from *cis*-retinal to *trans*-retinal form. In the absence of light, the *trans* form coverts back to the *cis* form, assisted by an enzyme. The dominance of the *cis* or the *trans* form signals the light level.

Another example of charge delocalization and its associated color can be seen in the case of *charge transfer* from one ion to another—for example, in blue sapphire. This material is Al_2O_3 with Fe^{2+} and Ti^{4+}. An excited electronic state is formed when adjacent Fe^{2+} and Ti^{4+} ions exchange an electron to give Fe^{3+} and Ti^{3+} by a photochemically induced oxidation-reduction reaction. The energy required to reach this excited state is about $2\,eV$, which corresponds to light of wavelength 620 nm. This light is yellow, and since the material is absorbing yellow from the incident white light, sapphire appears blue. Only a few hundredths of 1% of Fe^{2+} and Ti^{4+} are required to achieve a deep blue color in sapphire. The deep colors of many mixed-valence transition metal oxides (e.g., Fe_3O_4, which can be represented as $FeO\cdot Fe_2O_3$) arise from homonuclear charge transfer mechanisms.

Some molecules with delocalized electrons also are capable of *luminescence* (i.e., light emission from a cool body[*]), which includes fluorescence, phosphorescence, and chemiluminescence. In *fluorescence*, an excited state decays to an intermediate state that differs from the ground state by an energy that corresponds to rapid emission of light in the visible range of the spectrum, as shown in Figure 2.14. Compounds that are added to detergents as fabric brighteners absorb UV components of daylight (to get to an excited state) and then fluoresce, giving off blue light. This blue light makes the material appear bright by combining with the yellow of dirt (blue and yellow are complementary colors) to give white.

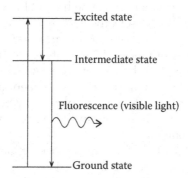

Excited state

Intermediate state

Fluorescence (visible light)

Ground state

FIGURE 2.14
Schematic representation of fluorescence.

---

[*] The lower temperature of luminescence distinguishes it from incandescence associated with black-body radiation.

Phosphorescence is the slow emission of light following excitation to a higher-energy state. Phosphorescence is discussed further in Chapter 3.

Chemiluminescence is fluorescence initiated by chemical reactions that give products in an excited state. The material might decay to its ground state energy level by a fluorescent pathway, as described earlier. An example is a glow stick. When chemiluminescence takes place in a biological system, it is referred to as *bioluminescence*; examples include the glow of fireflies and some deep-sea fish.

COMMENT: CARBONLESS COPY PAPER

Carbonless copy paper, used to make countless forms, is a 10^9-kg business annually, based on simple chemistry in which organic reactants change color on protonation. Copies are made when the pressure of a pen ruptures microcapsules that are coated on the lower surface of the top page. These microcapsules contain organic dye precursor molecules that, after rupture of the microcapsules, can react with a proton-donating reagent layer that is coated on the upper surface of the lower paper. In the acidic conditions, the organic dye precursor becomes colored, making a copy of the top page's written message onto the bottom page.

COMMENT: COLOR BLINDNESS

In 1794, the great chemist John Dalton (1766–1844), gave the first formal account of color blindness: his own. He believed that he was unable to distinguish red from green because he had a blue tint in the vitreous humor of his eyes that could selectively absorb the long wavelengths of light. He had ordered that his eyes be examined after his death, but no blue tint was found. More recent DNA tests have shown that Dalton's color blindness was caused by alteration of the red-responsive visual pigment. Although Dalton would not have been able to distinguish the color red, he possessed the red receptor and would have perceived light throughout the visible region, including red, as confirmed from the writings of his contemporaries.

More recently, lenses have been developed to aid people with red-green color blindness, resulting from red- and green-sensitive photopigments in their eyes that have more overlap than normal. The corrective lenses have filters that cut out narrow wavelength regions of the spectrum, to enhance specific colors, helping improve red-green color vision.

COMMENT: LASERS

Lasers derive their name from light amplification by stimulated emission of radiation. The laser was invented by Charles Townes (1915–2015, co-recipient of the 1964 Nobel Prize in Physics). The light produced can be very intense, and it has the special property of *coherence*, i.e., the waves produced are all in phase.

The action of a laser can be understood in terms of the energy levels of the system. A laser requires excitation (pumping) from the ground state energy to an excited state (absorption level). This excitation can be electrical, optical, or chemical; an example is shown schematically in Figure 2.15. From this absorption level, the energy rapidly decreases as the system goes to a long-lived excited state. From this state, the energy can decrease further as the system returns to the ground state; this last process can be stimulated by light, and more light is emitted in this process.

In the laser cavity, which is a tube with perfectly aligned end mirrors, the emitted light that is directed exactly along the length of the cavity will cause additional stimulations along the length of the cavity. Emission will also take place in other directions, but, in contrast to the emission along the length of the cavity, emission in other directions will not be reflected by the mirrors. The emission in the direction along the cavity will be coherent and very intense. One of the mirrors is slightly transparent, and typically 1% of the coherent stimulated emission escapes from this end.

In a helium–neon laser, direct current or radio frequency radiation is used to excite the helium atoms, and then these collide with neon atoms, exciting them into a long-lived excited state. The emission from this state gives rise to the familiar red light of a helium–neon laser.

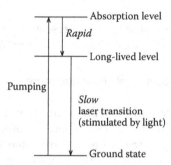

FIGURE 2.15
Schematic energy level diagram for a laser.

2.8 Light of Our Lives: A Tutorial

For many decades, the most common lights in our homes were incandescent lights. These have a wire inside them that glows when current is passed through it.

a. The early developers of light bulbs had to struggle to get bulbs that would last for a long time. Suggest some of their difficulties, and how they have been overcome.

b. Tungsten is now the preferred material for the filaments of incandescent bulbs, whereas carbon was initially used. Why is tungsten superior?

c. Incandescent light bulbs get hot when they have been operating for some time. Why?

d. If a drop of water falls on an operating light bulb, the bulb will crack. Within a few minutes or less, the bulb will dim and sputter and then stop working. Describe the process of this failure.

Fluorescent lights are otherwise known as *gas discharge tubes*. Electrical discharge along the tube excites the Hg atoms in the vapor phase in the tube. These atoms emit some visible light and considerable light in the UV region, which we cannot see with our eyes. The UV light causes excitations in the fluorescent coating (e.g., magnesium tungstate, zinc silicate, cadmium borate, cadmium phosphates) on the inside of the tube; this coating then emits additional light in the visible range.

e. Draw energy level diagrams to depict the fluorescent light processes.

f. Discuss factors that are important in choosing materials for the vapor phase. Consider the full life cycle (manufacturing to disposal) of the fluorescent light.

g. Discuss factors that are important in choosing materials for the fluorescent coating.

h. How does the temperature of a fluorescent light compare with that of an incandescent light during use? Comment on relative energy efficiencies.

i. How can incandescent lights be made more energy efficient?

A halogen (or quartz halogen) lamp has a tungsten filament housed inside a smaller quartz casing, inside the bulb. There is halogen gas, such as iodine or bromine, inside the quartz casing. The tungsten filament is heated to a very high temperature, and some of the tungsten atoms vaporize and react with the halogen. In a subsequent step, they redeposit as tungsten metal on the filament. This recycling process increases the durability of the tungsten filament and allows the lamp to produce more light for less energy input than normal incandescent lamps.

j. Why does the inner casing need to be quartz, not ordinary glass?

In the 1990s, high-intensity discharge lights began to be commonly used for car headlights, giving an intense bluish light. These lights contain high-pressure mercury and metal halides such as sodium and scandium iodides and sometimes also xenon (which shortens start-up times). The process by which they work is similar to fluorescent lamps except that the light is produced directly with no need for phosphors. A great deal of UV light also is produced.

k. How would the lighting process of high-intensity discharge lights influence the material choice for the outer bulb?

Neon lights are useful because the glass tubes can be bent into virtually any shape. Within the tube, neon atoms are excited by electrons knocked off other neon atoms by electrical discharge across the tube. The excited neon atoms decay back to their ground state, emitting red light.

l. Would other discharge gases be expected to emit the same color of light?

Some of the most energy efficient lights are light-emitting diodes (LEDs; see Chapter 3 for the principles of their operation). Most of the LEDs that are used to provide white light contain an active element that emits at short wavelength in the visible region (e.g., blue emission) and another material that absorbs the blue light and converts it to longer wavelength light. The most common converters are phosphors (see Chapter 3 for further discussion). Whereas the conversion of electrical energy to light in an incandescent light is about 15% efficient, the conversion of electrical energy to light for an LED is about 80% efficient.

m. If the active element of the LED emits blue light, what color(s) of light should be emitted by the phosphor in order to make the overall LED produce light that appears white?

n. A problem was encountered in the use of LED traffic lights in the midwestern United States in the winter. The lights last longer and are more efficient than the previously used incandescent lights, but snow can accumulate on the LED lights and block the view of them. Why is this a problem with the LEDs and not with incandescent lights?

o. Energy is conserved in all processes, so where does the "lost" energy go from the inefficient conversion of electrical energy to light for an incandescent bulb?

2.9 Learning Goals

- Electromagnetic spectrum
- Transmission, reflection, absorption, emission
- Energy levels
- Complementary colors
- Atomic transitions
- Black-body radiation
- Vibrational transitions
- Crystal field
- Point and line defects in crystals
- Color centers
- Holes
- Charge delocalization and molecular orbitals
- Luminescence (fluorescence, phosphorescence, bioluminescence, chemiluminescence)
- Lasers
- Types of home and office lighting

2.10 Problems

2.1 If six paints (red, orange, yellow, green, blue, and violet) are mixed in equal portions, the resulting paint is black. However, if six lights (red, orange, yellow, green, blue, and violet) of equal intensity shine at the same spot, the resultant spot is white. Explain why two different "colors" (white versus black) arise.

2.2 The jadeite form of jade is composed of $NaAl(SiO_3)_2$, and its green color results from the presence of a small amount of iron.

a. It would appear that the color in jade could be caused by a color center or a transition metal absorption. Describe the source and physical processes involved in each.

b. Suggest an experiment that would distinguish between the above possibilities. Explain what you would do, what you would expect to observe in each case (and why), and how this would distinguish between the two possibilities.

2.3 The equation describing radiant power (*M*) for black-body radiation as a function of temperature (*T*) and wavelength (λ) is given by the Planck energy distribution:

$$M = \frac{2\pi hc^2}{\lambda^5} \left\{ \frac{1}{e^{\left(\frac{hc}{kT\lambda}\right)} - 1} \right\} \tag{2.3}$$

where *h* is Planck's constant, *k* is Boltzmann's constant, and *c* is the speed of light.

a. Calculate *M* for *T*=1000 K at the following wavelengths (in m): 10^{-4}, 10^{-5}, 7×10^{-6}, 5×10^{-6}, 3×10^{-6}, 2×10^{-6}, 1.3×10^{-6}, 10^{-6}, 9×10^{-7}, 8×10^{-7}, 10^{-7}. Show a sample calculation (complete with working out of units, three significant figures in *M*) and put the results in a table that includes $\log_{10}(\lambda/\text{m})$.

b. Calculate *M* for *T*=2000 K at the following wavelengths (in m): 10^{-4}, 10^{-5}, 3×10^{-6}, 2×10^{-6}, 1.5×10^{-6}, 10^{-6}, 9×10^{-7}, 8×10^{-7}, 7×10^{-7}, 6×10^{-7}, 5×10^{-7}, 3×10^{-7}. Put the results in a table that includes $\log_{10}(\lambda/\text{m})$.

c. Calculate *M* for *T*=4000 K at the following wavelengths (in m): 10^{-4}, 10^{-5}, 3×10^{-6}, 2×10^{-6}, 1.5×10^{-6}, 10^{-6}, 9×10^{-7}, 8×10^{-7}, 7×10^{-7}, 6×10^{-7}, 5×10^{-7}, 3×10^{-7}, 2×10^{-7}. Put the results in a table that includes $\log_{10}(\lambda/\text{m})$.

d. On the same graph, plot the results of (a), (b), and (c). Plot the graph as *M* versus $\log_{10}(\lambda/\text{m})$. Use different symbols for the three sets of results and draw smooth curves. Use three sets of *y*-axes, one for (a), one for (b), and one for (c), so that each curve nearly fills the page. The zero line on the *y*-axis should be the same for all curves.

e. Mark the visible light region on your graph.

f. Comment, based on your graph, on the relative areas under the three peaks. If they are different, explain why.

g. Comment, based on your graph, on the color(s) of black-body radiation from an object at *T*=1000, 2000, and 4000 K.

2.4 Rubies are red because of the absorption of light by Cr^{3+} ions. The color of emeralds also is because of absorption of light by Cr^{3+} ions; emerald is $Be_3Al_2Si_6O_{18}$ doped with Cr^{3+}. Given that emeralds are green, explain whether the crystal field experienced by Cr^{3+} is less in emeralds or in ruby. Refer to Figure 2.9.

2.5 Some types of old glass contain iron and manganese. After many years of intense exposure to sunlight, this type of glass turns violet through the development of color centers. Why is exposure to light needed?

2.6 Some materials develop color from exposure to radiation. This coloration is due to the formation of color centers, and the number formed is proportional to the radiation dosage received. Suggest a method that could be used to quantify the radiation exposure, leaving the material in "new condition" to be used again.

2.7 When heated, some materials, such as diamond, give off a faint glow that decays with time. This effect is called *thermoluminescence*. Suggest a plausible reason for it. Explain how thermoluminescence could be used to determine the age and location of geological materials.

2.8 An old "magic" trick involves taking a seemingly normal piece of pale pink cloth and holding it over a glowing incandescent light bulb. While the audience watches, the cloth turns blue. The "magician" can return its color to pale pink by blowing on it. The cloth has been treated by dipping it in cobalt (II) chloride. $CoCl_2$ is a blue salt, and $CoCl_2 \cdot 2H_2O$ and $CoCl_2 \cdot 6H_2O$ are red salts. Suggest an explanation for the "magic" trick, including the physical basis for the origins of the colors.

2.9 A large organic molecule with an acidic group can be used as an acid/base indicator. It has an equilibrium between its acid form (call it HA for RCOOH) and its conjugate base (call it A⁻ for RCOO⁻), as follows:

$$R-\overset{\displaystyle O}{\overset{\displaystyle ||}{C}}-OH \rightleftarrows H^+ + R-\overset{\displaystyle O}{\overset{\displaystyle ||}{C}}-O^-$$

In the presence of acid, the HA form is favored. In the presence of base, the A⁻ form is favored (as H⁺ is consumed, pulling the equilibrium to the right). The color depends on pH such that in one pH range the color is red and in another it is blue.

a. In the red form, what color is absorbed from white light?

b. In the blue form, what color is absorbed from white light?

c. Are electrons in organic acids more delocalized in the acid form or in the base form? Why?

d. When the electrons are more delocalized, does this cause the absorption of visible light to be at lower or higher energy? Explain briefly.

e. On the basis of your answers to (a), (b), (c), and (d), which is red and which is blue (of HA and A⁻)? Explain.

2.10 It is possible to make a dill pickle "glow" by attaching both ends to leads to a 110-V power source. (Safety note: This experiment requires safe handling procedures; see *Journal of Chemical Education*, 1993, 70,

250, and *Journal of Chemical Education*, 1996, 73, 456, for details of a safe demonstration.) The glow is yellow. What is the origin of the color?

2.11 What color would a flower that is white in daylight appear to be when observed in red light? Explain.

2.12 Most clothing stores have fluorescent lighting. In this light, two items can appear to be color-matched but, when viewed in sunlight, they are not well matched. Explain.

2.13 The window panes in some very old houses have a violet tint. Faraday noticed that this color was because of the effect of sunlight on the glass, and it has been determined that this effect requires a small amount of manganese. If you heat the glass, the color disappears, and it does not reappear on cooling. What is the cause of the color?

2.14 Stars with different surface temperatures have different colors (e.g., bluish-white, orange, red, yellow, white). Arrange these star colors in order of increasing star surface temperature.

2.15 By differentiating the expression for $M(\lambda)$ given in Equation 2.3, show that λ_{max}, the wavelength of the most intense black-body radiation, is inversely proportional to the temperature, that is

$$\lambda_{max} = \frac{0.0029 \text{ m K}}{T} \tag{2.4}$$

where T is in kelvin and λ_{max} is in m. Equation 2.4 is known as Wien's law.

2.16 Some types of white plastic turn from white to yellow over time. Explain the physical origin of the color, including an explanation of the yellow color and why it has turned yellow rather than, for example, red.

2.17 A convenient type of thermometer determines body temperature by placement of a probe in the ear. The device must be placed accurately in the ear canal as the thermometer measures infrared energy from the eardrum and converts this into a temperature reading. Using the Stefan–Boltzmann law, the radiant power (M) for black-body radiation is given by

$$M = \sigma T^4 \tag{2.5}$$

where T is the temperature in kelvin, and σ, the Stefan constant, has a value of 5.67×10^{-8} W m^{-2} K^{-4}, calculate the increase in energy flux with an increase in temperature from 37 °C to 38 °C.

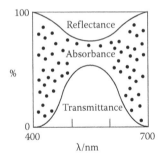

FIGURE 2.16
Proportions of light that are absorbed, transmitted, and reflected as functions of wavelength.

2.18 Some polymers, such as polycarbonate, change color when exposed to gamma radiation. Explain the physical principle that gives rise to the color change and suggest an application of this property.

2.19 A glass head representing a Pharaoh, made about 3400 years ago, is presently housed in an American museum. The work of art was once bright blue but now is faded. Discuss possible origins of the color and reasons why it has changed.

2.20 From the data given in Figure 2.16 for a particular object, what is the observed color of the object (i.e., observed in reflected light; assume that the surface is smooth so that scattering of light can be neglected)? Explain your reasoning.

2.21 Some brown or yellow topaz (aluminum fluorosilicate) gemstones have their color created artificially by irradiation, and this treatment has been controversial because some of the resulting gemstones have been left highly radioactive. Suggest the origin of the color.

2.22 Would a mercury-based fluorescent light be expected to work at very low temperature? Explain.

2.23 On the basis of Equation 2.2 and the data in Section 2.4, what percentage of red light would be absorbed by a 1-cm path length of water?

2.24 The flame from a glass-blowing torch is yellow-orange. As a safety measure to reduce the intensity of light that reaches a glassblower's eyes, what color should the glassblower's glasses be? Explain your reasoning briefly.

2.25 A commercial handheld digital infrared thermometer uses emissivity to measure surface temperatures of materials such as asphalt, ceramics, and highly oxidized metals. The thermometer does not work on aluminum or other shiny metals. Why not?

2.26 Some mail is now irradiated with intense electromagnetic radiation to kill biologically active substances. It is said that some precious gems are colored after this radiation and should not be sent through the mail. (Not that you should send a precious gem in the mail anyhow!)

a. Explain the origin of the color.

b. Suggest a possible means of removing the color.

2.27 A fleck of ruby can be added as a calibrant to high-pressure spectroscopic experiments to determine the pressure. Refer to Figure 2.9 to explain why the fluorescence of ruby depends on the pressure. Does the fluorescence of ruby increase or decrease in wavelength as the pressure increases? Explain.

2.28 Thermochromic materials change color with temperature. Many commercially available thermochromic mixtures are composed of a mixture: A dye that can

> See the video "Thermochromic Ink" under Student Resources at PhysicalPropertiesOfMaterials.com

change color depending on its surroundings, and a developer, and a solvent. The solvent is the main component by mass, and when it melts, the developer can interact in a different way than in the solid state. If the developer interacts preferentially with the dye, then the material is colored. If the developer interacts preferentially with the solvent, then there is no color. The color change can be reversible or irreversible. The former mechanism is used to make toner or ink that can be erased by heating. A typical dye is crystal violet lactone, which goes from a spirolactam ring-closed form (uncolored) to a ring-open zwitterionic form (intensely blue), where the latter is stabilized by the presence of developer. In which form do you expect the dye to have its electrons more delocalized? Explain your reasoning.

2.29 A typical LED light consumes 10 W, whereas a typical incandescent lamp uses 60 W to provide the same amount of light. Given that the energy used for lighting is about 10% of the worldwide energy consumption, and the latter is about 6×10^{20} J per year, how many J can be saved in a year by converting all lamps to LEDs from incandescent? What other factors should you consider before deciding on such a conversion?

2.30 The 2008 Nobel Prize in Chemistry was awarded to Osamu Shimomura (1928–), Martin Chalfie (1947–), and Roger Tsien (1952–2016) for the discovery and development of green fluorescent protein. This protein is important because it can be linked to other proteins to study biological processes in cells. Draw a simple energy level diagram for a material that exhibits green fluorescence.

Further Reading

General References

Most introductory physical chemistry textbooks describe aspects of electromagnetic radiation and black-body radiation. Books on rocks and minerals often contain information concerning color and composition of geological materials.

P. Bamfield and M. G. Hutchings, 2010. *Chromic Phenomena*, 2nd ed. Royal Society of Chemistry, Cambridge.

R. S. Berns, 2000. *Billmeyer and Saltzman's Principles of Colour Technology*, 3rd ed. John Wiley & Sons, Hoboken, NJ.

C. L. Braun and S. N. Smirnov, 1993. Why is water blue? *Journal of Chemical Education*, 70, 612.

W. D. Callister, Jr. and D. G. Rethwisch, 2013. *Materials Science and Engineering: An Introduction*, 9th ed. John Wiley & Sons, Hoboken, NJ.

R. M. Evans, 1948. *An Introduction to Color*. John Wiley & Sons, Hoboken, NJ.

M. Gouterman, 1997. Oxygen quenching of luminescence of pressure sensitive paint for wind tunnel research. *Journal of Chemical Education*, 74, 697.

P. Gregory, 2000. Colouring the jet set. *Chemistry in Britain*, August, 39.

M. Jacoby, 2000. Bright prospects for organic lasers. *Chemical and Engineering News*, July 31, 10.

A. Javan, 1967. The optical properties of materials. *Scientific American*, September, 239.

N. M. Johnson, A. V. Nurmikko, and S. P. DenBaars, 2000. Blue diode lasers. *Physics Today*, October, 31.

D. C. MacLaren and M. A. White, 2005. Design rules for reversible thermochromic mixtures. *Journal of Materials Science*, 40, 669.

M. J. McKelvy, P. Mitian, K. Hintze, E. Patrick, K. Allagadda, B. L. Ramakrishna, C. Denny, B. Pryor, A. V. G. Chizmeshya, and V. Pizziconi, 2000. Why does a light bulb burn out? *Journal of Materials Education*, 22, 5.

C. G. Mueller, M. Rudolph, and the Editors of LIFE, 1966. *Light and Vision*. Life Science Library, Time Inc.

K. Nassau, 1980. The causes of colour. *Scientific American*, October, 124.

K. Nassau, 2001. *The Physics and Chemistry of Colour*, 2nd ed. Wiley-Interscience, Hoboken, NJ.

T. Quickenden and A. Hanlon, 2000. The colours of water and ice. *Chemistry in Britain*, December, 37.

T. D. Rossing and C. J. Chiaverina, 1999. *Light Science: Physics and the Visual Arts*. Springer, New York.

H. Rossotti, 1983. *Colour*. Princeton University Press, Princeton, NJ.

B. Schwarzschild, 2006. Mather and smoot share nobel physics prize for measuring cosmic microwave background. *Physics Today*, December, 18.

R. Tilley, 2011. *Colour and the Optical Properties of Materials*, 2nd ed. John Wiley & Sons, Hoboken, NJ.

D. R. Tyler, 1997. Organometallic photochemistry: Basic principles and applications to materials science. *Journal of Chemical Education*, 74, 668.

V. F. Weisskopf, 1968. How light interacts with matter. *Scientific American*, September, 60.

M. A. White, 1998. The chemistry behind carbonless copy paper. *Journal of Chemical Education*, 75, 1119.

M. A. White and M. LeBlanc, 1999. Thermochromism in commercial products. *Journal of Chemical Education*, 76, 1201.

C. Wolinsky, 1999. The quest for color. *National Geographic*, July, 72.

G. Wyszecki and W. S. Stiles, 2000. *Color Science*, 2nd ed. Wiley-Interscience, Hoboken, NJ.

Color of Water

C. L. Braun and S. N. Smirnov, 1993. Why is water blue? *Journal of Chemical Education*, 70, 612.

T. Quickenden and A. Hanlon, 2000. The colours of water and ice. *Chemistry in Britain*, December, 37.

Carbonless Copy Paper

M. A. White, 1998. The chemistry behind carbonless copy paper. *Journal of Chemical Education*, 75, 1119.

Lasers

N. M. Johnson, A. V. Nurmikko, and S. P. DenBaars, 2000. Blue diode lasers. *Physics Today*, October, 31.

Thermochromism

D. C. MacLaren and M. A. White, 2005. Design rules for reversible thermochromic mixtures. *Journal of Materials Science*, 40, 669.

M. A. White and M. LeBlanc, 1999. Thermochromism in commercial products. *Journal of Chemical Education*, 76, 1201.

Websites

For links to relevant websites, see PhysicalPropertiesOfMaterials.com.

3

Color in Metals and Semiconductors

3.1 Introduction

In the previous chapter, we looked at atomic and molecular origins of color, emphasizing insulating materials. For the most part, the principles involved only a single atom (e.g., an electronic transition) or a small group of atoms (e.g., colors due to electronic transitions in molecules and crystal field effects). In this chapter we look at colors and luster in metals and origins of color in semiconductors. In both metals and semiconductors, very large numbers of atoms are usually required for the presence of color.

3.2 Metallic Luster

Electrons have their maximum possible delocalization in metals, where the *free electron gas model* (i.e., electrons modeled as virtually floating in a sea of positive charges) has been used to describe many features.

In a metal, all the conduction electrons are essentially equivalent because they can freely exchange places with one another, but they do not all have the same energy, largely due to the quantum mechanical limitation that no two electrons in the same system (meaning the metal, in this case) can have the same set of quantum numbers. Therefore, for a chunk of metal, there will be of the order of 10^{23} electrons and nearly as many energy levels. (If all the levels were filled, there would be half as many levels as electrons since each level can accommodate only two electrons, one of each spin. This is in keeping with the *Pauli* exclusion principle, which states that no two electrons in a system can occupy the same quantum state simultaneously.) With this many energy levels in a metal, its energy ladder is not discrete and can be considered to be a virtual continuum.

---

* Wolfgang Pauli (1900–1958) was born in Vienna and studied under Arnold Sommerfeld, Max Born, and Niels Bohr. Pauli helped to lay the foundations of quantum theory and was awarded the Nobel Prize in Physics in 1945 for expounding the exclusion principle.

FIGURE 3.1
Boltzmann's gravesite in Vienna. (Photo by
Robert L. White. With permission.)

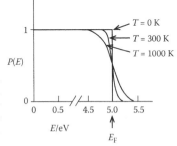

FIGURE 3.2
Probability distribution for electrons in a metal,
according to the Fermi–Dirac distribution function
(Equation 3.1), at various temperatures. E_F is the Fermi
energy, where E_F separates the filled levels from the
unoccupied levels at $T = 0$ K. Note that at all temperatures, the probability of occupation is ½ at $E = E_F$. Here,
$E_F = 5.0$ eV.

Fermi[*] and Dirac[†] worked out that $P(E)$, the probability that a state with
energy E will be occupied in a free electron gas, is

$$P(E) = \frac{1}{e^{(E-E_F)/kT} + 1},\tag{3.1}$$

where E_F, the *Fermi energy*, is the energy at which $P(E) = ½$, and k is the
Boltzmann[‡] constant, and T is temperature. Equation 3.1 is known as the
Fermi–Dirac distribution function.

This function looks "square" at $T = 0$ K, as one would expect since all the
energy levels are filled up one by one at $T = 0$ K, up to the Fermi energy. This
occupation is shown in Figure 3.2.

---

[*] Enrico Fermi (1901–1954) was an Italian-born physicist who made major contributions to the
foundations of quantum mechanics and the first controlled self-sustaining nuclear reactor;
he was the winner of the 1938 Nobel Prize in Physics.

[†] Paul A. M. Dirac (1902–1984) was a British theoretical physicist noted for his contributions to
quantum mechanics. Dirac shared the 1933 Nobel Prize in Physics with Schrödinger. From
1932 to 1969, Dirac was the Lucasian Professor of Mathematics at Cambridge University a
position previously held by Sir Isaac Newton.

[‡] Ludwig Boltzmann (1844–1906) was an Austrian theoretical physicist. Boltzmann was
an engaging lecturer who demanded discipline of his students, and, in return, they were
devoted to him. Although revered today for his work in the areas of thermodynamics, statistical mechanics, and kinetic theory of gases, Boltzmann's work was not well received by his
colleagues in his lifetime. This was largely due to his espousal of the unfashionable corpus-
cular theory of matter. Boltzmann took his own life in 1906. Today his grave site, in the same
Viennese cemetery where Beethoven lies, bears Boltzmann's famous equation: $S = k \log \omega$
(Figure 3.1).

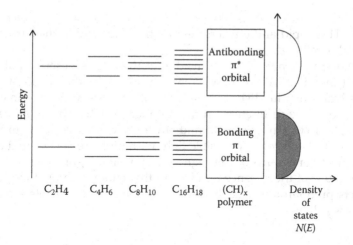

FIGURE 3.3

In going from localized molecular orbitals (smaller molecules, left side of diagram) to delocalized molecular orbitals (larger molecules, right side of diagram), electronic energy levels become more numerous until they become a virtual continuum, as shown schematically here. The *density of states*, $N(E)$, is defined as $N(E)\ dE$, i.e., the number of allowed energy levels per unit volume in the range between E and $E+dE$. $N(E)$ is zero in the *band gap*, the forbidden region between the bonding and antibonding orbitals in this case. Metals have energy bands but no band gap (see Figure 3.2).

COMMENT: BAND STRUCTURE

In going from localized molecular orbitals to delocalized molecular orbitals, the increased number of electrons leads to increasingly numerous electronic energy levels until they become a virtual continuum, as shown schematically as bands in Figure 3.3.

At higher temperatures, some energies above the Fermi energy are occupied. The spillover (to states beyond E_F) depends on T, as shown in Figure 3.2.

Because a metal has a near continuum of excited energy levels (energies above the Fermi energy), a metal can absorb light of any wavelength, including visible wavelengths. If the absorption were the only effect operable here, all metals would appear black because all visible light would be absorbed. However, when an electron in a metal absorbs a photon of light, it is promoted to an excited state that has many other energy levels available for de-excitation (due to the virtual continuum of levels). Light can be absorbed easily, and since light is an electromagnetic wave, and a metal is a good conductor of electricity, the absorbed light induces alternating electrical currents on the metal surface. These currents rapidly emit light out of the metal. This rapid and efficient re-radiation means that the surface of a metal is *reflective*.

Although all metals are shiny, due to the effects described above, not all metals have the same color. For example, the colors of gold and silver are quite distinct from each other. This difference in color is due to subtle differences in the number of states above the Fermi level. The reflectances of gold and silver are compared in Figure 3.4. Silver reflects all colors of visible light with high efficiency; gold does not reflect very much of the high-energy visible light (blue and violet) because of the absence of levels in this energy region. Since gold reflects predominantly at the low-energy end of the visible range, it appears yellow.

COMMENT: POLISHING CHANGES COLOR

Almost every material has some degree of luster, as some incident light can be reflected back to the viewer. The degree of luster can depend on such matters as surface smoothness and electronic energy levels.

The change in the look of a piece of wood when it has been polished with wax or varnish is familiar. Two processes are taking place here to increase the luster and color intensity of the polished wood: a reduction in diffuse refection from the outer surface and introduction of multiple reflections under the polish.

The bare wood surface is not smooth, so much of the incident light is *diffusely scattered* (i.e., in all directions). However, the polished surface is much flatter, so it will reflect primarily at a particular angle (the same as the angle of incidence; this is called *specular reflection*) giving the wood its luster. The color of the polished surface also is more intense because the colored light from the rough wood surface will not be polluted with diffusely scattered light.

The polish layer also allows multiple internal scattering of the light (i.e., incident light reflected several times from wood and polish surfaces in the polish layer before escaping), as shown schematically in Figure 3.5. Multiple reflections narrow the wavelength range of the observed color, and this again intensifies the richness of a particular hue.

It is possible for metals to transmit light if the sample is sufficiently thin. The transmission will vary with the wavelength of the light, and such

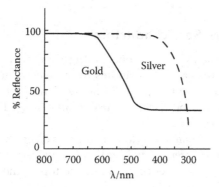

FIGURE 3.4

Reflectance spectra of gold and silver as functions of wavelength. Gold reflects red, orange, and yellow strongly but gives less reflection of blue and violet, and therefore it appears yellow. In contrast, silver reflects incident light over most of the visible region of the spectrum (wavelengths 400–700 nm), and appears white. (Adapted from A. Javan. *Scientific American*, September 1967, 244.)

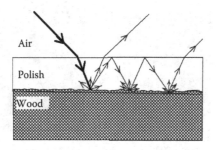

FIGURE 3.5

A schematic view showing some of the paths of a ray of light striking a polished wood surface. Specular refection and multiple internal scattering are emphasized in this diagram, and diffuse reflection at the polish-wood interface is also shown.

transmission can give rise to color. A thin gold film (100 nm or less) will transmit blue-violet light, again because blue-violet is not of the appropriate energy to be reflected.

On the other hand, if very small particles of gold, typically 40–140 nm in size, are suspended in glass, the electrical currents required for high reflectivity in a bulk metal cannot develop, and the resulting red color is due to the absorption of green light in a process called *Mie scattering*. This material is commonly called "ruby glass" due to its red color, although it is chemically distinct from the gem ruby.

Therefore, for a metal, the color observed depends on both composition and sample dimensions.

3.3 Colors of Pure Semiconductors

If an electron in a metal is at an energy higher than the Fermi energy, it is said to be conducting as it contributes to the electrical and thermal conductivities of a metal (see Chapters 8 and 12). The probability diagram of Figure 3.2 can be turned on its side to reveal an energy diagram for a metal as shown in Figure 3.6.

Materials called *semiconductors* do not conduct electricity very well due to the gap between their *valence band* and their *conduction band*, as shown in Figure 3.7. The energy difference between the top of the valence band and

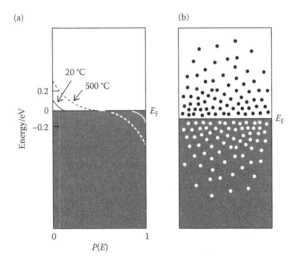

FIGURE 3.6
Other ways to view the Fermi–Dirac distribution for electrons in a metal: (a) occupational probability, $P(E)$, for energy levels in a typical metal at various temperatures; (b) occupational probability schematically for $T > 0$ K, with ● representing excited electrons and ○ representing holes.

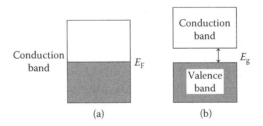

FIGURE 3.7
Energy bands in (a) a metal, and (b) a pure semiconductor. E_F is the Fermi energy and E_g is the energy gap.

the bottom of the conduction band, designated E_g, is the *energy gap* (or *band gap*); there are no energy levels in this region in pure semiconductors. A similar gap appeared in the energy range between the bonding and antibonding orbitals (see Figure 3.3). The semiconductor energy gap changes from element to element; for example, it increases on going through the series Sn, Ge, Si, C. This order correlates with increasing bond interaction, and shorter, stronger bonds lead to a larger band gap energy. For a given element, application of pressure or dramatic lowering of temperature can shorten the interatomic distance and increase the band gap.

If there is sufficient thermal energy available, large numbers of electrons in a semiconductor could be promoted from the valence band to the conduction band, but this thermal energy might correspond to a temperature in excess of 1000 K, and the material could very well melt before achieving electrical conduction comparable to that in a metal.

However, visible light may provide sufficient energy to promote electrons from the valence band to the conduction band of a semiconductor, and this absorption gives semiconductors their colors.

For a pure semiconductor, the color observed depends on the value of E_g. If E_g is less than the lowest energy of visible light (i.e., lower energy than red light, $\lambda \sim 700$ nm, $E \sim 1.7$ eV), then any wavelength of visible light will be absorbed by this semiconductor. (Excess energy beyond E_g will be used to promote the electrons higher in the conduction band.) Because all visible light is absorbed, this semiconductor will appear black. If reemission is rapid and efficient (this depends on the energy levels), then this material will have a metal-like luster; Si is an example. If reemission is not rapid, then this semiconductor will appear black and lusterless (e.g., CdSe).

If E_g is greater than the highest energy of visible light (i.e., violet light, $\lambda \sim 400$ nm, $E \sim 3.0$ eV), then no visible light is absorbed and the material is colorless. Pure diamond, with E_g of 5.4 eV (corresponding to light of wavelength 230 nm, i.e., UV), is an example.

When a semiconductor has E_g falling in the energy range of visible light, the color of the material depends on the exact value of E_g. The value of E_g depends on two factors: the strength of the interaction that separates bonding and antibonding orbitals, and the spread in energy of each band. Periodic trends in lattice parameters, bond dissociation energies, and band gap are summarized for one isostructural group of the periodic table in Table 3.1. The shorter, stronger bonds can be seen to give rise to larger values of E_g.

Let us now consider the observed colors of some pure semiconductors. For example, HgS (called by the mineral name *cinnabar*, or the pigment name *vermilion*), has E_g of 2.1 eV, which corresponds to $\lambda = 590$ nm, that is, yellow light. Therefore, in HgS all light with energies greater than 2.1 eV (wavelengths shorter than 590 nm, i.e., green, blue, and violet) is absorbed. This means that, in white light, only wavelengths greater than 590 nm are transmitted, and HgS appears red.

TABLE 3.1

Periodic Trends in Diamond Structures in Group 14

Element	Lattice Parameter/Å	Bond Dissociation Energy/(kJ mol⁻¹)	E_g/eV
C (diamond)	3.57	346	5.4
Si	5.43	222	1.1
Ge	5.66	188	0.66
α-Sn	6.49	146	0.1

TABLE 3.2

Some Examples of Colors and Band Gaps in Pure Semiconductors

Material	Color	E_g/eV
C (diamond)	Colorless	5.4
ZnS	Colorless	3.6
ZnO	Colorless	3.2
CdS	Yellow-orange	2.4
HgS	Red	2.1
GaAs	Black	1.43
Si	Metallic gray	1.11

As another example, consider CdS. Here the band gap energy is 2.4 eV, which corresponds to light of wavelength 520 nm (blue light). Since blue and higher-energy (i.e., violet) visible light are absorbed, this leaves behind a region of the visible spectrum that is centered on yellow, and CdS appears yellow-orange.

To generalize, the colors of pure semiconductors in order of increasing E_g, are black, red, orange, yellow, and colorless. Some examples are given in Table 3.2.

If the semiconductor has a large gap, it will be colorless if it is pure, but the presence of impurities can introduce color in the semiconductor, as we will see in the next section.

3.4 Colors of Doped Semiconductors

There are two categories of impurities in semiconductors, and these are defined here. These definitions are also important to discussions of electrical properties of semiconductors (see Chapter 12).

An impurity that donates electrons to the conduction band of a semiconductor creates what is known as an *n-type semiconductor*. The "n" stands

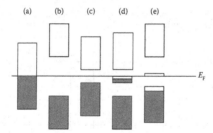

FIGURE 3.8
Simple view of energy bands in (a) a metal, (b) an insulator, (c) a pure semiconductor, (d) an n-type semiconductor, and (e) a p-type semiconductor. E_F is the Fermi energy.

for negative as this impurity gives rise to negative charge carriers (electrons in the conduction band). This doping (i.e., purposeful addition of an impurity) is also known as a *donor impurity*.

Conversely, an impurity that produces electron vacancies that behave as positive charge centers (holes) in the valence band leads to a *p-type semiconductor*. The "p" stands for positive as the holes act as positive charge carriers. Because the impurity accepts an electron (giving rise to the hole), this also is known as an *acceptor impurity*.

The impurity in a semiconductor often can introduce a set of energy levels intermediate between the valence band and the conduction band, i.e., in the gap, as shown in Figure 3.8. Because the transition to such impurity levels takes less energy than the transition across the entire band gap, this transition can influence the color of a wide-band-gap semiconductor.

For example, if a small number of nitrogen atoms are substituted into the diamond lattice, the extra electrons (nitrogen has one more electron than carbon) can form a donor impurity level within the wide band gap of diamond. Excitation from this impurity band absorbs violet light, depleting white light of violet and making nitrogen-doped diamonds appear yellow at nitrogen levels as low as 1 atom in 10^5. As discussed above, pure diamond has a wide band gap and is colorless because visible light is not absorbed.

Conversely, boron has one electron less than carbon, and doping boron into diamond gives rise to acceptor (hole) levels in the band gap. These intermediate levels can be reached by absorption of red light (but not shorter wavelength light—do you know why?) so boron-doped diamonds appear blue. The Hope diamond is an example.

Doped semiconductors also can act as *phosphors*, which are materials that give off light with high efficiency when stimulated, for example, by an electrical pulse. (*Phosphorescence* results when light is emitted some time after the initial excitation; the delay is caused by the change in spin between the excited state and the ground state. In *fluorescence*, which is a faster process, the spins of the excited and ground states are the same.) On the surface of the cathode ray tube (CRT) of a traditional color television, there are three phosphors: one emitting red light, one emitting green light, and one emitting

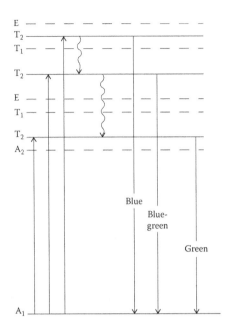

FIGURE 3.9
Schematic view of phosphorescence due to emission from excited states in the blue phosphor ZnS doped with Ag^+. (Adapted from Y. Uehara, 1975. *Journal of Chemical Physics*, 62, 2983.)

blue light; emission is stimulated by excitation with an electron beam. The energy level diagram for the blue phosphor ZnS doped with Ag^+ is shown in Figure 3.9.

COMMENT: TRIBOLUMINESCENCE

Some materials emit light when they are stimulated mechanically; this phenomenon is known as *triboluminescence*. One of the best-known examples is Wint-O-Green Lifesavers®, which can give rise to flashes of blue-green light when crushed. The sugar leads to triboluminescence due to the production of large electric fields during the formation of a newly charged crystal surface, the acceleration of electrons in this field, and the subsequent production of *cathodoluminescence* (fluorescence induced by cathode rays that energize a phosphor to produce luminescence) from the nitrogen atoms (from the air) adsorbed on the surface. The blue luminescence is associated with N_2^+. Some of the light emitted during this process is in the UV range, which then stimulates the wintergreen (methyl salicylate) molecules to fluoresce in the visible range. Therefore, the presence of wintergreen flavoring enhances the observed triboluminescence.

When an n-type semiconductor is placed next to a p-type semiconductor, interesting features can result. Here we concentrate on color, but in Chapter 12 we will look at the importance of *p,n-junctions* to electronic properties of semiconductors. When an electrical potential is applied to a p,n-junction such that electrons are supplied through an external circuit to the n-side of the junction, the extra electrons in the conduction band of the n-type semiconductor drop down in energy (to the valence band) and neutralize the holes in the valence band of the p-type semiconductor, as shown in Figure 3.10. As the electrons go to the lower energy state, they can

> See the video "Phosphorescent Paper" under Student Resources at PhysicalPropertiesOfMaterials.com

emit energy in the form of light. This is the basis of the electronic device known as the *light-emitting diode* (LED), and it is $GaAs_{0.6}P_{0.4}$ that emits the red light that is commonly associated with these devices in applications such as digital displays. An LED shows *electroluminescence*, that is, the production of light from electrical stimulus. Various colors can be emitted depending on the materials; see Table 3.3.

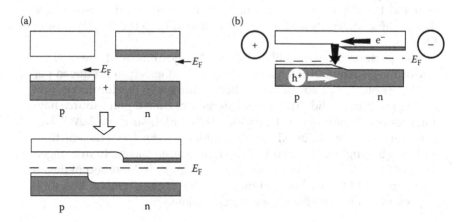

FIGURE 3.10
The electronic band structure of a p,n-type junction forms a light-emitting diode (LED). (a) When a p-type semiconductor is placed beside an n-type semiconductor, the Fermi levels equalize. (For simplicity, the donor and acceptor levels are assumed to be negligible here.) (b) When a field is applied to a p,n-junction as shown, electrons in the n-conduction band migrate to the positive potential on the p-side and fall down to the valence band, recombining with the holes in the p-semiconductor. Similarly, the holes in the p-type semiconductor migrate toward the negative n-side and recombine with electrons. The electrons dropping from the conduction band to the valence band emit light corresponding to the energy of the band gap. For example, the emission from $GaAs_{0.6}P_{0.4}$ is red.

COMMENT: NEW DIRECTIONS IN LIGHT-EMITTING DIODES

Relatively recent advances in LEDs include emission of white light nearly as bright as a fluorescent lamp. This particular device uses layers of polymers, with blue, green, or red light, depending on the layer. The combination of blue, green, and red makes the emission appear white, extending the utility of LEDs. With readily available red and green LEDs, the stumbling block to white LEDs was the need for efficient blue LEDs, as invented by Nobelists Isamu Akasaki,[*] Hiroshi Amano,[†] and Shuji Nakamura.[‡]

Some other new LEDs are not limited to one color of emitted light. An LED composed of layers of cadmium selenide nanocrystals between layers of poly(*p*-phenylenevinylene) can have different colors, depending on the voltage and the preparation. At low voltage, CdSe is the light emitter, and the color can be changed from red to yellow by varying the size of the nanocrystals. (The latter is a consequence of quantum effects.) At higher voltages, the polymer is the emitter, and the emitted light is green. Materials like these are now used for display devices in which the color of the *pixel* (dot of light on a screen) is determined by the applied voltage.

Much excitement about LEDs involves the use of organic materials. *Organic light-emitting diodes* (OLEDs) are reduced power consumption and thereby improve efficiency relative to inorganic-based LEDs. The organic materials add mechanical flexibility to the displays using OLEDs.

The performance of LEDs has certainly improved over time. Performance is quantified as the number of lumens (lm) produced per watt (W) of energy consumed, where a lumen is the luminous flux emitted within a solid unit angle (1 steradian) from a point source having a uniform intensity of 1 candela. Early LEDs, such as GaAsP, had a performance of about 0.1 lm W^{-1}, making them less efficient than Edison's first light bulb (2 lm W^{-1}), and much less efficient than halogen lamps (20 lm W^{-1}) and fluorescent lights (80 lm W^{-1}). More recent developments in LEDs have brought them into the range of 200 lm W^{-1}, with expectations of further improvements to come.

---

[*] Isamu Akasaki (1929–) is a Professor at Nagoya University in Japan and co-winner of the 2014 Nobel Prize in Physics as co-inventor of a high-efficiency blue LED.
[†] Hiroshi Amano (1960–) is a Professor at Nagoya University in Japan, and co-winner of the 2014 Nobel Prize in Physics as co-inventor of a high-efficiency blue LED.
[‡] Shuji Nakamura (1954–) is a Professor at University of California, Santa Barbara, who, as a researcher at Nichia Chemicals in Japan, was co-inventor of a high-efficiency blue LED, for which he was co-winner of the 2014 Nobel Prize in Physics.

TABLE 3.3

Colors of LEDs

Semiconductor	$\lambda_{emission}$/nm	Color
GaAlAs/GaAs	880	IR
AlGaInP	663	Red
GaAsP/GaP	605	Orange
GaP	555	Green
GaN/SiC	470	Blue
InGaN/SiC	395	UV

COMMENT: PHOTOVOLTAIC SOLAR CELLS

The sun provides energy flux of about 300 W m^{-2} at the earth's surface, providing about 2×10^{24} J annually, i.e., much more than the world's energy consumption of about 600 EJ (*Note*: 1 EJ=1 exajoule=10^{18}J). Therefore, there is great interest in harvesting solar energy. Photovoltaic (PV) cells directly convert sunlight into electrical energy. Photovoltaics are based on semiconductors, and solar energy promotes electrons from the valence band to the conduction band. For example, phosphorous-doped silicon can be the n-type semiconductor, with boron-doped silicon as the p-type semiconductor. An electric field is formed when the n-doped and p-doped semiconductors come together to form a diode (see Chapter 12), and when light hits the device, electrons move to the n-side and holes move to the p-side, providing a current. Another important optical aspect of solar PVs is the antireflective coating needed to increase light absorbance by reducing losses due to reflection. Considerable materials research is focused on increased efficiency and improved material properties of the solar cell materials, at reduced costs. Single-crystalline silicon materials have efficiencies > 75% of the theoretical maximum; other PV materials include GaAs, GaInP, CdTe, perovskites, organic and polymeric materials, and quantum dot materials.

3.5 The Photocopying Process: A Tutorial

Semiconductors also are important in the photoelectric process that we know as photocopying. The photocopying process, or xerography (which means "dry copying," literally from the Greek *xeros graphy*), relies on a few basic physical principles and a fair bit of ingenuity. The process was developed by Chester Carlson in the mid-1930s.

One of the two main parts of a photocopy machine is a corona discharge wire, which has such a strong electric field, > 30,000 V cm^{-1}, that it ionizes air to produce O_2^+ and makes the air in its immediate vicinity glow green due to photons emitted by excited electrons. The other main part is a semiconductor. Traditionally, this was amorphous selenium, a semiconductor with a band gap of 1.8 eV. (Although new photocopiers use organic semiconductors, the principles are the same.)

a. What color is a bulk sample of amorphous selenium?

b. Monoclinic Se has a band gap of 2.05 eV. What color is it?

c. After you run a photocopying machine for quite a while, why does the surrounding air smell like ozone (O_3)?

d. After the corona wire passes by the drum that is coated with amorphous selenium, the Se is charged positively. Why?

e. In the next step, a very strong light shines on the document to be copied, and this light is then reflected onto the Se drum. Does the light uniformly strike the drum, or is there an "image" from the document on the drum? Explain.

f. Where the light "image" hits the Se drum, electron/hole pairs are formed. Why? (This process results in this area becoming electrically neutral since the drum is grounded, and the holes move to the ground and the electrons move to the surface where they locally neutralize the positive charge.)

g. Where the dark "image" hits the drum, the surface stays positively charged. Why is it not neutral everywhere?

h. Devise a way to convert *the latent electrical image* (i.e., electrical image that is present but not yet visible) on the Se drum to an image on paper.

i. Yet another form of Se has a band gap of 2.6 eV. Will this be useful as a direct replacement for amorphous selenium? What other factors could be important?

j. What areas of the xerographic process would you suggest for improvement or advancement?

A laser printer works on a very similar principle to a photocopier. In the printing process, the light, from a laser or a bank of LEDs, hits the drum in the printer. However, there is a major difference in the creation of an electrostatic image for photocopiers and most laser printers. In the photocopier, light reflects off the paper onto the photoreceptor corresponding to the white areas, but not the dark areas. The background is discharged, and the positive charge remains in the areas of the drum corresponding to the print. This method is called "write-white" since the light "whitens" the background. On the other hand, most laser printers use the reverse process, referred to as "write-black," in which the laser gives a pulse of light for every dot to be printed, and no pulse for a dot to be left unprinted. The areas to be left unprinted therefore have a positive charge, while the areas to be printed have a negative charge.

k. Do you expect the charge of the toner for a write-black laser printer to be the same as for a photocopier? Explain.

l. Why is the write-black process more efficient for a laser printer than the write-white process?

m. Although selenium was used in the traditional photocopier, now organic semiconductors are more commonly used for the photo-receptive element. Which do you expect is easier to use to make a photosensitive drum: selenium or an organic material? Explain.

3.6 Photographic Processes: A Tutorial

The basis of classic "wet" film photography is the reactivity of silver halides, especially silver bromide and silver chloride, in the presence of light.

A photographic film contains silver halide particles (typically 50–2000 nm in size) embedded in an emulsion. When the silver halide particles are illuminated by a light of sufficient energy, conduction band electrons and valence band holes are created efficiently. The electrons move through the silver halide particle until, eventually, one becomes localized at a defect site. This trapped electron can combine with an interstitial silver ion (Ag^+) to form a neutral silver atom (Ag). Because a single silver atom is unstable in the lattice, it is energetically favorable for additional electrons and interstitial Ag^+ ions to arrive at the trapped site, and their combination causes more silver atoms to form from Ag^+. Eventually, a stable silver cluster of three to five silver atoms forms; this step takes somewhere between 6 and 30 absorbed photons. At this site, there is said to be a stable latent image in the film.

a. When a photographic film is exposed to a white object, is the corresponding area of the film more or less subject to Ag cluster formation than the area corresponding to exposure to a black object?

b. The band gap in silver halides is about 2.7 eV. What is the corresponding wavelength of light? Does this explain why silver halides are sensitive to visible light?

c. After a film is exposed to a subject, why can it be handled in red light but not daylight?

d. After the latent image has been formed, the exposed film is placed in a developer that contains a reducing agent such as hydroquinone.

The regions of the latent image that have Ag clusters undergo further reduction of Ag^+ in their immediate region, "amplifying" the number of Ag clusters in highly exposed regions of the film by as much as a factor of 10^9. The unexposed grains of silver halide, unchanged by photographic development, are then removed by dissolution. Explain why the image formed on the film is a "negative" (dark subject appears light and vice versa).

e. How could a "positive" print be produced from the negative film? Consider the black and white process only at this stage.

f. Chemical and spectral sensitizing agents, often containing sulfur and/or gold and/or organic dye molecules, are present with the silver halide in the film. These create sensitive sites on certain surfaces of silver halide microcrystals, thereby sensitizing the film to sub-band-gap light. Films of different "speeds" (i.e., meeting different light-level requirements) exist; speculate on how films of different speeds can be made.

g. Formation of a color image is more complex than for black-and-white, although it still relies on silver halides. Simple color film contains three layers of emulsion, each containing silver halide and a coupler that is sensitive to particular colors. In the order in which light hits the film, the first layer is sensitive to blue, and next there is a filter that stops blue and violet. The second emulsion is sensitive to green; the third emulsion is sensitive to red. (In a good-quality color film, there will be many more emulsions to get truer colors, but the principle is the same.) What color is the filter that stops the blue?

h. In the development of color film, the oxidized developer reacts with each coupler to produce a precursor to an organic dye molecule. The colors of the dye molecules produced in the development process are the complements of the light that exposed the emulsions (i.e., orange color forms in the first layer since orange is the complement to blue; magenta [red-violet], the complement to green, forms in the second layer; cyan [bluish-green], the complement to red, forms in the third layer). Why is synthetic organic chemistry important in making color films?

i. Discuss the role of materials researchers in measurement and calculation of properties such as the effect of impurities and lattice defects on the mobility of ions and holes in silver halides, and topics such as the surface energies of different faces of silver halide crystals.

j. How could the printing process for color film work?

k. Speculate on how the Polaroid® (instant processing) process works.

l. The Polaroid® process depends rather critically on temperature. Why?

m. A digital camera does not use film. Instead, it has a semiconductor-based sensor that converts light into electrical charges. The sensor is a 2D array with many, many individual sensors (pixels). What type of material likely is used to convert light to electrical charges?

n. Many digital cameras use a charge-coupled device (CCD).* The CCD transports the charge across the chip, and an analog-to-digital converter then turns each pixel's value into a number corresponding to the amount of charge at each photosite. Why does a digital camera require a lot of memory?

o. Some cameras use a different type of sensor called a complementary metal oxide semiconductor (CMOS). CMOS devices use several transistors at each pixel to amplify and move the charge using more traditional wires. Since the CMOS signal is digital, analog-to-digital conversion is not required. CCD sensors give high-quality, low-noise images, while CMOS sensors are generally more susceptible to noise. Furthermore, each pixel on a CMOS sensor has several transistors located next to it, and many of the photons hit the transistors instead of the photodiode, making the light sensitivity of a CMOS chip lower than that of the CCD. Which technology, CMOS or CCD, do you expect to have lower power consumption?

3.7 Learning Goals

- Metallic luster
- Electronic energy distribution in metals (Fermi–Dirac distribution)
- Colors and reflectance of metals
- Electronic energy distribution in semiconductors (band theory)
- Colors of semiconductors
- Band structure
- Colors of doped semiconductors
- n-type dopants (donor impurity)
- p-type dopants (acceptor impurity)
- Light-emitting diode
- PV solar cells
- Photocopier and laser printer processes
- Photographic processes

---

* The 2009 Nobel Prize in Physics was awarded to Willard S. Boyle (1924–2011) and George E. Smith (1930–) for their invention of the CCD sensor.

3.8 Problems

3.1 The color of the heating coil in a toaster dies off slowly when the toaster is unplugged. However, the color of the LED in a clock dies off quickly when the clock is unplugged. Comment on the sources of the difference.

3.2 Two very different materials both absorb light of wavelength 580 nm, which corresponds to yellow light.
   a. One material is a pure semiconductor. What color is it? Explain your reasoning.
   b. The absorption in the other material (which is an insulator) is due to a transition metal impurity. What color is this material? Explain your reasoning.

3.3 Most camera film works on the silver halide process. Most silver halides are semiconductors with band gaps of around 2.7 eV ($\lambda = 460$ nm). There are films available that can be used for "infrared" photographs. Briefly explain how the silver halide process would have to be modified to give an infrared film.

3.4 Photographic films come in different speeds. High-speed film requires less light to obtain an image, whereas low-speed film requires more light. Some people worry about passing their film camera through the x-ray process at security counters in airports in case the x-rays damage the film. Is high-speed or low-speed film more susceptible to x-ray damage? Explain.

3.5 The blood of invertebrates in the phyla *Arthropoda* and *Mollusca* (e.g., crab, crayfish, lobster, snail, slug, octopus) contains hemocyanin as its oxygen carrier. Hemocyanin contains copper, which is bound directly to a protein. The oxygenated form, oxy-hemocyanin, contains bound oxygen, and this species absorbs light. There are three regions of maximum absorbance: one at a wavelength of 280 nm (due to the protein), and two others due to the copper one at 346 nm and one at 580 nm.
   a. Suggest the physical source of the color of their blood. Give your reasoning.
   b. What color is the blood of species with hemocyanin? Give your reasoning.

3.6 Explain how the optical properties of a semiconducting chromophore based on crystalline organic compounds can be influenced by different packing arrangements of the molecules in the solid state.

3.7 Lasers produce light by stimulated emission. Some lasers use semiconductors, with the light emission corresponding to the energy of the band gap. This makes the light for a given system fixed at one energy.

a. Suggest how the addition of impurities to a semiconductor laser material can be used to change the wavelength of the emitted light.

b. Can the wavelength be both shortened and lengthened using this approach? Explain.

c. Stimulated emission leading to lasing requires a high density of lasing centers. How will this affect the intensity of lasing from a doped semiconductor, compared with an intrinsic semiconductor with the same effective energy gap?

3.8 The compound yttrium vanadate with 5% of the Y^{3+} ions replaced by Eu^{3+} gives a very efficient red phosphor. This is the red phosphor that is commonly used as a red emitter in a CRT color television screen (the other emitters are blue and green); its emission is perceptively pure red radiation at 612 nm from its $^5D_0$ electronic state upon excitation. In such a television screen, what excites the phosphor causing it to emit light?

3.9 Photochromic glass changes its transparency depending on the light level. The principle involved is similar to the influence of light on silver halides in photography. This glass contains copper-doped silver halide crystallites. When UV light hits the glass, copper gives up an electron, which is captured by the silver ion, leading to aggregates of silver atoms that darken the glass. This glass becomes transparent again when the UV source is removed (e.g., on moving indoors). Propose a mechanism that would allow for this reversion to the original state.

3.10 Two walls are painted with orange paint; one is "plain" orange and the other is fluorescent orange. Use energy level diagrams to show the difference in the two paints.

3.11 A science museum is building a new exhibit concerning the photocopying machine. A museum employee suggested that part of the exhibit be a working photocopier with the front panel made of glass so that visitors can see the interior workings of the copier when in use. Explain whether the photocopier would work if you could see inside.

3.12 How would the color of a light-emitting diode be expected to change on cooling from room temperature (where it emits red light) to the temperature of the boiling point of liquid nitrogen ($T=77$ K)? Consider that the lattice will have contracted and how this will influence E_g.

See the video "LED in Liquid Nitrogen" under Student Resources at PhysicalPropertiesOfMaterials.com

3.13 J. D. Bernal wrote, "All that glisters ["glistens"] may not be gold, but at least it contains free electrons." Give an example that refutes this statement.

3.14 From the data in Table 3.1, what color is Ge? Explain your reasoning.

3.15 The pigment zinc white (ZnO) is white at room temperature (i.e., colorless with a white appearance due to scattering from the powder) and bright yellow when heated. Explain the color change.

3.16 Blue color can be created in diamond by addition of boron as a dopant, or by irradiation to create lattice defects. How could measurement of the electrical conductivity distinguish between these sources of color in a blue diamond?

3.17 Use Equation 3.1 to calculate the probability of occupation of energy levels from 1.5 eV to 2.5 eV (in 0.05 eV increments) for a free electron gas with a Fermi energy of 2.0 eV, at (a) $T = 300$ K and (b) $T = 1000$ K. Show a sample calculation and put your results in a table. Graph your results, with results for both temperatures on the same graph.

3.18 Diamonds that are synthesized in laboratory conditions are often very strongly colored yellow. This is said to be due to an elemental contaminant from the atmosphere.

a. Since pure diamond is colorless, explain the origin of the yellow color. Begin by explaining why pure diamond is colorless and how the presence of the impurity affects the color.

b. Naturally produced diamonds also have the same impurity present, but over the very long time of their formation, these impurity atoms have migrated and clumped together. Explain why natural diamonds, then, are usually colorless (i.e., why this impurity does not lead to color when the impurity atoms are clumped together).

3.19 Lasers have been developed to give blue light. These make use of semiconductors. Is the band gap wider or narrower for a blue lasing semiconductor compared with a red lasing semiconductor material? Explain your reasoning.

3.20 It can be difficult to photocopy a document that has been written in pencil. The pencil color is dark and the light should be strongly absorbed, but it is not. Why not? (Consider the layered structure of graphite.)

3.21 The TELSTAR satellite, launched in 1962, suffered irreversible damage to its solar cells from an unexpectedly high number of electrons in its path in the Van Allen Belt surrounding the earth as a result of nuclear testing just before the TELSTAR launch. How did the extra electrons change the solar cells? Begin by explaining what category of material you expect to give a solar cell the property of converting sunlight to electrical energy, and your reason for that choice.

3.22 From Table 1.2, the embodied energy for high-purity Si for use in electronics is 6000 MJ kg^{-1}. About 3 kg of Si is required to manufacture 1 kW of conventional solar modules. How long must the modules operate (assume 8 h per day of solar use) to pay back the embodied energy of the Si?

3.23 Alexander Graham Bell worked on the invention of a "photophone" at his summer home lab in Baddeck, Nova Scotia, Canada. The idea of the photophone is to carry sound signals over space as light waves. The display on the photophone at the Bell Museum there reads: "The key to the photophone was the rare chemical element called selenium whose resistance to electrical current varied according to the amount of light which falls on it." Why does the electrical conduction of Se depend on the light it receives? Use an energy diagram to aid your explanation.

3.24 A pure semiconductor and another material for which color arises from a crystal-field effect both appear orange. Sketch the absorbance as function of wavelength across the visible range for both on the same graph in which you also indicate the colors corresponding to that wavelength.

Further Reading

General References

Many introductory physical chemistry and solid-state physics textbooks contain information concerning metals and semiconductors.

W. D. Callister, Jr. and D. G. Rethwisch, 2013. *Materials Science and Engineering: An Introduction*, 9th ed. John Wiley & Sons, Hoboken, NJ.

R. Cotterill, 2008. *The Material World*. Cambridge University Press, Cambridge.

M. de Podesta, 1996. *Understanding the Properties of Matter*. Taylor & Francis, Washington.

R. M. Evans, 1948. *An Introduction to Color*. John Wiley & Sons, Hoboken, NJ.

G. G. Hall, 1991. *Molecular Solid State Physics*. Springer-Verlag, New York.

A. Javan, 1967. The optical properties of materials. *Scientific American*, September, 239.

C. Kittel and H. Kroemer, 1980. *Thermal Physics*, 2nd ed. Freeman, New York.

G. C. Lisensky, R. Penn, M. J. Geslbracht, and A. B. Ellis, 1992. Periodic trends in a family of common semiconductors. *Journal of Chemical Education*, 69, 151.

G. J. Meyer, 1997. Efficient light-to-electrical energy conversion: Nanocrystalline TiO_2 films modified with inorganic sensitizers. *Journal of Chemical Education*, 74, 652.

K. Nassau, 1980. The causes of colour. *Scientific American*, October, 124.

K. Nassau, 2001. *The Physics and Chemistry of Color*, 2nd ed. John Wiley & Sons, Hoboken, NJ.

R. J. D. Tilley, 2011. *Colour and the Optical Properties of Materials*, 2nd ed. John Wiley & Sons, Hoboken, NJ.

R. J. D. Tilley, 2013. *Understanding Solids: The Science of Materials*, 2nd ed. John Wiley & Sons, Hoboken, NJ.

E. A. Wood, 1977. *Crystals and Light*. Dover Publications, New York.

G. Wyszecki and W. S. Stiles, 2002. *Color Science*, 2nd ed. Wiley-Interscience, Hoboken, NJ.
A. Zukauskas and M. S. Shur, 2001. Light-emitting diodes: Progress in solid-state lighting. *MRS Bulletin*, October, 764.

Color in Gems

W. Schumann, 2013. *Gemstones of the World*, 5th ed. Sterling, New York.

Lasers

M. G. D. Baumann, J. C. Wright, A. B. Ellis, T. Kuech, and G. C. Lisensky, 1992. Diode lasers. *Journal of Chemical Education*, 69, 89.

LEDs, Photovoltaics and Solar Cells

A. Bergh, G. Craford, A. Duggal, and R. Haitz, 2001. The promise and challenge of solid state lighting. *Physics Today*, December, 42.
R. T. Collins, P. M. Fauchet, and M. A. Tischler, 1997. Porous silicon: From luminescence to LEDs. *Physics Today*, January, 24.
C. Day, 2006. Light-emitting diodes reach for the ultraviolet. *Physics Today*, July, 15.
M. Freemantle, 2001. Color control for organic LEDs. *Chemical and Engineering News*, August 27, 17.
M. Gunther, 2017. Molecular movie exposes perovskite solar cell's inner workings. *Chemistry World*, September, 38.
A. Polman, M. Knight, E. C. Garnett, B. Ehrler and W. C. Sinke, 2016. Photovoltaic materials: Present efficiencies and future challenges. *Science*, 352, 307.
E. F. Schubert, 2003. *Light-Emitting Diodes*. Cambridge University Press, Cambridge, UK.

Light-Sensitive Glass

B. Erickson, 2009. Self-darkening eyeglasses. *Chemical and Engineering News*, April 13, 54.
D. M. Trotter Jr., 1991. Photochromic and photosensitive glass. *Scientific American*, April, 124.

Photocopying Process

October 22, 1938: Invention of xerography. *APS News*, October 2003, 2.
A. D. Moore, 1972. Electrostatics. *Scientific American*, March, 47.
J. Mort, 1994. Xerography: A study in innovation and economic competitiveness. *Physics Today*, April, 32.
D. Owen, 1986. Copies in seconds. *The Atlantic Monthly*, 257(2), 64.

Photographic Materials and Processes

Silver halides in photography. *MRS Bulletin*, Vol. XIV, Number 5, May 1989 (five research papers concerning this subject).

A. B. Bocarsly, C. C. Chang, and Y. Wu, 1997. Inorganic photolithography: Interfacial multicomponent pattern generation. *Journal of Chemical Education*, 74, 663.

C. A. Fleischer, C. L. Bauer, D. J. Massa, and J. F. Taylor, 1996. Film as a composite material. *MRS Bulletin*, July, 14.

M. Freemantle, 2000. Doubling film speed, radically. *Chemical and Engineering News*, December 11, 9.

W. C. Guida and D. J. Raber, 1975. The chemistry of color photography. *Journal of Chemical Education*, 52, 622.

W. L. Jolly, 1991. Solarization: The photographic Sabatier effect. *Journal of Chemical Education*, 68, 3.

J. F. Hamilton, 1988. The silver halide photographic process. *Advances in Physics*, 37, 359.

J. Kapecki and J. Rodgers, 1993. *Color photography. Kirk–Othmer Encyclopedia of Chemical Technology.* John Wiley & Sons, Hoboken, NJ.

M. S. Simon, 1994. New developments in instant photography. *Journal of Chemical Education*, 71, 132.

Triboluminescence

R. Angelos, J. I. Zink, and G. E. Hardy, 1979. Triboluminescence spectroscopy of common candies. *Journal of Chemical Education*, 56, 413.

L. M. Sweeting, A. L. Rheingold, J. M. Gingerich, A. W. Rutter, R. A. Spence, C. D. Cox, and T. J. Kim, 1997. Crystal structure and triboluminescence. *Chemistry of Materials*, 9, 1103.

Websites

For links to relevant websites, see PhysicalPropertiesOfMaterials.com

4

Color from Interactions of Light Waves with Bulk Matter

4.1 Introduction

In Chapter 2, we saw that color can originate from single atoms or small numbers of atoms. Chapter 3 showed that very large numbers of atoms of metallic or semiconductor materials can give rise to energy bands that in turn can provide color when visible light is absorbed.

Light can interact in other ways with bulk matter (i.e., matter with of the order of the Avogadro* number of atoms), and this can also lead to color. In this chapter, we explore some of these sources of color.

4.2 Refraction

Thus far, the processes giving rise to color have involved absorption or emission of light. However, there are other physical interactions that give rise to color by change in the direction of the light, and one of these is considered here.

Refraction results from the change in speed of light when passing from one medium to another. When a uniform, nonabsorbing medium, such as air or glass or salt, transmits light, the photons are absorbed and immediately re-emitted in turn by all the atoms in the path of the ray, slowing down the light. Light travels fastest in a vacuum, and more slowly in all other materials. Light travels faster in air than in a solid.

---

* Amedeo Avogadro (1776–1856) was an Italian lawyer and scientist. In 1811, in a paper that should have settled the dispute concerning the relationship between atomic weights and properties of gases, Avogadro proposed that equal volumes of gases under the same conditions should contain equal numbers of molecules. This work was ignored for about 50 years, but its importance is recognized today in naming Avogadro's number.

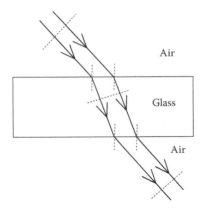

FIGURE 4.1
As monochromatic light passes from air to glass, it is slowed. When the light hits the glass at a glancing angle, one side of the light beam is slowed first, causing the light to be bent toward the perpendicular of the air-glass interface on entering the glass and away from the perpendicular of the glass-air interface on leaving the glass.

Consider the case of *monochromatic* (i.e., single wavelength) light entering a piece of glass, as shown in Figure 4.1. One edge of the beam enters the prism and is slowed before the other edge hits the glass. This bends the light toward the direction perpendicular to the edge of the glass. When exiting the prism, the side of the beam that exits first "pulls" the light in the direction that takes it away from the perpendicular direction. This bending of light as its velocity changes with a change in medium can be considered to be a consequence of *Fermat's* principle of least time* which states that light rays will always take the path that requires the least time. Bending of light on changing media is shown by analogy in Figure 4.2.

The *index of refraction* (also known as the *refractive index*), n, is defined as[†]

$$n = \frac{v}{v'} \tag{4.1}$$

where v is the speed of light in vacuum and v' is the speed of light in the medium under consideration; n depends on the material, the wavelength of the light used, and the temperature. Values of n for a range of materials (for light of the wavelength of the Na yellow line) are given in Table 4.1.

Although we are perhaps not aware of it, many optical phenomena are associated with refractive index (see Figure 4.3). For example, if we drop a piece of glass in water, we see the edges of the glass due to bending of light

---

* Pierre de Fermat (1601–1665) was a French lawyer who studied mathematics as a hobby. He made several important contributions in the areas of number theory, analytic geometry (including the equation for a straight line), and calculus.
[†] The most important equations in this chapter are designated with ■ to the left of the equation.

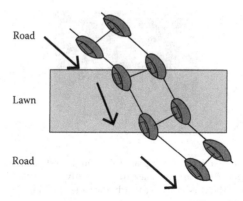

FIGURE 4.2
An analogy showing how light is bent as it changes speed on passing from one medium to another. When a pair of wheels from a vehicle hits a patch of soft lawn at a glancing angle, the tire on the left is slowed first, and this causes bending of the light toward the perpendicular of the road-lawn interface; the path bends away from the perpendicular of the lawn-road interface on leaving the lawn.

TABLE 4.1

Refractive Index Values for Some Common Materials at 25 °C Except as Noted[a]

Material	n
Air	1.00029
Ice	1.305
Water	1.33
Acetone	1.357
n-Heptane	1.385
Tetrahydrofuran	1.404
Fused quartz	1.456
Benzene	1.498
Aniline	1.584
Light flint glass (average)	1.59
Methylene iodide	1.738
Dense flint glass (average)	1.75
Diamond (at 0 °C)	2.42

[a] All values were determined using light from the sodium-D line (yellow light).

because the refractive index of glass is different from that of water. If alcohol and water are mixed, their mixing can be traced by the non-uniformity of the light's path, due to refraction associated with the small differences in their refractive indices. Similarly, the change in refractive index of air as it is heated is apparent in the shimmering effect seen in an air mass above hot asphalt.

Refractive index measurement can be used in identifying materials. In forensic work, a very small chip of colorless glass from a crime scene can

FIGURE 4.3
Two vials, each containing a liquid and a glass tube. On the left, the liquid and the glass tube have different refractive indices, and the glass tube is clearly visible. On the right, the refractive indices of the liquid and the glass tube are nearly the same, making it very difficult to see the glass tube immersed in the liquid.

COMMENT: MIRAGES

The refractive index for air depends on its density, which in turn depends on temperature and pressure. Lower density air has a lower refractive index. Under conditions of density gradients, light bends toward the denser air. For example, in a strong vertical temperature gradient, such as warm air above cold water in the ocean, the refractive index can change so rapidly that it can induce curvature in the light rays exceeding the curvature of the earth. This refractive-based curvature makes objects at or below the horizon appear in the field of vision as a *mirage*, as shown in Figure 4.4.

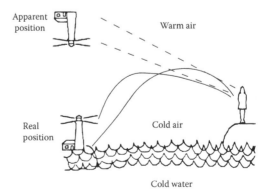

FIGURE 4.4
A mirage results when light is bent toward the denser air in a strong vertical temperature gradient. Because light normally travels in straight line, we perceive a mirage, greatly exaggerated in elevation here.

be identified and traced by its refractive index. The refractive index can be determined as follows. This chip of glass is placed in a liquid of a particular refractive index and another liquid of different refractive index is added to the first until the blended refractive index matches that of the glass chip, making the chip "invisible" at this point due to the matching of the refractive index of the glass chip and the solution. When the refractive index of the solution has been determined with a *refractometer*, knowledge of the numerical value of the refractive index of the glass can be used to trace its origin, often to a specific make and model of vehicle.

Another rather peculiar example of a refractive index oddity results when using liquid helium. Most liquids have refractive indices very different from the gas above them, and this is why we can see their surfaces distinctly. However, in the case of helium (which boils at $T = 4.2\,\text{K}$), the liquid and the gas have very similar refractive indices. Furthermore, the gas above liquid helium is just pure helium, as any other gas would condense to its solid state. When using liquid helium, which must be transferred carefully from one closed container to another in an insulated siphon, it can sometimes be difficult to see how much liquid has been transferred because the surface is almost "invisible."

Figure 4.1 shows refraction for monochromatic light, but refraction can lead to beautiful effects when dealing with white light. For example, refraction is responsible for the beautiful colors in a rainbow. In Cambridge, England, in 1666, at the age of 23, Isaac Newton* described an experiment with a prism "to try therewith the phenomena of the colors." He wrote,[†] "In a very dark Chamber, at a round hole... I placed a Glass Prism, whereby the Beam of the Sun's Light, which came in at that hole, might be refracted upwards toward the opposite wall of the Chamber, and there form a color'd Image of the Sun." We now know that the source of the spectrum of colors is the refraction property of glass, as first described correctly by Newton, using the term *refrangibility* (known today as refraction).

When white light, which we now know to be composed of the light of all visible colors, hits a prism, some colors are more refracted than others. In particular, higher-energy (shorter-wavelength) light is more refracted. This variation of refraction with wavelength causes white light to be dispersed by a prism, as shown in Figure 4.5.

Dispersion of white light by water droplets or ice crystals is the source of the colors of the rainbow. Similarly, the flashes of color characteristic of facetted gemstones (see Comment: Sparkling Like Diamonds) and high-quality "crystal" glassware are due to dispersion effects associated with the high refractive index of the lead-containing glass.

---

* Sir Isaac Newton (1642–1727) was an English natural philosopher and mathematician. Owing to the Black Plague, Newton spent 1665–1666 in the English countryside where he laid the foundations for his great contributions in optics, dynamics, and mathematics.
[†] I. Newton, 1730. *Opticks*, W. Innys, London.

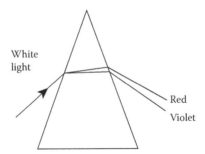

FIGURE 4.5
Refraction causes white light to be dispersed by a prism because different colors (wavelengths) of light have different refractive indices. Shorter-wavelength light (violet here) is refracted more than longer-wavelength light (red).

Dispersion is not always a desired feature, however. One example of a situation where it would be preferable to avoid such dispersion is the *chromatic aberration* (i.e., unwanted color patterns) sometimes found in optical systems such as telescope lenses and camera lenses.

COMMENT: SPARKLING LIKE DIAMONDS

Light leaving a material of refractive index n and passing into air with an incident light angle i (see Figure 4.6a), gives light with a refracted light angle, r, where i and r are related by Snell's* law:

$$\frac{\sin r}{\sin i} = n, \tag{4.2}$$

where the refractive index of air has been assumed to be 1. From Equation 4.2, the maximum value of i is $\sin^{-1}(n^{-1})$, and this value of i is called the *critical angle*. For i greater than the critical angle, there is no refraction, and total internal reflection results (see Figure 4.6b).

For diamond, $n = 2.42$, which leads to a critical angle of 24.4°. Therefore, a light ray that makes an angle of 65.6° (i.e., 90° − 24.4°) or less with the outer surface of a diamond cannot pass through the diamond surface to air but instead is totally reflected, as shown for monochromatic light in a diamond in Figure 4.7a. An appropriately cut diamond will appear opaque when viewed from below because no light from the upper surface will exit the bottom (Figure 4.7a). When white light hits the diamond's surface, it is dispersed into its colors. The dispersion, illustrated in Figure 4.7b, gives diamonds (and other high-refractive index materials) their sparkle.

---

* Willebrod Snell van Royen (1581–1626) was a Dutch mathematician and scientist.

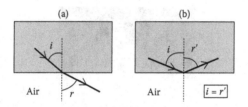

FIGURE 4.6
(a) When light leaves a medium to go to air, the angle of incidence, *i*, is related to the angle of refraction, *r*, by Snell's law (Equation 4.2). (b) If the angle of incidence is greater than the critical angle, there is no refraction but only internal reflection. Note that the angle of incidence in case (b) equals the angle of reflection, *i=r'* (i.e., specular reflection results).

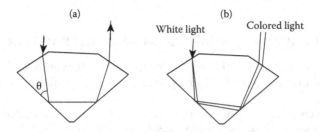

FIGURE 4.7
(a) Monochromatic light within a diamond that makes an angle θ of 65.6° or less with the surface will be totally reflected at the diamond–air interface. An appropriately cut diamond can appear opaque when viewed from the bottom due to total internal reflection of the impingent light, as shown. (b) White light enters a cut diamond and is dispersed. With an appropriate cut, colored light exits from the top face, giving dispersion colors (sparkle), and no light exits from the bottom face.

COMMENT: THE FRESNEL LENS

A *Fresnel* lens* is a flat lens (can be glass or more commonly plastic such as a copolymer of polyvinyl acetate and polyvinyl butyrate) that can act as a lens (Figure 4.8). It does so using the principle of refraction. Series of small concentric stepped prisms are molded into the lens, each step only a few thousandths of a centimeter wide. The steps extend concentrically from the center of the lens to its outer edges and can easily be felt with a finger. Each prism causes refraction and bends the light. Taken together, they form a lens. Fresnel lenses are relatively inexpensive to produce and much thinner than conventional lenses. They have many applications, including concentration of light for use in solar ovens, in solar collectors to heat domestic hot water, and in inexpensive eye glass lenses for developing countries.

---

* Augustin Jean Fresnel (1788–1827) was a French physicist. He invented his lens design to allow efficient and effective light signaling between lighthouses.

FIGURE 4.8
Piano keys viewed through a Fresnel lens. These lenses derive their optical properties from small concentric stepped prisms that refract light.

COMMENT: NEGATIVE INDEX OF REFRACTION

A material with a negative refractive index, $n < 0$, would lead to both refracted and incident light on the same side of normal, as shown in Figure 4.9.

Negative index of refraction materials were first pondered theoretically in 1968 and are now being realized through *metamaterials*, which are materials that derive their properties from their manipulated structure rather than the intrinsic properties of their components. (Sometimes referred to as "artificial materials," metamaterials are indeed real!) In principle, materials with $n = -1$ can focus light perfectly without any curved surfaces. Sometimes called "left-handed materials" because the wave vector is antiparallel to the usual right-hand cross product of the magnetic and electric fields, negative refractive index materials are now being used in many applications, especially in the microwave region.

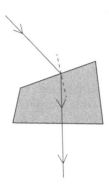

FIGURE 4.9
The path of light entering a negative refractive index material bends away from the perpendicular to the air-material interface.

4.3 Interference

Newton also was the first to document an understanding of the sources of color due to *interference*.

To understand interference patterns in light, it is instructive to first consider two extreme ways in which waves can interact. These interactions apply equally to light waves or any other waves such as sound or water waves.

When two waves of precisely the same wavelength intersect, the resulting wave depends very much on the *phases* of the waves. For example, if two identical waves exactly in phase (peaks and troughs at same time and location, as shown in Figure 4.10) interact, the resulting wave will be of the same wavelength, and its amplitude will be the sum of the amplitudes of the incident waves, as shown in Figure 4.10. These waves are said to undergo *constructive interference*.

Although there is only one way in which waves can be in phase, there are many ways for the waves to be "not in phase." The most extreme case is two identical waves that are exactly out of phase with one another: When one has a peak, the other has a trough. Two such waves of exactly the same wavelength and exactly the same amplitude will come together to result in total destruction: no wave results (see Figure 4.11). This effect is called *destructive interference*.

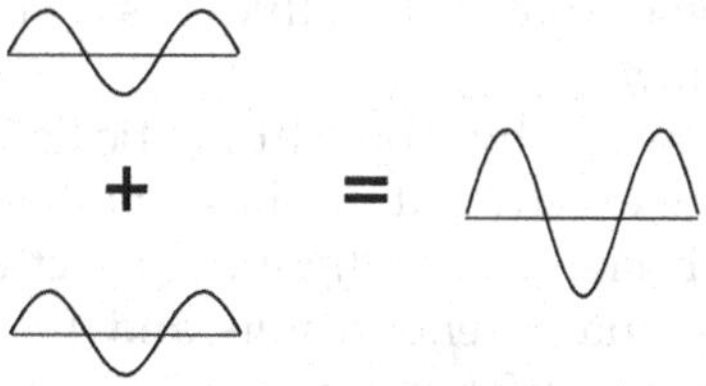

FIGURE 4.10
Constructive interference of two identical waves precisely in phase (peak aligned with peak, trough aligned with trough) gives a resultant wave of the same wavelength and twice the amplitude of the original waves.

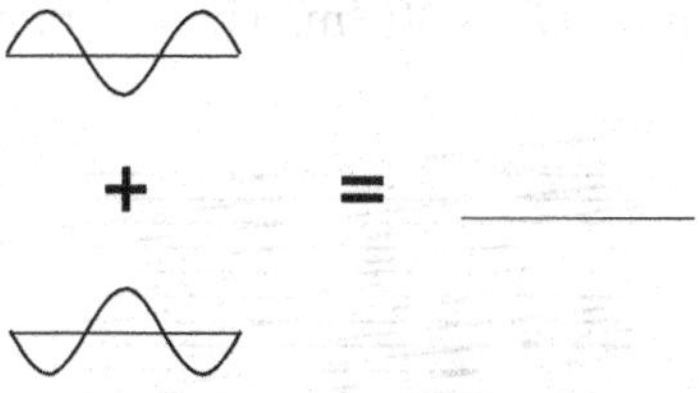

FIGURE 4.11
Destructive interference: Two identical waves exactly out of phase (peak aligned with trough) give no resultant wave.

COMMENT: SOUND INTERFERENCE

Interference effects apply to any kind of wave, including both light and sound. Destructive interference effects are used in some earphones that remove background noise: They work by generating sound waves that exactly destructively interfere with the sound waves of the background.

COMMENT: MOIRÉ PATTERNS

Two sets of equally spaced lines overlaid with a slight displacement angle, as shown in Figure 4.12, give rise to an interference pattern, called a *moiré pattern*, from the French word *moiré*, meaning "water-spotted." The presence of moiré patterns has been used for more than a century to detect aberrations due to irregularities in optical components, by looking for such unwanted patterns.

Interference effects are especially prominent when light interacts with thin films. In such cases, light can be reflected from the both front face of the film and also the back face of the film. Whether the front-reflected and back-reflected light cancels (destructive interference) or reinforces (constructive interference) depends on the nature and thickness of the film, the color of the light, and the viewing angle.

If the extra distance traveled by monochromatic light causes the front- and back-reflected light to be exactly out of phase, no light of this color will be seen. If the front- and back-reflected light are exactly in phase for a given wavelength, constructive interference results, and this color will be observed.

Therefore, in the case of interference due to front- and back-reflected white light from thin films, the color and intensity of light observed will depend on the viewing angle. We are familiar with this situation in such everyday experiences as the colors in a bubble or the colors of oil on water. Even thin slices of inorganic materials can derive their colors from interference of front- and back-reflected light. For example, the colors of silicon oxide wafers range from blue-green at a thickness of 0.92×10^{-6} m to maroon-red at 0.85×10^{-6} m, back to blue-green again at 0.72×10^{-6} m. The colors repeat as the thickness

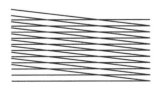

FIGURE 4.12
A moiré pattern.

changes, as additional layers add integer multiples of the wavelength to the extra path, allowing the same color to interfere constructively at different film thicknesses.

See the video "Liquid Crystal Displays" under Student Resources at PhysicalPropertiesOfMaterials.com

Interference colors in soap films bear further examination. Soap works as a cleansing agent because of its chemical structure: one end is polar, and the other is nonpolar. This structure is shown schematically in Figure 4.13 for a "typical" soap.

The hydrocarbon "tail" of the soap molecule is hydrophobic and acts as a solvent for oils, while the polar end is attracted to water (hydrophilic). This property of being both oil-loving and water-loving makes soap *amphiphilic* and allows soap to attract both water and oils, thereby allowing cleansing action.

A soap film is composed of soap molecules and water molecules. It is more energetically favorable for the soap molecules to have their hydrophobic ends away from the water pointing out into the air, which can, at appropriate concentration, give a bilayer structure, as shown schematically in Figure 4.14.

FIGURE 4.13
Schematic structure of a "typical" soap molecule, showing the alkyl (nonpolar) tail and polar head group.

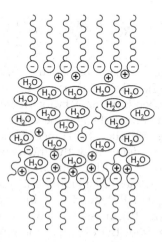

FIGURE 4.14
Schematic bilayer structure of a typical soap film. The alkyl chains of the soap molecules point into the air, with the polar head groups of the soap molecules pointing into the water. Much more water would be present in a real soap film, and different thicknesses of soap films will result from different numbers of water molecules, with the soap still aligned on the outside of the film.

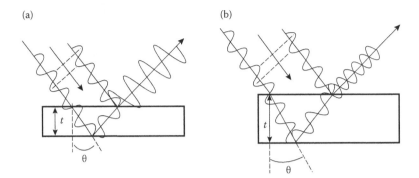

FIGURE 4.15
Schematic view of front- and back-surface (top- and bottom-) reflection of light from a thin film, such as a soap film: (a) constructive interference (the exiting waves add to give a resultant wave of larger amplitude); (b) destructive interference (the exiting waves exactly cancel to give no light out). Note that front reflection leads to a change in phase, whereas back reflection does not.

The main difference between a thick soap film and a thin one is the number of water molecules between the soap layers.

The soap film can be modeled schematically as a front and back surface, each reflecting light, as shown in Figure 4.15. Views of constructive and destructive interference from front- and back-reflection are also shown.

For light of wavelength λ and incident intensity I_i hitting the thin layer (soap film) of refractive index n, consideration of wave scattering gives the intensity of the reflected light, I_r, as*

$$I_r = 4I_i R \cos^2\left(\frac{2\pi nt}{\lambda}\cos\theta + \frac{\pi}{2}\right),\tag{4.3}$$

where t is the thickness of the soap film, θ is the angle defined in Figure 4.15, and R is the fraction of light reflected. Equation 4.3 shows that the intensity of reflected light observed depends on many factors, including the thickness of the soap film, the wavelength of the light, and the viewing angle. This situation is immediately familiar to us as the beautiful range of colors seen in a soap bubble.

Let us consider in more detail how the intensity of reflected light will depend on the thickness of the film. Due to gravity, a vertical soap film will not be of uniform thickness (see Figure 4.16), and t varies with height. Since the intensity of light will depend on t, in monochromatic light there will be regions where the thickness is such that maximum constructive interference will be observed (the extra path length traveled by the back-reflected light combined with the phase change of the front-reflected light

---

* See C. Isenberg (1992). *The Science of Soap and Soap Films*. Dover Publications, p. 36, for a derivation.

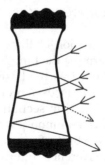

FIGURE 4.16
Cross section of a vertical soap film showing a nonuniform thickness and constructive (solid line) interference and destructive (dashed line) interference of monochromatic light from different regions of the film.

keeps them exactly in phase on exiting), and the reflected light will be very intense. There will be other regions (other thicknesses) in the soap film where maximum destructive interference will be observed (the extra path length traveled by the back-reflected light combined with the phase change of the front-reflected light puts the exiting beams exactly out of phase), and there will be no net reflected light. There also will be intermediates between these two extremes, again depending on the thickness of the film. Therefore, for a wedge-shaped film observed in monochromatic light, there will be an interference pattern of fringes of light and dark areas, depending on the thickness of the film.

For white light, colored fringes are observed since the emerging beam will be the sum of all the contributions from each wavelength, and each wavelength has I_r given by Equation 4.3. This situation gives rise to the familiar pattern of color that we know from soap films and from other interference effects, such as the interference colors of oil on water.

When a soap film is allowed to sit in very stable conditions and drain over a period of time, much of the water between the bilayers drains away. If, over time, the thickness of the film becomes such that $t \ll \lambda$, Equation 4.3 leads to an interesting result. Since, for $t \ll \lambda$

$$\cos^2\left(\frac{2\pi nt}{\lambda}\cos\theta + \frac{\pi}{2}\right) = \cos^2\left(0 + \frac{\pi}{2}\right) = \cos^2\left(\frac{\pi}{2}\right) = 0, \qquad (4.4)$$

for very thin films ($t \ll \lambda$), there will be no observed reflected intensity. In such a case, all light that interacts with the film is absorbed, and the film appears black in reflected light. This phenomenon is in fact observed, although it really does require very still surroundings to achieve such thinning of the film without it breaking! Black films have typical thicknesses in the range 5–30 nm.

It should be apparent from Equation 4.3 that an analysis of color fringes in a soap film can be used to determine the film thickness.

Although soap films were used as the basis for the discussion above, the approach applies equally to discussion of any thin transparent film. The colors of a beveled edge of a thin section of a mineral, gasoline on a puddle and a soap bubble have common physical origins in interference effects. Colors of a fish scale or a butterfly wing or CD have a similar basis and are due to multilayer interference. Interference coloration is usually *iridescent*, that is, displaying all the colors of the rainbow. Furthermore, the color observed changes with the viewing angle, and changes on immersion in a fluid.

COMMENT: HOLOGRAMS

The hologram was invented in 1947 by Dennis Gabor*; he was awarded the 1971 Nobel Prize in Physics for this invention.

The hologram is an example of the use of interference effects to produce color. It also produces a three-dimensional image.

The production of a high-quality hologram requires film, a light source (usually a laser), and a small three-dimensional model for the object to be captured on the hologram. To make a master hologram, the laser beam is split into two beams: the illuminating beam and the reference beam. The illuminating beam is aimed at the object, and then its reflection strikes a light-sensitive film. The reference beam meets the reflected beam at the film, and the interference between the reflected and reference beam leaves a pattern on the film, representative of the object scanned.

The master holographic image is converted to a physical image of microscopic pits in a fine layer of nickel. A special plate is then created to emboss these microscopic pits into a clear plastic film. Then the film is coated with a thin reflective layer of aluminum. The master hologram can be used to produce many such holograms.

Viewing the hologram requires ambient light from the same direction as the reference beam. This light interacts with the pattern on the film to reconstruct the three-dimensional image of the original object.

X-ray holograms with atomic resolution have been developed for use as data storage devices.

---

* Dennis Gabor (1900–1979) was a Hungarian-born electrical engineer who developed the holographic principles while carrying out research aimed to improve the resolution of electron microscope images.

4.4 Scattering of Light

Light is bent or scattered when it hits the edge of an opaque object. This is usually a minor effect when the object is large but becomes a major effect when the size of the object approaches the wavelength of light involved.

Scattering of light was put on a firm basis by the studies of Rayleigh[*] and Tyndall,[†] and their names are often associated with light-scattering processes.

For a small object, about 10 nm–300 nm in diameter, that scatters light of incident intensity I_i and wavelength λ, the intensity of light observed at a distance d from the scatterer and angle θ between incident and detected intensity, I_θ, is given by[‡]:

$$\frac{I_\theta}{I_i} = \frac{8\pi^4 \alpha^2 \left(1 + \cos^2 \theta\right)}{d^2 \, \lambda^4} \tag{4.5}$$

where α is the *polarizability* of the sample (i.e., a measure of how easy it is to displace electrons in the sample). Polarizability is important for interactions with light because light is electromagnetic radiation. Equation 4.5 describes *Rayleigh scattering*.

Since Equation 4.5 shows that I_θ is proportional to λ^{-4} in Rayleigh scattering, the most intensely scattered light is that which is at the shortest wavelengths. In the visible range, blue light is much more scattered than red light.

Rayleigh scattering is responsible for the blue color of the daytime sky. The light from the sun contains all colors; we know that because we can see it dispersed into the whole spectrum when there is a rainbow. Of the visible wavelengths of light emitted by the sun, violet light will be most scattered by particles in our atmosphere, but the intensity of violet light from the sun is much less than that of blue light. When we stand beneath the noon sun and look at the sky, most of the color that we see in the sky, not looking directly at the sun, is therefore blue (Figure 4.17a).

On the other hand, we are all familiar with the beautiful red sky at sunset. In the direction of the sun, much of the light at shorter wavelengths has been scattered by the atmosphere, particularly at sunset (and sunrise) because the sun's light has a longer path to travel through the atmosphere before getting to us. The scattering away of the higher-energy light leaves behind predominantly red color in the direction of the sun at sunset (Figure 4.17b).

---

[*] Lord Rayleigh (born John William Strutt; 1842–1919) was a British mathematical physicist and winner of the 1904 Nobel Prize in Physics for accurate measurements of the density of the atmosphere and its component gases. This work led to the discovery of argon and other members of the previously unknown family of noble gases.

[†] John Tyndall (1820–1893) was an Irish-born physicist who carried out precise experiments on heat transfer and also on scattering of light by fine particles in air, known today as the Tyndall effect.

[‡] See, for example, R. J. Hunter (1993), *Introduction to Modern Colloid Science*, Oxford, p. 42.

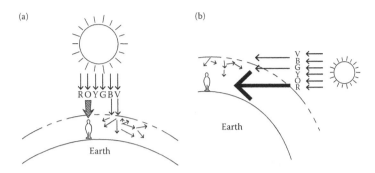

FIGURE 4.17
(a) The sky appears blue at noon because the shortest wavelengths of sunlight (corresponding to blue-violet light) are most highly scattered. The sun appears yellow-orange because red to yellow light is much less scattered. (b) At sunset, the light from the sun travels a longer path through the light-scattering atmosphere to reach us. The scattering away of even more of the short-wavelength light than at noon makes the sun appear red at sunset.

The white color of clouds also is due to light scattering, but now the particle size (>1000 nm) is much larger than for Rayleigh scattering (10–300 nm) and is more accurately described by equations for Tyndall scattering (particles >50 times the wavelength of light). The clouds are made of water droplets that can scatter light. The scattering is due to surface roughening and is nearly equal at all wavelengths (as for a rough surface of paper or ground glass; see the Tutorial in Chapter 5), and clouds appear white.

Indeed, without an atmosphere, the sky would appear black even in daytime, as it does in space or on the moon.

COMMENT: COLOR-CHANGING SPECIES

The color-matching of the chameleon to its environment is largely a result of light scattering. The skin of a chameleon contains many layers—an outer yellow carotenoid layer, followed by a red chromophore layer, then a granular guanine layer, and finally an inner granular melanin layer. The ambient light level alters the stimulation of the nerves in the skin, which, in turn, changes the degree of aggregation of the guanine layer. In low light levels, granules aggregate, and the layer appears dark. In higher light levels, the granules are more dispersed, and the incoming light is scattered by the silvery blue guanine layer, making the skin appear more brightly colored. The degree to which the red and yellow layers are activated is in direct response to the stimulation of light from the surroundings.

The Hercules beetle changes color in response to humidity. The upper layer of its skin is spongy and, when dry, it scatters yellow light.

In moist conditions, this layer becomes translucent, and the black underlayer dominates, giving a black appearance. Thus, the beetle looks yellow in low-humidity daylight conditions and black in the humid night air or amid rotting fruit. Once again, light scattering can provide camouflage.

4.5 Diffraction Grating

Interference and angular dependence can act together to produce colors in the case of a *diffraction grating*, which is a regular structure with very small slits (or grooves), with spacing on the order of the wavelength of light. When the light passes through such a slit, or is reflected from such grooves, optical effects can arise from constructive and destructive interference.

For a single slit, the exiting light intensity varies with distance from the slit and angle. For many slits, the light coming from adjacent slits can interfere either constructively or destructively. For a given wavelength of light, there will be angles at which there is complete constructive interference, and other angles for which there is complete destructive interference (Figure 4.18; the Young* experiment). The exact angles will depend both on λ and on the spacing between the slits.

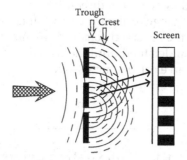

FIGURE 4.18

Light passing through two very small slits can interact in such a way that both constructive and destructive interference is possible. This gives "fringe patterns" (white for constructive interference from crests meeting crests and troughs meeting troughs, black for destructive interference from crests meeting troughs) as shown. This experiment was designed and first carried out by Thomas Young, and was the first proof of the wave nature of light.

---

* Thomas Young (1773–1829) was an English natural philosopher who made many contributions to physical sciences, including the principle of interference and principles of elasticity (memorialized in the term "Young's modulus"; see Chapter 14.)

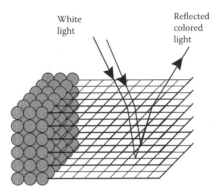

FIGURE 4.19
Spectral colors can arise from the interaction of light with a regular lattice array with a spacing close to or shorter than the wavelength of light. The reflected light interferes constructively, enhancing some colors, and also interferes destructively, depleting other colors. The color observed depends on the viewing angle, and is the origin of color in opals.

When white light is passed through a diffraction grating (or is reflected from it), each wavelength is enhanced in different sets of directions; therefore, as the diffraction grating is rotated in different directions, different wavelengths come into view. This is the principle by which a grating *monochromator* (which means "single-color producer") works. The diffraction elements can be made, for example, by engraving fine rulings on a glass plate.

One example of a natural diffraction grating is the gemstone opal (Figure 4.19). Opal is composed of spheres of SiO_2 with a small proportion of water packed in a three-dimensional lattice with a repeat distance of about 25 nm. When white light enters this three-dimensional grating, pure colors appear, but the colors change as the eye or the stone move, as this changes the viewing angle and the observed constructive interference. The source of color in opals was not understood until the structure of opal was elucidated in 1964 with the use of an electron microscope.

4.6 An Example of Diffraction Grating Colors: Liquid Crystals

Liquid crystals can provide another example of diffraction-grating colors. Liquid crystals were first reported by botanist Reinitzer* in 1888. He discovered that cholesteryl benzoate had two melting points. On heating, at 145 °C it goes from a solid to a cloudy liquid; at 179 °C it transforms from a cloudy

---

* Friedrich Reinitzer (1857–1927) was a Prague-born botanist who spent much of his working life in Austria. His insightful observation of a "double melting point" led to the discovery of liquid crystals.

$$CH_3(CH_2)_nCOO^-M^+$$

DNA

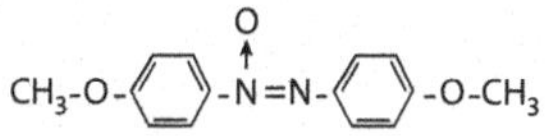

FIGURE 4.20
Some molecules that can form calamitic liquid crystals. Note that all have rod-like shapes.

FIGURE 4.21
Liquid crystals are both pourable (liquid) and ordered (crystalline) because of their molecular shape. The molecular size is exaggerated here.

liquid to a clear liquid. It is now known that the intermediate state (cloudy liquid) is liquid crystalline.

Liquid crystals derive their appropriate name from the facts that they are pourable (liquid) yet structured (crystal). They owe these seemingly conflicting properties to their molecular shapes: materials that form liquid crystals are long and thin (*calamitic liquid crystals*) or disc-shaped (*discotic liquid crystals*). Some typical molecular structures are shown in Figure 4.20.

Liquid crystals are *mesophase* (i.e., intermediate phase) materials. The liquid crystalline phase can be produced only in a certain temperature range (*thermotropic* liquid crystal) or by addition of an ordinary solvent (*lyotropic* liquid crystal, e.g., concentrated aqueous soap solution). The molecular shape of a pure liquid-crystal-forming molecule allows liquid crystals to be pourable yet show some order (Figure 4.21). There are many types of liquid crystals and much nomenclature on this subject; the categories are defined by the molecular arrangements.

Nematic (Greek for "threadlike") liquid crystals have their molecules aligned in the direction of the long axis, but they are interspersed, not in planes. This is shown schematically in Figure 4.22.

Smectic (Greek for "soap") liquid crystals have layers of molecules aligned in planes, as shown in Figure 4.22. A soap bubble is an example of a smectic liquid crystal, with the number of layers reduced as the bubble expands (see Figure 4.14).

Liquid crystals can be used to produce colors from the principle of a diffraction grating. The coloration is due to the regularity of their stacking, which can act as a diffraction grating, much as in the case of opal. As the

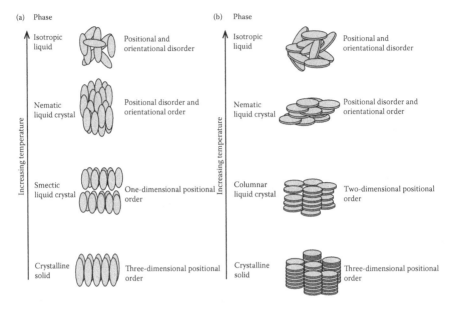

FIGURE 4.22
Liquid crystal structures showing the wide variety of phases exhibited by liquid crystals. (a) Phases of calamitic (rod-shaped) liquid crystals, showing nematic and smectic phases. (b) Phases of discotic (disc-shaped) liquid crystals, showing nematic and columnar phases. In both cases, the liquid crystalline mesophases exist at temperatures intermediate between the isotropic (ordinary) liquid and the crystalline solid.

temperature changes, it changes the spacing in the liquid crystal due to thermal expansion (at higher temperatures the molecules require a little more room to jostle around). Such a change in spacing causes a change in color. The change in color as a result of change in temperature, which is not limited to liquid crystals (see Chapter 2), is an example of thermochromism. In general, on cooling of a diffraction-grating liquid crystal, the reflectance maximum moves from violet to blue to green to yellow to orange to red to colorless. So-called "mood" rings make use of thermochromism in liquid crystals; their color, which indicates hand temperature and possibly correlates with mood, is due to the thin film of liquid crystalline material behind a bead of colorless transparent plastic. In some thermochromic liquid crystal devices, temperature changes as small as 0.001 K can be detected.

Both the color of a liquid crystal and the temperature of its color change can be modified chemically. For example, 20% cholesteryl chloride with 80% cholesteryl nanoate is green, but an added solvent can change the color: benzene changes the color to blue, and chloroform changes the color to red.

One important factor concerning thermotropic liquid crystalline phases is that they are intermediate in the temperature range between low-temperature crystalline forms and high-temperature "normal" liquid phases (see Figure 4.22). The latter is often referred to as an *isotropic* liquid phase, meaning that there is no preferential direction for molecular orientation in this phase. For materials that form liquid crystals, the isotropic liquid phase exists when the temperature is sufficiently high to allow the molecules to tumble freely in all directions. As the temperature is lowered, the molecules preferentially align (e.g., in a smectic or nematic structure), but they are still relatively free to translate. As the temperature is lowered further, there is not enough thermal energy for the molecules to translate, and the structure changes to a crystalline form. For this reason, liquid crystal displays (LCDs; see Chapter 5 for a discussion of their operating principles) have a limited temperature range of operation, namely that of liquid crystal stability.

COMMENT: COLORS OF LIQUID CRYSTALS

An early application of liquid crystals was in the temperature-dependence of the color of thin films of *cholesteric* liquid crystals. Cholesteric liquid crystals (also called *twisted nematic liquid crystals*) have layers of chiral molecules arranged such that each layer has a different orientation, as shown schematically in Figure 4.23. The repeat distance, called the *pitch*, changes with temperature. These types of liquid crystals have been used to sensitively detect temperature changes in applications as varied as detection of irregularities in blood circulation, metabolic changes due to breast cancer, and measurements of stress and strain in building components.

See the video "Thermochromic Liquid Crystal" under Student Resources at PhysicalPropertiesOfMaterials.com

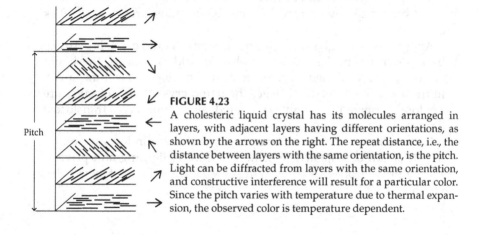

FIGURE 4.23
A cholesteric liquid crystal has its molecules arranged in layers, with adjacent layers having different orientations, as shown by the arrows on the right. The repeat distance, i.e., the distance between layers with the same orientation, is the pitch. Light can be diffracted from layers with the same orientation, and constructive interference will result for a particular color. Since the pitch varies with temperature due to thermal expansion, the observed color is temperature dependent.

Pitch

COMMENT: COLORED LIQUID CRYSTAL DISPLAYS

Variable-colored liquid crystalline materials can be produced using *pleochroic* dyes in a nematic phase. A pleochroic material has different light absorptions and therefore different colors along different axes due to its anisotropic nature (i.e., it has different properties in different directions). The dye molecules will tend to align themselves along the axis of alignment of the liquid crystals, and the color observed will depend on the orientation of the liquid crystal molecules with respect to the viewer. When an electric field is applied across this phase, the liquid crystal molecules can reorient, moving the dye molecules to another orientation, changing the observed color of the film. An example of the molecular structure of a pleochroic dye is shown in Figure 4.24.

COMMENT: "SMART" GLASS

While we are all familiar with transparent glass, we would probably find it convenient at times to be able to make the glass opaque. Two devices do just that; in each, the glass changes transparency when a small electric field is applied.

The *electrochromic* device, so named to indicate that the material changes color due to changes in electric field, is typically 2×10^{-6} m thick, with outer layers of a mixture of indium and tin oxides acting as transparent electrodes, with a middle layer of tungsten oxide. When voltage is applied, the tungsten oxide is reduced to metallic tungsten, which is bronze-colored and thereby reduces the amount of transmitted light. The process can be reversed with a change in the direction of the electric field, giving reversible change in transparency at the flick of a switch.

Another "smart" glass contains a thin film of liquid crystals between transparent electrodes. When an electric field is applied, the liquid crystals become aligned with the field, and the material is transparent. In the absence of an electric field, the liquid crystalline molecules are randomly oriented, and light is scattered, giving the glass a translucent (i.e., partly opaque) appearance.

Both these devices have many uses, such as windows on buildings that can be adjusted to achieve optimum lighting and privacy. As the price drops, we likely will see widespread use of switchable windows.

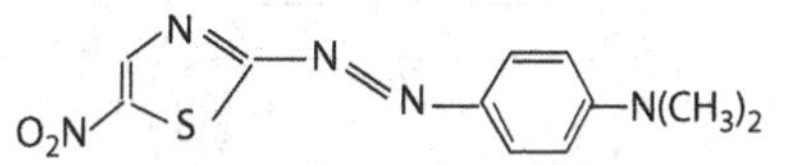

FIGURE 4.24
An example of a pleochroic dye for which the dominant color is blue.

COMMENT: PHOTONIC CRYSTALS

The periodic arrangement of atoms gives rise to crystal structures (Chapter 1), whereas periodic variation in refractive index can control how photons can be transmitted through a *photonic crystal*. For example, a periodic array of 1 mm sized voids in a high-refractive index material can make it totally absorbing for microwave radiation. These manipulated structures can provide the optical equivalent of a semiconductor band-gap where photons in the photonic material have only an allowed range of energies. The structures need not be laboratory made; some natural materials such as butterfly wings have structures that exhibit photonic properties. New photonic materials will play an important role in development of ever faster computers and optical communication devices.

4.7 Fiber Optics: A Tutorial

One of the most important devices using optical properties of materials is the *optical fiber*.* These fibers can transmit light signals for long distances with high efficiency. Internal reflection within the fiber can be used to propagate the light. A typical radius is of the order of $100\,\mu m$ (= 10^{-4}m, i.e., the diameter of a hair).

a. To decrease the scattering of light along the fiber, will it be better to use shorter or longer wavelengths of light? Explain.

b. The core of an inorganic optical fiber is typically made of a drawn fiber of SiO_2. What sorts of chemical impurities can be particularly harmful to the efficiency of transmission?

c. One advantage of *fiber optics* is that the fibers can transmit light of many different energies (wavelengths).

   i. How is this an advantage over transmission along metal wires?

   ii. Are there any limitations on possible energies for transmission? Explain.

---

* Charles Kao (1933–) was awarded the 2009 Nobel Prize in Physics "for groundbreaking achievements concerning the transmission of light in fibers for optical communication."

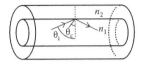

FIGURE 4.25
The indices of refraction of the optical fiber and its cladding are n_1 and n_2, respectively. The
critical angle is θ_c, where $\theta_i > \theta_c$ leads to total internal reflection.

    d. Efficient transmission is possible so long as sufficient care is taken
concerning the coating on the fiber. In particular, the coating (also
called *cladding*) should allow for essentially complete internal reflection of the light that hits it. For the optical fiber shown in Figure
4.25, with a refractive index of n_1 and a coating of refractive index n_2,
there is a critical angle, θ_c, for which any incident angle, $\theta_i > \theta_c$, leads
to internal reflection. The internal reflection in this fiber is shown in
Figure 4.25. Extension of Equation 4.2 leads to the following relationship among θ_c, n_1, and n_2:

$$\sin \theta_c = \frac{n_2}{n_1}. \tag{4.6}$$

What should the relative values of n_1 and n_2 be to have maximum
internal reflection?

    e. An optical fiber can be either single-mode or multimode, depend-
ing on whether it carries just one or many light rays. Multimode
fibers are used to transmit a greater amount of information but
usually over shorter distances than single-mode fibers, which are
used for long-distance transmission of telephone and cable television signals. Why are multimode fibers used only for shorter
distances?

    f. What other properties of the optical fiber are important for its
application?

4.8 Learning Goals

- Refraction
- Refractive index
- Total internal reflection
- Dispersion of light
- Interference effects

- Constructive and destructive interference
- Origins of color in thin films
- Rayleigh scattering
- Diffraction grating
- Liquid crystals
- Photonic crystals
- Fiber optics

4.9 Problems

4.1 a. The depth of the water is not responsible for the difference in hue (color) between lake or ocean water close to the shore and water offshore. What difference does the depth make to the optical properties of water?

   b. The difference in the color of water near the shore and offshore is related to the abundance of suspended particles in the water near the shore due to the action of waves on the shoreline. Explain briefly how the presence of suspended particles might affect the color of the water.

4.2 There is a commercial product available for fever detection. It is a strip of material that is placed on the forehead, and it changes color if the person is feverish. Comment on the principles on which such a device could be based.

4.3 If you add a few drops of milk to a glass of water, darken the room, shine a flashlight directly at the glass, and observe the glass at right angles to the flashlight beam, what color will the milky suspension appear? What color will be observed looking through the glass into the flashlight beam? Explain.

4.4 In a consumer column in a newspaper published in Ottawa, Canada, a reader reported trouble with the liquid crystal display of a car's dashboard features in the winter. The display was initially inoperable, but the problem was rectified after the car ran for some time. What do you think the problem was?

4.5 Toys that change color on exposure to sunlight are available commercially. Suggest an explanation for this phenomenon.

4.6 Toys that change color on exposure to water are available commercially. Suggest an explanation for this phenomenon.

4.7 Toys that change color on cooling are available commercially. Suggest an explanation for this phenomenon.

4.8   *The Invisible Man* is a famous science fiction novel by H. G. Wells. According to our understanding of scientific principles, should the invisible man be able to see? Explain.

4.9   Ammolite is a gemstone, part of the crushed, fossilized mother-of-pearl shell of the prehistoric ammonite. These rare gemstones are found near the banks of the St. Mary and Bow Rivers in southern Alberta, Canada. They can display dazzling primary colors, especially red. Ammolite was formed from mother-of-pearl when some elements were added or taken away. An important factor in the production is that the ammonite shell was crushed by ancient aquatic reptiles, leaving each fragment tilted at a slightly different angle. What physical process gives rise to the colors of ammolite?

4.10 If you were required to design a material for use in eye shadow, you might base your selection on light absorbance (for color) and light interference effects (for sparkle). One such material used in eye shadow is titanium-dioxide-coated mica. Would you expect this to meet the requirements?

4.11 Hypercolor™ T-shirts change color (e.g., from pink to violet) when heated. It has been said that the color change cannot be based on thermochromic liquid crystals. Why not? Explain. Suggest another explanation for this effect.

4.12 Explain how one could use a diffraction grating to produce a monochromator. Use a diagram.

4.13 Consider the interference of light that is both front-reflected and back-reflected from a thin film. The thickness of the film is d. Consider the refractive index of the film to be the same as in the air surrounding the film. (Although a difference in refractive index is required to get reflection, this assumption will simplify the mathematical model.) The angle of incidence of the light to the film surface is θ, as shown in Figure 4.26.

   a. Derive a general expression for the condition of constructive interference of the front- and back-reflected beams for this system in terms of $m\lambda$, where m is the extra number of wavelengths for the back-reflected beam and λ is the wavelength of light under

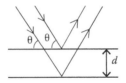

FIGURE 4.26
Light scattering from a thin film.

consideration. Keep in mind that the front-reflected beam will have a change in phase, whereas the back-reflected beam will not.

b. For light of $\lambda = 500$ nm, $\theta = 30°$, find d (in nm) for $m = 1, 2, 3$, and 10.

c. For light of $\lambda = 500$ nm and $d = 250$ nm, calculate the values of θ (in degrees) between $10°$ and $90°$ that give constructive interference.

d. If change in the refractive index on entering the film was considered, draw the back-reflected light compared with the case described earlier.

4.14 Why do stars twinkle? (The light source is continuous, not pulsating at the frequency of the twinkling.)

4.15 During a lunar eclipse (earth positioned between the sun and the moon), the moon appears red. Why?

4.16 An announcement by a Japanese car company presented "insect-inspired technologies," with colors of cars based on a South African butterfly. These insects appear to be brilliantly colored, but their fragile wings, which are covered with microscopic indentations, carry no pigment. The manufacturer produces colors by making the car's surface with irregularities spaced only hundreds of nanometers apart.

a. What is the principle on which the color is produced?

b. Discuss advantages and disadvantages of this type of color compared with the more usual car color due to pigments in paint.

4.17 After the eruption of a volcano, the sunset can have even more spectacular color than usual. Why?

4.18 A particular polymer is thermochromic. When heated from room temperature to $240°C$, the optical absorption peak blue shifts (i.e., goes to shorter wavelength). This change has been associated with a change in the degree of conjugation in the system.

a. Is the HOMO–LUMO gap larger or smaller at higher temperature? Explain your reasoning.

b. Is the system more or less conjugated at higher temperature? Explain your reasoning.

c. Propose a physical explanation for the change in conjugation.

4.19 Liquid crystals can be used to orient molecules (e.g., to examine direction-dependent properties, such as IR, UV-vis, and NMR parameters). How are liquid crystals useful for these studies? Suggest the advantages and limitations of this method.

4.20 Why are there no white gases?

4.21 A coffee mug has been advertised as having a "disappearing STAR TREK® crew." The crew appear on the mug when it is empty, but when the mug is filled with a hot beverage, the crew is no longer visible ("probably beaming down to some planet in distress"). Explain the principle of the disappearing crew.

4.22 The Pāua shell, native to New Zealand waters, has beautiful colors, mostly in the blue-violet range. Part of its attraction is that the observed hue changes with angle of viewing. Suggest a plausible physical origin for the source of the color. Explain your reasoning.

4.23 Sketch the path of a beam of light focused by a Fresnel lens. Show details of the light path through two adjacent stepped prisms in the lens as the light enters, passes through, and exits. To simplify matters, consider the light to be monochromatic.

4.24 Ice crystals can cause a rainbow halo (circle of light with the full rainbow of colors) to appear around the sun. Would red or violet of the rainbow halo appear closer to the sun? Explain.

4.25 Titanium, niobium, and tantalum are all used by metalwork artists as materials that can have iridescent colors introduced by electrochemical oxidation of the surface. Explain why the color observed depends on the extent of the surface oxidation.

4.26 What color is the Earth's sky when viewed from outer space? Explain any difference from the color of the sky as viewed from Earth.

4.27 A new mineral composed of Al, Ca, Mg, Fe, and O has a structure based on fibers somewhat like asbestos. This mineral has the remarkable quality of a striking blue color that changes to violet and cream on rotation. Draw a diagram that illustrates the interaction of light with this material and the resulting observed colors.

4.28 Air above hot pavement or hot desert sand has a steep temperature gradient, with cooler air at the higher elevation. Sketch the resultant mirage effect under these conditions, comparable with the diagram in Figure 4.4. Consider the real image to be a vertical arrow (head at the top) and show both it and the apparent image (mirage).

4.29 Titania (TiO_2) is sometimes added to laundry detergents to replace the TiO_2 that has washed away from synthetic fibers. The TiO_2 was added to the synthetic fibers in the manufacturing process to make them appear white instead of transparent or translucent.

   a. Describe the interaction of light with TiO_2. It has a very high refractive index (about 2.6).

   b. TiO_2 is somewhat photoactive, in that it can absorb a high-energy photon (e.g., UV light) and release an electron. Would this limit textile applications of TiO_2? Explain.

4.30 Shampoos come in many colors but their foams are always white. Why?

4.31 At first glance, ordinary glass and diamond look quite a bit alike. If both were cut to the classic diamond cut (see Figure 4.7), how would the dispersion of light compare for cut glass and diamond? Note the indices of refraction: n(glass)=1.4, n(diamond)=2.4.

4.32 Is the refractive index for glass ($n \approx 1.4$) greater or less for light of wavelength 400 nm compared with light of wavelength 700 nm?

4.33 The blue coloration of some bird feathers is due to scattering of light from barbs, which have a colorless transparent outer layer about 10^{-5} m thick over a layer of melanin-containing cells with irregularly shaped air cavities, <30 nm to about 300 nm across. Explain whether or not the blue color is expected to remain if the feather is crushed, destroying the air cavities.

4.34 Fog lights are manufactured to produce yellow light, rather than white light, so that less scattering of the light occurs from fine droplets of water. Explain why the color results in a difference in the light scattering.

4.35 It is well known in the television business that plaids and small-checked patterns should not be worn on camera as they do not reproduce well on television. Explain why.

4.36 A CD works by having the light of a laser hit the CD and, from a given spot, the light is either reflected to the detector or not.

    a. In a commercial CD, this situation is produced by having very small pits imprinted on the CD in the pressing process. Explain briefly why the presence of pits can lead to different light path for the monochromatic laser beam.

    b. In a writable CD, the information is "burned" on the CD by local heating of some sections and not others. The heating process turns an organic layer from transparent (allowing the light to hit the metal layer beneath and bounce back) to opaque. Explain how the opacity leads to an outcome similar to that in (a).

4.37 If the sun were cooler than it is, would the color of our sky change as we perceive it? Explain your reasoning. Use one or more diagrams. (Assume that nothing else on earth was changed and that we are still around!)

4.38 The moon has almost no atmosphere. If you were standing on the sunlit side of the moon, what color would the moon's "sky" be? Explain.

4.39 Moissanite is a naturally occurring mineral. Its chemical composition is silicon carbide, SiC. Moissanite is a very hard material, with a high refractive index (ca. 2.7). Synthetic moissanite costs about one-tenth the cost of diamonds, so it is a less expensive alternative for jewelry. How does moissanite compare with diamond for dispersion of light (the so-called brilliance of the gem)? Explain.

4.40 Chromatic aberration in photography is due to different colors of light having different indices of refraction, and therefore, after passing through the lens, different colors land at different locations on the sensor. This situation can give a color fringe around objects in the resulting image. Chromatic aberration can be reduced by using a smaller lens aperture, giving a greater depth of field by focusing the light rays closer together when they hit the sensor. However, if the aperture is too small, diffraction can result. Why is diffraction a problem for the resulting photographic image?

4.41 Some rocks show strong colors when observed in UV light; these colors are not observed in ordinary light. Explain the origin of the color in UV light. Use an energy level diagram.

4.42 There are walkways under the city streets in Seattle that receive daylight from the street above, by passing through a solid piece of glass that is embedded in the sidewalk. The top surface of the glass is flat, even with the sidewalk. Why does the glass have a curved underside, not a flat underside?

4.43 Film from old movies is often scratched. For example, a cross section of such a film might have a "V"-shaped notch in it.

   a. Draw a diagram to indicate the path of light through this damaged section of the film, compared with the light path through an undamaged section.

   b. If the notch is filled in during film restoration, what are the important optical properties to consider for the filler material?

Further Reading

General References

Many introductory physics and physical chemistry textbooks contain information concerning physical interactions of light with bulk matter.

All that glitters, 1999. *Chemistry in Britain*, November, 20.

Negative index materials. Special issue. W. Park and J. Kim, Eds. *MRS Bulletin*, October 2008, 907–934.

D. J. Blumenthal, 2001. Routing packets with light. *Scientific American*, January, 96.

W. D. Callister, Jr. and D. G. Rethwisch, 2013. *Materials Science and Engineering: An Introduction*, 9th ed. John Wiley & Sons, Hoboken, NJ.

G. P. Collins, 2003. Heat and light. *Scientific American*, January, 26.

G. P. Collins, 2003. Holographic control. *Scientific American*, July, 22.

C. Day, 2002. Why do lobsters change color when cooked? *Physics Today*, November, 22.

R. M. Evans, 1948. *An Introduction to Color*. John Wiley & Sons, Hoboken, NJ.

R. E. Hummel, 2004. *Understanding Materials Science*, 2nd ed. Springer, New York.

A. Javan, 1967. The optical properties of materials. *Scientific American*, September, 239.

M. G. J. Minnaert, 1993. *Light and Colour in the Outdoors*, 5th ed. Springer-Verlag, New York.

C. G. Mueller, M. Rudolph, and the Editors of LIFE, 1966. *Light and Vision*. Life Science Library, Time Inc, New York.

K. Nassau, 1980. The causes of colour. *Scientific American*, October, 124.

K. Nassau, 2001. *The Physics and Chemistry of Color*, 2nd ed. John Wiley & Sons, Hoboken, NJ.

I. P. Parkin and T. D. Manning, 2006. Intelligent thermochromic windows. *Journal of Chemical Education*, March, 393.

T. D. Rossing and C. J. Chiaverina, 1999. *Light Science: Physics and the Visual Arts*. Springer, New York.

H. Rossotti, 1983. *Colour*. Princeton University Press, Princeton, NJ.

F. C. Sauls, 2006. A Demonstration of refractive index matching using isopropyl alcohol and MgF_2. *Journal of Chemical Education* 83, 1170.

R. J. D. Tilley, 2004. *Understanding Solids: The Science of Materials*. John Wiley & Sons, Hoboken, NJ.

F. Vögtle, Ed., 1991. Chapter 8: "Liquid crystals." In *Supramolecular Chemistry: An Introduction*. John Wiley & Sons, Hoboken, NJ.

V. F. Weisskopf, 1968. How light interacts with matter. *Scientific American*, September, 60.

E. A. Wood, 1977. *Crystals and Light: An Introduction to Optical Crystallography*. Dover, New York.

G. Wyszecki and W. S. Stiles, 1967. *Color Science*. John Wiley & Sons, Hoboken, NJ.

Fiber Optics

A. M. Glass, 1993. Fiber optics. *Physics Today*, October, 34.

D. Hewak, 1997. Travelling light. *Chemistry in Britain*, May, 26.

Films (including Soap Films)

F. J. Almgren and J. E. Taylor, 1976. The geometry of soap films and soap bubbles. *Scientific American*, July, 82.

C. V. Boys, 1959. *Soap Bubbles*. Dover, New York.

C. Isenberg, 1992. *The Science of Soap Films and Soap Bubbles*. Dover, New York.

Interference Effects

S. K. Blau, 2004. Light as a feather: Structural elements give peacock plumes their color. *Physics Today*, January, 18.

A. Goldberg-Gist, 2003. Opal. *Chemical and Engineering News*, January 27, 58.

R. Greenler, 1994. Sunlight, ice crystals, and sky archeology. In *The Candle Revisited*, P. Day and R. Catlow, Eds. Oxford University Press, Oxford.

D. Hawaleshka, 2001. Fingerprinting diamonds. *Maclean's Magazine*, February 26, 54.

A. J. Kinneging, 1993. Demonstrating the optic principles of Bragg's law with moiré patterns. *Journal of Chemical Education*, 70, 451.

G. Oster and Y. Nishijima, 1963. Moiré patterns. *Scientific American*, May, 54.

D. Psaltis and F. Mok, 1995. Holographic memories. *Scientific American*, November, 70.

P. Vukusic, J. R. Sambles, C. R. Lawrence, and R. J. Wootton, 2001. Structural colour: Now you see it—Now you don't. *Nature*, 410, 36.

Liquid Crystals

Materials for flat panel displays. Special Issue. J. S. Im and A. Chiang, Eds. *MRS Bulletin*, March 1996, 27–68.

J. D. Brock, R. J. Birgeneau, J. D. Litser, and A. Aharony, 1989. Liquids, crystals and liquid crystals. *Physics Today*, July, 52.

G. H. Brown, 1972. Liquid crystals and their roles in inanimate and animate systems. *American Scientist*, 60, 64.

P. G. Collings, 2002. *Liquid Crystals: Nature's Delicate Phase of Matter*, 2nd ed. Princeton University Press, Princeton, NJ.

P. G. de Gennes and J. Prost, 1993. *The Physics of Liquid Crystals*, 2nd ed. Clarendon Press, Oxford.

E. Demirbas and R. Devonshire, 1996. A computer experiment in physical chemistry: Linear dichroism in nematic liquid crystals. *Journal of Chemical Education*, 73, 586.

J. L. Fergason, 1964. Liquid crystals. *Scientific American*, August, 76.

M. Freemantle, 1996. Polishing LCDs. *Chemical and Engineering News*, December 16, 33.

M. Freemantle, 2002. Paintable liquid-crystal displays. *Chemical and Engineering News*, May 6, 11.

G. H. Heilmeier, 1970. Liquid-crystalline display devices. *Scientific American*, April, 100.

R. H. Hurt and Z.-Y. Chen, 2000. Liquid crystals and carbon materials. *Physics Today*, March, 39.

M. Jacoby, 2005. Liquid-crystal collage. *Chemical and Engineering News*, January, 7.

S. Kumar, Ed., 2000. *Liquid Crystals: Experimental Study of Physical Properties and Phase Transitions*. Cambridge University Press, Cambridge.

A. J. Leadbetter, 1990. Solid liquids and liquid crystals. *Proceedings of the Royal Institute of Great Britain*, 62, 61.

S. G. Steinberg, 1996. Liquid crystal displays. *Wired*, January, 68.

R. Templer and G. Attard, 1991. The world of liquid crystals. *The New Scientist*, May 4, 25.

J.-F. Tremblay, 2005. Thinning flat panels. *Chemical and Engineering News*, June 27, 20.

J.-F. Tremblay, 2006. Riding on flat panels. *Chemical and Engineering News*, June 26, 13.

G. R. Van Hecke, K. K. Karukstis, H. Li, H. C. Hendargo, A. J. Cosand, and M. M. Fox, 2005. Synthesis and physical properties of liquid crystals: An interdisciplinary experiment. *Journal of Chemical Education*, 82, 1349.

E. Wilson, 1999. Liquid crystals find their voice. *Chemical and Engineering News*, September 27, 13.

Opals

P. J. Darragh, A. J. Gaskin, and J. V. Saunders, 1976. Opals. *Scientific American*, April, 84.

Photonic Crystals

J. D. Joannopoulos, S. G. Johnson, J. N. Winn, and Robert D. Meade, 2008. *Photonic Crystals: Molding the Flow of Light*, 2nd ed. Princeton University Press, Princeton, NJ.

Y. Liu and X. Zhang, 2011. Metamaterials: A new frontier of science and technology, *Chemical Society Reviews*, 40, 2494.

C. M. Soukoulis and M. Wegener, 2011. Past achievements and future challenges in the development of three-dimensional photonic metamaterials. *Nature Photonics*, 5, 523.

P. Vukusic and J. R. Sambles, 2003. Photonic structures in biology. *Nature*, 424, 852.

E. Yablonovitch, 2001. Photonic crystals: Semiconductors of light. *Scientific American*, December, 46.

Thermochromism

D. Lavabre, J. C. Micheau, and G. Levy, 1988. Comparison of thermochromic equilibria of Co(II) and Ni(II) complexes, *Journal of Chemical Education*, 65, 274.

D. C. MacLaren and M. A. White, 2005. Design rules for reversible thermochromic mixtures. *Journal of Materials Science*, 40, 669.

M. A. White and M. LeBlanc, 1999. Thermochromism in commercial products. *Journal of Chemical Education*, 76, 1201.

Undergraduate Laboratory Experiment

B. Millier and G. Aleman Milán, 2014. A Cost-Effective Optical Device for the Characterization of Liquid Crystals. *Journal of Chemical Education*, 91, 518.

Websites

For links to relevant websites, see PhysicalPropertiesOfMaterials.com

5

Other Optical Effects

5.1 Introduction

Although color is an aesthetically pleasing optical effect and can also have many uses, there are other interesting optical properties of materials that do not necessarily involve color. Some of these effects are described in this chapter.

5.2 Optical Activity and Related Effects

Normally, light is *unpolarized*, that is, it contains waves with electromagnetic fields oscillating in all directions perpendicular to the direction of the light's motion, as shown schematically in Figure 5.1.

When light passes through a filter called a *polarizer*, only one oscillation direction of the electromagnetic field is allowed through. On emerging from the polarizer, the light is said to be *polarized* (or, more specifically plane or linearly polarized; see Figure 5.1).

Some materials have the property of rotating the oscillation direction of the electromagnetic field of linearly polarized light; this property is called *optical activity*. Molecules or crystals that are *chiral* (i.e., not superimposable on their mirror image) are optically active, but the degree of optical activity depends on the particular material. It also depends on the state of matter; a crystal can derive its optical activity from its chiral packing arrangement or from the chirality of the constituent molecules. In the latter case, there will still be optical activity even if the crystal is melted or dissolved in solution. A 1-mm path length of quartz (which is an optically active crystal) can rotate the plane of polarization by about 20°.

Cholesteric liquid crystals, which we met in Chapter 4, have layers of chiral molecules, arranged so that each layer has a different orientation. As shown schematically in Figure 5.2, the orientation of the plane of polarization rotates as it passes through a cholesteric liquid crystal, and the optical activity can be

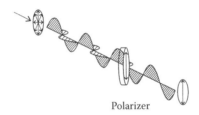

Polarizer

FIGURE 5.1
Normal unpolarized light contains waves with oscillating electromagnetic fields in all directions perpendicular to the direction of the light's motion as indicated by the wheel at the upper left (start of the light wave). For clarity, only two wave oscillation directions are shown here. Plane polarized light contains light of only one oscillation direction, and this oscillation can be selected from unpolarized (ordinary) light by a polarizer, giving linearly (or plane) polarized light.

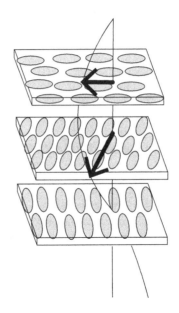

FIGURE 5.2
A schematic structure of a cholesteric liquid crystal. The arrows indicate the direction of the polarization vector, which is rotated as it passes through the material.

among the highest known, as much as 50 full rotations (i.e., $50 \times 360° = 18,000°$) for a 1-mm path length! This optical activity can be used to advantage in a liquid crystal display (LCD) device, as detailed below.

In a *field-effect LCD*, each digit of the display device is composed of seven pairs of electrode sandwiches (see Figure 5.3). The two outer layers, like the bread of a sandwich, are composed of a transparent material

See the video "Polarizers" under Student Resources at PhysicalPropertiesOfMaterials.com

FIGURE 5.3
Seven pairs of electrodes are required for each digit in a liquid crystal display.

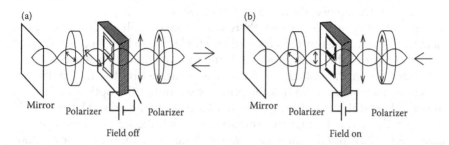

(a)   (b)

Mirror   Polarizer   Polarizer   Mirror   Polarizer   Polarizer

Field off   Field on

FIGURE 5.4
A schematic view of a field-effect liquid crystal display (LCD) device. (a) When the field across the electrode containing the cholesteric liquid crystals is off, the light enters, is linearly polarized (the direction is indicated by the double-headed arrow), has its plane of polarization rotated by the liquid crystal, passes through a matched polarizer, is reflected by a mirror, and reverses the entire process on the way back to the outside world. Since no light is absorbed, the LCD digit has no color. (b) When an electric field is applied across the electrode containing the cholesteric liquid crystal (switch closed), its optical activity is changed. This means that the entering polarized light is rotated by an amount different from the field-off case, and the light polarization no longer matches the second polarizer. Therefore, the light cannot pass through this polarizer and cannot reach the mirror to be reflected out. Since light is absorbed in this process, the LCD digit appears black in the presence of the electric field.

(e.g., glass) coated with a transparent, electrically conducting material, typically a thin layer of indium tin oxide; between these layers there is a layer of cholesteric liquid crystal. Activation of each of the seven elements of the digit is caused by the application of an electric field; independent activation or deactivation of the seven electrodes can allow display of the digits from 0 to 9.

Another important factor in this display involves plane polarized light. Outside light passing into a LCD device first passes through a polarizer (Figure 5.4). It then travels into the cholesteric liquid crystal cell where the plane of polarization is rotated by the optically active liquid crystal. The display device is constructed so that this new direction of the plane of polarization exactly matches the direction of polarization of the next polarizer in the light path. Because the polarization directions match, the light passes through this second polarizer. When the light hits a mirror at the back of the display device, it is reflected back along the original path and toward the outside world. This light is still polarized in the direction of the innermost polarizer, and it continues outward. When this polarized light encounters the liquid crystal cell, the polarization is changed by an amount equal but opposite in direction to the change on passing into the cell from the other side. For example, as shown in Figure 5.4a, if the change in polarization was +90° on entering the cell, it is −90° on leaving the cell. This rotation allows the polarization of the light exiting the cell to match that of the outermost polarizer so that the light can exit the device. Since light entered the device and it also can exit the device, the cell appears transparent.

In the situation just described, there was no field applied across the seven electrode sandwiches in the cell. If a small electric field is applied (Figure 5.4b), the color of the electrodes is modified for the following reasons. Light still enters from the outside world and is polarized by the first polarizer. However, the electric field has the effect of changing the optical activity of the liquid crystal by modifying the structure of the material. The change in optical activity means that the amount by which the liquid crystal rotates polarized light is not the same as it was without the field, and the new polarization of the light emerging from the liquid crystal in the electric field no longer matches the innermost polarizer. Since the polarizer only passes light of the matched polarization, the light that hits this polarizer cannot go further. Therefore, light is absorbed with no reemission, and the electrode elements appear black. Each of the seven elements actually has its own electric field control, so each can be on or off, allowing the full range of digits from 0 to 9 to be selectively produced.

This simple LCD device was made possible technically by the development of thin coatings of transparent electrodes, usually indium tin oxide. Without this development, light could not pass into the liquid crystal cell.

The type of LCD just described requires an external light source, and these are referred to as *reflective*. The external light is reflected back, or not, depending on the state of the electrodes. Ambient light is required to see the output.

Backlit LCDs use an internal light source. For example, most LCD computer monitors are lit with built-in fluorescent lights around and behind the LCD. The fluorescent light is diffused and scattered by a panel behind the LCD to provide uniform lighting. However, about half of this light is lost by absorption in the device and eventually converted to heat.

The terms *passive-matrix* and *active-matrix* refer to the way in which each pixel in the LCD is activated. In the passive-matrix mode, a row and a column in a grid supply voltage to the pixel at their intersection point. Active-matrix LCDs address a particular pixel by switching on the row and then sending a charge down the correct column; at the intersection, a thin-film capacitor can be carefully charged and the pixel activated. Although the passive-matrix device is simpler, it has the disadvantage of slow response time and ghost images on neighboring pixels, reducing resolution. These disadvantages are overcome with the active-matrix approach, but they are somewhat more complex (and expensive) to produce.

A colored LCD has three subpixels with red, green, and blue color filters for each pixel. Making use of light absorption properties, the resulting optical effect can be color controlled. For an active-matrix LCD for a laptop computer, more than 3 million thin-film transistors need to be present. This is impressive, as is the fact that more than 15 million different shades can be produced at each pixel.

5.3 Birefringence

For materials that are *anisotropic* (*not* the same in all directions, the opposite of isotropic), an unusual optical property called *birefringence* (also known as *double refraction*) can occur. Birefringence was first described in 1669 by Bartholin,* and explained by Huygens† in 1678 using the wave theory of light. Examples of anisotropic materials are liquid crystals (e.g., the smectic and nematic structures with the molecules aligned along the long axis) and also noncubic solids. In a liquid, the atomic arrangements are, on average, the same in the x-, y-, and z-directions, and therefore they are isotropic, but such is not the case for an anisotropic solid.

Birefringence in an anisotropic material arises because unpolarized incident light is split into polarized components (two components for tetragonal and hexagonal crystals; three components for less symmetric crystals) because the speed of light is different in different directions (i.e., the refractive indices are different in different directions; see Figure 5.5). Thus, parallel but displaced beams of polarized light emerge from the anisotropic material (Figure 5.5). Birefringence in a colorless transparent crystal of Iceland spar is shown in Figure 5.6.

See the video "Polarizers and Birefringence" under Student Resources at PhysicalPropertiesOfMaterials.com

---

* Erasmus Bartholin (1625–1698) was a Danish scientist who made contributions in mathematics and astronomy. His greatest contribution was his study of double refraction in Icelandic spar (collected by an expedition to Iceland in 1668), but his explanation was faulty: He assumed there was a double set of "pores" that led to two refraction rays.

† Christiaan Huygens (1629–1695) was a Dutch scientist who contributed to many areas, including mathematics, statics, dynamics, astronomy, and optics. In addition, Huygens discovered Titan (a satellite of Saturn), was the first to realize that Saturn's odd shape was due to its rings, and he also invented the pendulum clock. His wave theory of light was used to explain optical phenomena, including double refraction.

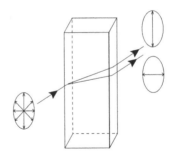

FIGURE 5.5
The effect of double refraction, also known as birefringence, in a hexagonal material. Because the material is anisotropic, the refractive index is not the same in all directions. The different refractive index in different directions causes unpolarized incident light to be split into two polarized components with different speeds, giving rise to two emerging beams of light.

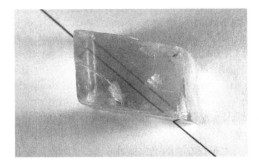

FIGURE 5.6
A sample of Iceland spar (also known as calcite, $CaCO_3$) showing birefringence, which makes a double image appear when looking (here at a single line) through the crystal.

The presence of birefringence is a straightforward way to detect anisotropy in a transparent material. In addition, birefringence can lead to *dichroism*, that is, two different colors of material when viewed in different directions.

5.4 Circular Dichroism and Optical Rotatory Dispersion

We have seen in Figure 5.1 that light can be polarized in a particular direction, giving linearly (or plane) polarized light. Light also can be polarized such that the direction of polarization rotates in the plane perpendicular to the direction of propagation of the light wave. Such a rotation can be clockwise or anticlockwise, giving rise to right- and left-*circularly polarized light*, respectively, as shown in Figure 5.7.

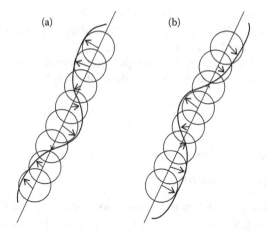

FIGURE 5.7
(a) Right- and (b) left-circularly polarized light. The polarization vectors (shown as arrows) rotate in opposite directions for the two forms of circularly polarized light.

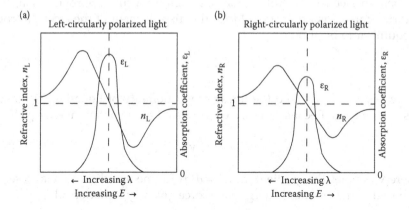

FIGURE 5.8
The variation of refractive index, n, and molar absorption coefficient, ε, as functions of wavelength, as measured with left- and right-circularly polarized light (subscripts L and R, respectively). When $\Delta\varepsilon = \varepsilon_L - \varepsilon_R \neq 0$, the material exhibits circular dichroism. When $\Delta n = n_L - n_R \neq 0$, the material exhibits optical rotatory dispersion.

The refractive index, n, and also the absorption coefficient, ε, can depend on both the wavelength of light and whether the light is right- or left-circularly polarized. This result is illustrated schematically in Figure 5.8.

The difference in the absorption coefficient for right- and left-circularly polarized light is referred to as *circular dichroism* (CD). The origin of this name is in the fact that the different absorption coefficients for left- and right-circularly polarized light lead to different colors (i.e., dichroism), depending on the circular polarization of the light used to make the observation.

A nonzero difference in refractive index for right- and left-circularly polar-ized light is called *optical rotatory dispersion* (ORD). If both CD and ORD are present, this is referred to as the *Cotton effect*.

Investigation of CD and ORD can provide information concerning the structure, configuration, and conformation of complicated optically active molecules.

5.5 Nonlinear Optical Effects

The term *nonlinear effect* refers to the nonlinear proportionality of a response to an input. It is a general concept, and most physical processes, when exam-ined closely, are nonlinear in detail although they may be approximately linear.* Let us consider the case of an electron bound to an atom and its potential energy, as this situation is important in considering the interaction of light with matter, especially when more than one photon is involved.

When an electron is pulled from an atom, to a first approximation, the restoring force F is proportional to x, the displacement of the electron from its equilibrium position[†]:

$$F = -kx \tag{5.1}$$

where k is a constant, and the linear proportionality of Equation 5.1 makes this a *linear effect*. Because any force F is related to the potential energy, V, as

$$F = -\nabla V \tag{5.2}$$

where ∇ denotes the gradient (i.e., $\nabla = \partial/\partial x$ for the one-dimensional case), it follows that the potential energy for a force given by Equation 5.1 is

$$V = \frac{1}{2}k\,x^2 \tag{5.3}$$

where the potential energy has been chosen to be zero at $x = 0$. Equation 5.3 corresponds to a harmonic potential, as shown in Figure 5.9.

When an electric field E (e.g., light) interacts with electrons in a material, the electron density can be provoked to be displaced. When the optical effect

---

* Any effect is nonlinear if the response to an input (i.e., the output) changes the process itself. In the case of nonlinear optical effects, the interaction of light with a material causes the material's properties to change such that the next photon that arrives will have a different interaction with the material.

† The most important equations in this chapter are designated with ▮ to the left of the equation.

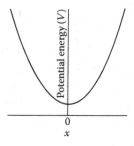

FIGURE 5.9
Potential energy as a function of x, the displacement of an electron from an atom in the harmonic approximation.

is linear, this situation leads to an *induced* dipole (or *polarization*) of magnitude μ_i:

$$\mu_i = \alpha E \tag{5.4}$$

where α is the *linear polarizability*. This relationship is illustrated in Figure 5.10.

Two important optical properties arise directly from *linear optical effects*—the dielectric constant (ε; note that this is the same symbol as used above for absorption coefficient but these properties are distinct) and the refractive index (n). The *dielectric constant* is defined as the ratio of the total electric field within the material, D (also called the displaced field), to the applied electric field of the light, E:

$$\varepsilon = \frac{D}{E} \tag{5.5}$$

and, in the absence of absorption of light, the refractive index is related to the dielectric constant ε by the following equation:

$$\varepsilon(v) = n^2(v) \tag{5.6}$$

where the dependence on the frequency of the light, v, is given explicitly. Both Equations 5.5 and 5.6 are written in their isotropic forms, but we know that refractive index can be anisotropic (as observed by birefringence, for example), and Equations 5.5 and 5.6 could be written in tensor form to account for directional dependence explicitly.

The harmonic restoring potential in a linear optical material gives rise to a motion of charge (i.e., electric field oscillation) that is of the same frequency as the light (Figure 5.10). From a quantum mechanical perspective, the oscillating applied electric field induces a time-dependent induced polarization of the molecule that can be expressed as a mixing of the ground state and polarized excited states. (These are called *virtual transitions* because there is excited-state character but no long-lived population of the excited states.)

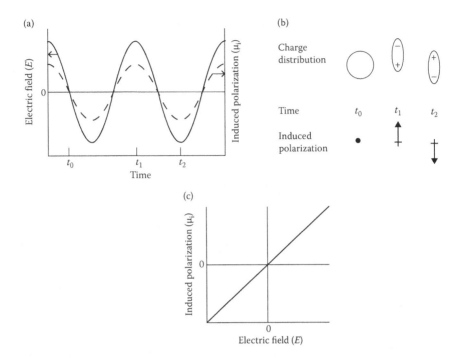

FIGURE 5.10
Linear optical effects. (a) The electric field (solid line) and the induced polarization wave (dotted line) both vary as sine waves in time for an optically linear material. (b) Schematic of the induced polarization of an optically linear material interacting with light as a function of time. (c) The induced (linear) polarization as a function of applied field for an optically linear material. (Figure styled after S. R. Marder et al., Eds., 1991. Materials for nonlinear optics: Chemical perspectives. *American Chemical Society Symposium Series 455*, ACS.)

The motion of charge (or, equivalently, mixing of excited states) leads to re-emission of the radiation at the same frequency as the polarizing light.

Many times in physical science the harmonic model is a good first approximation but closer examination indicates the importance of other terms in the potential. If the restoring force of Equation 5.1 can more accurately be written as

$$F = -kx - \frac{1}{2}k'x^2 \tag{5.7}$$

this form is now nonlinear. Equation 5.7 leads to the following expression for potential energy:

$$V = \frac{1}{2}kx^2 + \frac{1}{6}k'x^3 \tag{5.8}$$

that is, there is an additional anharmonic term compared to Equation 5.3. This potential is illustrated in Figure 5.11.

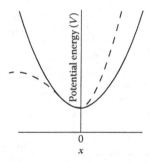

FIGURE 5.11
Potential energy as a function of displacement coordinate for a material with a harmonic potential (solid line, Equation 5.3) and for a material with an additional anharmonic term (broken line, Equation 5.8).

The anharmonic potential depends on the direction of the displacement of the electron (whereas Equation 5.3 gives the same value of V for x and $-x$, Equation 5.8 gives values of V that depend on the sign of x), and therefore the magnitude of the polarization also depends on direction. An example of such a material is one where there is an anisotropic charge distribution in the absence of an electric field (e.g., a noncentrosymmetric polar molecule in the gas phase or a noncentrosymmetric polar crystal). In such a case, the electric field of the light can displace the charge more in one direction than in the other, and the resulting polarization can be a *nonlinear optical (NLO) effect* (Figure 5.12).

In an NLO material, the polarization μ $(= \mu^0 + \mu_i$ where μ^0 is the *static dipole*, that is, in the absence of the field, and μ_i is the induced dipole) can be written as a Taylor series expansion:

$$\mu = \mu^0 + \alpha_{ij}E_j + \frac{1}{2}\beta_{ijk}E_jE_k + \frac{1}{6}\gamma_{ijkl}E_jE_kE_l + \cdots \qquad (5.9)$$

where this equation is written in the tensor form. The terms beyond αE (which is the *linear polarizability;* see Equation 5.4) are not linear in E; they are referred to as nonlinear polarization terms and give rise to NLO effects. The parameters β and γ are referred to as the first and second *hyperpolarizabilities,* respectively. For β to be nonzero, there must be no inversion center in the material. However, γ is nonzero for all materials although it induces weaker effects. Since $\alpha E \gg \beta E^2$ and γE^3, observation of NLO effects requires large electric fields, and lasers have been very important in developing this subject.

In a linear optical material, $\beta = 0$, the induced electric field was at the same frequency as the light, and the light emitted was purely of the same frequency as the exciting light. In an NLO material, $\beta \neq 0$, the induced polarization is not a pure sine wave at the frequency of the light (see Figure 5.12), but it can be considered to be composed of that frequency (the fundamental) and its harmonics, as well as an offset (DC component), as shown in Figure 5.13. The static polarization gives rise to the DC component, the linear

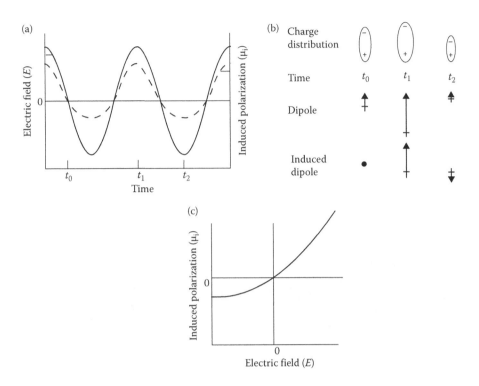

FIGURE 5.12
Nonlinear optical effects. (a) The electric field (solid line) and the induced polarization wave (dotted line) both vary as functions of time for an optically nonlinear material. Note that both the electric field and induced polarization have the same repeat time, but only the former is a pure sine wave. (b) Schematic of the dipole and induced dipole (which is the difference between the dipole at that time and at time t_0) of an optically nonlinear material interacting with light as a function of time. Note that the magnitude of the induced dipole depends on the direction of polarization. (c) The induced (nonlinear) polarization as a function of applied field for an optically nonlinear material. (Figure styled after S. R. Marder et al., Eds., 1991. Materials for nonlinear optics: chemical perspectives. *American Chemical Society Symposium Series 455*, ACS.)

polarizability gives rise to the fundamental (as in linear optical materials), and the hyperpolarizabilities give rise to the harmonics.

If light of frequency ν is incident on an NLO material, light of frequency ν and 2ν (and to a lesser extent 3ν, 4ν, etc.) will result. The production of light of frequency 2ν in this manner is known as *frequency doubling* by *second-harmonic generation* (SHG). This property is used in practice in NLO devices to achieve frequencies twice that of the incident light, for example using $Ba_2NaNb_5O_{15}$, KH_2PO_4, $LiNbO_3$ or HIO_3. (The associated DC component of the electric field is referred to as *optical rectification*.) SHG is considered to be a *three-wave mixing* process since two photons with frequency ν are required to combine to give a single photon with frequency 2ν (energy is conserved). By analogy, third-order harmonics, at frequency 3ν from the γE^3 term of Equation 5.9, require *four-wave mixing*.

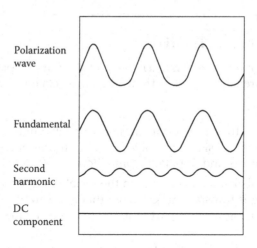

FIGURE 5.13

Fourier analysis of the asymmetric polarization wave of Figure 5.12a, showing its components: the fundamental (frequency v), the second harmonic (frequency $2v$), and the offset (DC component).

In the case of two beams of light of frequencies v_1 and v_2 interacting with an NLO material, it is also possible to have emitted light at the *sum frequency* $(v_1 + v_2)$ and at the *difference frequency* $(v_1 - v_2)$ in processes referred to generally as *sum frequency generation* (SFG).

Other NLO effects include the dependence of the refractive index of a NLO material on the applied voltage (*linear electrooptic* [LEO] or *Pockels effect*) where the applied voltage distorts the electron density within the material.

Nonlinear optical properties present exciting challenges to the materials researcher. Advances in this area have been greatly helped by the use of high-powered lasers. Current research in this area includes design of materials to have enhanced NLO properties, design of materials to examine the role of ground state and excited state structure in NLO properties, and further research in structure/property relations. One promising area involves polymeric NLO devices that could be used to carry more information than present optoelectronic devices.

COMMENT: PHOTOREFRACTIVE EFFECT

Another nonlinear optical property is the *photorefractive effect*, which describes the dependence of the index of refraction of the material on the local electric field. This property can change the refractive index of a material as light passes through it. The photorefractive effect allows the imprinting of an interference pattern on a material, and photorefractive materials will lead to more nearly perfect resolution, and retrieval of digitized images.

5.6 Transparency: A Tutorial

For application of an optical material, one of the most important considerations is transparency. Of course, transparency is related in part to the color of the material.

a. Is it possible that a material can absorb visible light but be transparent to light outside the visible range? Consider narrow band gap semiconductors and organic chromophores.

b. If a material absorbs negligibly in the visible range, does this guarantee that it is transparent? Consider that SiO_2 can be very transparent to visible light (e.g., quartz tubing) or quite opaque as in sand at the beach.

c. If the material in question does not absorb visible light when pure, it can be made to have interactions that lead to opacity by the introduction of second-phase particles or pores. The basis of the opacity is refraction at the pores. Is the size of the pore important? If so, approximately what size range would be most important? What other factors concerning the pores are important?

d. Window glass can be frosted or dimpled, rendering it translucent. What principles are involved here?

e. Some materials are designed to be transparent to only certain polarizations of light. Explain the principles involved here.

f. Although some materials are quite transparent, a significant portion of the light impingent on them can be reflected. This reflection might be unwanted as, in use, it could lead to glare. One way to reduce glare is to coat the material with another material. What factors will be important in selecting and applying the coating?

g. Early crystallographers knew that repeat units in crystal structures were smaller than 400 nm because otherwise crystals would appear frosted on the surface. Explain.

h. What does optical transparency of ordinary glass imply about the band gap in this medium?

5.7 Learning Goals

- Linearly polarized light
- Optical activity

- Field-effect liquid crystal display
- Birefringence
- Circular dichroism and optical rotatory dispersion
- Nonlinear optical effects
- Second-harmonic generation
- Sum frequency generation
- How to modify optical transparency

5.8 Problems

5.1 When linearly polarized light is passed through a normally isotropic transparent object that is under stress, and the transmitted light is viewed through another polarizer, colored fringe patterns can be observed. (This phenomenon is the origin of the color image on the cover of this book.) Suggest an explanation for this phenomenon.

See the video "Stress Between Polarizers" under Student Resources at PhysicalPropertiesOfMaterials.com

5.2 Polaroid™ sunglasses have lenses that are polarized to absorb the glare from surfaces in front of the wearer, such as a road, water, and snow, preventing the glare from reaching the eyes. What polarization orientation do the glasses absorb? Explain briefly.

5.3 The absence of a center of symmetry in a crystal is essential for it to exhibit (second-order) NLO properties. However, this condition is not sufficient. Why not?

5.4 a. Give an example of a material (it can be hypothetical), that is

   i. isotropic and centrosymmetric;

   ii. anisotropic and centrosymmetric;

   iii. anisotropic and noncentrosymmetric.

   b. Is it possible to give an example of a material that is isotropic and noncentrosymmetric? Explain.

5.5 The song "Anthem" by Leonard Cohen contains the following line in the refrain:

There is a crack, a crack in everything. That's how the light gets in.

Comment.

5.6 Dry sand is light brown in color, and water is virtually colorless. When water is added to dry sand, the resulting wet sand is dark brown. Why does the color of the sand depend on how wet it is?

5.7 Some polymer films are transparent but become translucent when stretched. Suggest an explanation. Consider the ordering of the polymer molecules in the original form compared with the stretched form.

5.8 Stained glass windows look rather dull from outside a building but quite beautifully colored when viewed from inside the building. Why is there an apparent difference in color?

5.9 The difference in color of a material in reflected light, compared with transmitted light, can be an important design consideration. For example, when a dye is bound to a transparent backing and light is passed through it, this leads to color due to transmitted light. The color may appear very different in reflected light. Explain why the color is different in the two cases.

5.10 Iridescent plastic films are composed of thin layers of two or more types of polymer. These materials have different colors when viewed in reflection and transmission. Furthermore, the colors viewed on reflection depend on the viewing angle. Sketch diagrams that show these effects when white light is incident on such a film. For simplicity, just consider two polymer layers.

5.11 Why is solid candle wax translucent, whereas molten candle wax is transparent?

5.12 When chiral molecules (i.e., those that are not superimposable on their mirror image) are incorporated into cholesteric liquid crystals, the different enantiomers (i.e., nonsuperimposable mirror image pairs) perturb the pitch of the liquid crystals (and hence the color of the system) in different ways. It is often difficult to distinguish enantiomers by usual chemical means as they can have the same melting points, boiling points, etc. Can you suggest an application of the different colors in distinguishing enantiomers? (This has been done successfully: T. Nishi, A. Ikeda, T. Matsuda, S. Shinkai, 1991. *Journal of the Chemical Society, Chemical Communications*, 339.)

5.13 Ruby can exhibit dichroism: violet-red versus orange-red. The $^4T_1$ and $^4T_2$ levels (see Figure 2.9) can each be split in two due to distortion of the octahedral symmetry at the Cr^{3+} site; if a ruby crystal is viewed in polarized light, it can be rotated to see the effect of absorption to one set of levels in one direction and another set in another direction. In the direction in which the absorption is between the ground state and the higher excited state, is the color violet-red or orange-red? Explain.

5.14 a. When a crystal that is birefringent but transparent is placed over a piece of paper on which a straight line has been drawn, explain what the resulting image looks like from above, and explain why.

b. If the same crystal has a polarizer placed on top of it, what could one expect to see looking from above as the polarizer is rotated? Explain briefly.

5.15 Polarizers are traditionally made by adding small amounts of mineral components, such as tourmaline or peridot, to molten glass and stretching the glass in one direction while it is in its flexible state. What does this process indicate about the structure of the mineral additive, and why is the glass stretched?

5.16 3D movies provide stereoscopic vision, usually by making separate images reach the viewers' left and right eyes. The two main approaches are the use of complementary colors and the use of polarized light.

a. In the case of anaglyph images, the projected image contains different aspects in different colors, and the movie is viewed through glasses in which the color of one lens is the complement to the other (most commonly red and cyan). Explain how the anaglyph lenses work to produce stereoscopic vision.

b. In the case of polarized images, polarized light of opposite polarizations is projected separately. The light can be either polarized linearly (the opposite polarizations would be perpendicular to or parallel to the vertical), or polarized circularly (the two images would be left- and right-circularly polarized). The viewer would wear glasses in which the lenses are of opposite polarizations (linearly for the linear method, circularly for the circular method). One drawback to this method is the need for a very flat silver screen to maintain the polarization on reflection from the screen to the viewers' eyes. Another drawback is the potential need for two projectors, one to project each image. Suggest some ways that the latter drawback could be overcome. Compare your suggestions with current technologies for 3D movies, such as RealD cinema and Dolby's 3D system.

5.17 As discussed in Chapter 1, the consumption of indium for indium tin oxide (ITO) transparent electrode materials has led to significant price increases and possibly a perilous world supply of the element.

a. If the layer of ITO is about 200 nm thick on each electrode, calculate the mass of indium in a display screen that is 30 cm × 20 cm. State and justify the assumptions in your calculation.

b. Based on your answer to (a), is it easy to recover the indium from the display screen at the end of its useful life? Discuss.

5.18 If a sheet of linearly polarized material is placed in the beam from some LCD projectors, the color viewed on the screen from the beam that passed through the polarizer depends

See the video "Polarized Light from a Liquid Crystal Display Projector" under Student Resources at PhysicalPropertiesOfMaterials.com

on the polarizer's orientation. For example, the projected image might appear green with one polarizer orientation and blue with another. What does this effect indicate about the polarization of the beam from the LCD projector?

Further Reading

General References

Many introductory physics and physical chemistry textbooks contain information concerning polarized light and its interaction with matter.

Materials for optical data storage. Special issue. H. Coufal and L. Dhar, Eds. *MRS Bulletin*, April 2006, 294–393.

S. Chang, 2014. Tailor-made surface swaps light polarization. *Physics Today*, August, 18.

R. Cotterill, 2008. *The Material World*. Cambridge University Press, Cambridge.B. Kahr and J. M. McBride, 1992. Optically anomalous crystals. *Angewandte Chemie International Edition in English*, 31, 1.

M. G. Lagorio, 2004. Why do marbles become paler on grinding? Reflectance, spectroscopy, color, and particle size. *Journal of Chemical Education*, 81, 1607.

J. M. Marentette and G. R. Brown, 1993. Polymer spherulites. I. Birefringence and morphology. *Journal of Chemical Education*, 70, 435.

K. Nassau, 2001. *The Physics and Chemistry of Colour*, 2nd ed. Wiley-Interscience, Hoboken, NJ.

T. D. Rossing and C. J. Chiaverina, 1999. *Light Science: Physics and the Visual Arts*. Springer, New York.

J. M. Rowell, 1986. Photonic materials. *Scientific American*, October, 146.

R. Tilley, 2011. *Colour and the Optical Properties of Materials*, 2nd ed. John Wiley & Sons, Hoboken, NJ.

V. F. Weisskopf, 1968. How light interacts with matter. *Scientific American*, September, 60.

E. A. Wood, 1977. *Crystals and Light: An Introduction to Optical Crystallography*, 2nd ed. Dover Publications, New York.

Coatings

T. Nguyen, 2017. Making glass disappear. *Chemical and Engineering News*, November 6, 9.

Devices

The physics of digital color. Special issue. P. Roetling, Ed. *Physics Today*, December 1992, 24–65.

R. H. Chen, 2011. *Liquid Crystal Displays: Fundamental Physics and Technology*. John Wiley & Sons, Hoboken, NJ.

S. W. Depp and W. E. Howard, 1993. Flat-panel displays. *Scientific American*, March, 90.

E. Wilson, 1985. Nonlinear optical polymer device may speed up information super-highway. *Chemical and Engineering News*, August 14, 27.

Liquid Crystals

G. H. Brown, 1983. Liquid crystals—the chameleon chemicals. *Journal of Chemical Education*, 60, 900.

P. G. Collings, 2002. *Liquid Crystals: Nature's Delicate Phase of Matter*, 2nd ed. Princeton University Press, Princeton, NJ.

J. Fünfscilling, 1991. Liquid crystals and liquid crystal displays. *Condensed Matter News* 1(1), 12.

Nonlinear EffectsNonlinear optics. Special issue. E. Garmire, Ed. *Physics Today*, May 1994, 23–57.

Photonic materials for optical communications. Special issue. H. Hillmer and R. Germann, Eds. *MRS Bulletin*, May 2003, 340–376.

Refractive index changes its sign. *Chemical and Engineering News*, April 9, 2001, 31.

R. T. Bailley, F. R. Cruickshank, P. Pavlides, D. Pugh, and J. N. Sherwood, 1991. Organic materials for non-linear optics: Inter-relationships between molecular properties, crystal structure and optical properties. *Journal of Physics D: Applied Physics*, 24, 135.

R. W. Boyd, 2008. *Nonlinear Optics* 2nd Edition. . Academic Press, Amsterdam.

R. Dagani, 1995. Photorefractive polymers poised to play key role in optical technologies. *Chemical and Engineering News*, February 20, 1995, 28.

J. Feinberg, 1988. Photorefractive nonlinear optics. *Physics Today*, October 1988, 46.

M. Freemantle, 2001. Opal chips: Photonic jewels. *Chemical and Engineering News*, January 22, 2001.

B. G. Levi, 1995. New compound brightens outlook for photorefractive polymers. *Physics Today*, January, 17.

S. R. Marder, 1992. Chapter 3: Metal containing materials for nonlinear optics. In *Inorganic Materials*, D.W. Bruce and D. O'Hare, Eds. John Wiley & Sons, Hoboken, NJ.

S. R. Marder, 2016. Materials for third-order nonlinear optics. *MRS Bulletin*, January 2016, 53.

S. R. Marder, J. E. Sohn, and G. D. Stucky, Eds., 1991. Materials for nonlinear optics: Chemical perspectives. In *American Chemical Society Symposium Series 455*, American Chemical Society.

S. Mukamel, 1995. *Principles of Nonlinear Optical Spectroscopy*. Oxford University Press, New York.

D. S. Rodgers, Ed. 2013. *Circular Dichroism: Theory and Spectroscopy*. Nova Science Publishers, Inc., Hauppauge, NY.

Transparency

J. E. Burke, 1996. Lucalox alumina: The ceramic that revolutionized outdoor lighting. *MRS Bulletin*, June, 61.

Websites

For links to relevant websites, see PhysicalPropertiesOfMaterials.com

Part III

Thermal Properties of Materials

Passed through the fiery furnace
As gold without alloy . . .

Philip Carrington
from a hymn, 1938

6

Heat Capacity, Heat Content, and Energy Storage

6.1 Introduction

The thermal properties of any material are among the most fundamental, whether directly or indirectly involved in the material's application. For example, it is almost always important to know how properties of materials change if the temperature is changed. In this section of the book, we explore the fundamentals of thermal properties, from basic thermodynamics to applications. This chapter emphasizes heat capacity, heat content, and energy storage. Heat capacity, which is the most fundamental of all thermal properties, is related to the strength of intermolecular interactions, phase stability, thermal conductivity, and energy storage capacity.

6.2 Equipartition of Energy

To understand the fundamentals of energy storage, it is useful to examine the *theory of equipartition of energy.*

First, let us consider the information needed to know where a molecule is. This will lead to considerations of how the energy of that material is partitioned.

For a single atom, three coordinate positions must be specified to know the atom's position. For example, this can be expressed as three Cartesian coordinates (x, y, z) in orthogonal axes. We express the freedom of position by saying that a single atom has three degrees of freedom.

For a polyatomic molecule with N atoms per molecule, $3N$ coordinates (3 for each of the N atoms) must be known to specify the total molecular position. An N-atom polyatomic molecule has $3N$ degrees of freedom.

The molecule's motion can be expressed in terms of three types of movement: translational (motion of whole molecule), rotational (rotation of whole molecule), and vibrational (internal vibrations within the molecule).

For a monatomic species, there is only one meaningful type of motion: translation. The translational motion can be considered to take place in three orthogonal directions (e.g., x, y, z), and any motion of the monatomic molecule can be described as a vector sum of the motions in these three directions. This statement is another way of expressing the three degrees of freedom of a monatomic species.

For a diatomic molecule, there are also three degrees of freedom of translational motion. In addition, there are two meaningful rotational motions for a diatomic molecule. (A third rotational motion, along the molecular axis, does not move the atoms and therefore does not contribute to the freedom of the molecule.) A diatomic molecule has one further vibrational degree of freedom due to an internal vibration. The sum of these degrees of freedom of motion (3 translational + 2 rotational + 1 vibrational = 6 total) is the same as the total number of degrees of freedom (for $N = 2$, $3N = 6$).

For a linear polyatomic molecule with N atoms, there are three translational degrees of freedom and two rotational degrees of freedom, leaving $(3N-5)$ vibrational degrees of freedom.

For a nonlinear polyatomic molecule, there can be three meaningful rotational motions. This gives, for N atoms per molecule, three translational degrees of freedom, three rotational degrees of freedom, and $(3N-6)$ vibrational degrees of freedom.

From quantum mechanical considerations, virtually any translational energy value is allowed. Of course, translational energies are quantized, but their spacing is so close together that it results in a virtual continuum of energy levels.

It takes more energy to excite rotations than it does to excite translations, which is reflected in the wider spacing of rotational energy levels relative to the translational ladder.

Vibrational energy levels are even further apart than rotational levels, and it takes considerable energy to excite them. (Recall from the discussion of color of matter that electronic energy levels are even more widely spaced than vibrational levels, as shown in Figure 2.2 in Chapter 2, so we do not consider electronic excitation here.)

The *principle of equipartition* states that for temperatures high enough that translational, rotational, and vibrational degrees of freedom are all "fully" excited,* each type of energy contributes $\frac{1}{2}kT$ (where k is the Boltzmann constant and T is the temperature in kelvin) to the internal energy, U, per degree of freedom.

This means that each translational degree of freedom contributes $\frac{1}{2}kT$ to U, and each rotational degree of freedom contributes $\frac{1}{2}kT$ to U, but each vibrational degree of freedom contributes $2 \times \frac{1}{2}kT = kT$ to U. The reason for the double contribution for vibration is that, whereas translation and rotation

---

* Strictly speaking, all the degrees of freedom are fully excited only at infinite temperature.

have only kinetic energies associated with them, vibrational motion has both kinetic and potential energy, each contributing $\frac{1}{2}kT$.

This simple theory of equipartition can be useful to work out the contributions of degrees of freedom to U and hence to the heat capacity. Two examples follow.

6.2.1 Heat Capacity of a Monatomic Gas

For a monatomic gaseous species, there are only three degrees of freedom in total, and all of these are translational degrees of freedom. Therefore, $U = 3 \times \frac{1}{2}kT$. This same result also is found from the kinetic theory of gases.

Considering that the *heat capacity at constant volume*, C_V, is defined by*:

$$C_V = \left(\frac{\partial U}{\partial T}\right)_V \tag{6.1}$$

then $C_V = 1.5k$ per molecule for a monatomic gas.

To scale up to a mole of gas, we need to multiply by the Avogadro constant, N_A, and this gives the molar heat capacity at constant volume, $C_{V,m}$:

$$C_{V,m} = C_V N_A \tag{6.2}$$

so, for a gaseous monatomic, $C_{V,m} = 1.5 N_A k = 1.5 R$ at any temperature, where $R (= N_A k)$ is the *gas constant*.

6.2.2 Heat Capacity of a Nonlinear Triatomic Gas

For a nonlinear triatomic molecule, such as H_2O, the high-temperature limit of the heat capacity can be calculated from the equipartition theory. Since $N = 3$, there are $3N = 9$ degrees of freedom in total, partitioned as shown in Table 6.1.

TABLE 6.1

Degrees of Freedom and their Contributions to U, C_V, and $C_{V,m}$ for a Nonlinear Triatomic Molecule in the Gas Phase

Motion	Degrees of Freedom	U	C_V	$C_{V,m}$
Translational	3	$3 \times \frac{1}{2}kT$	$3 \times \frac{1}{2}k$	$3 \times \frac{1}{2}R$
Rotational	3	$3 \times \frac{1}{2}kT$	$3 \times \frac{1}{2}k$	$3 \times \frac{1}{2}R$
Vibrational	3	$2 \times 3 \times \frac{1}{2}kT$	$3k$	$3R$
				Total $C_{V,m} = 6R$

---

* The chapters in this section on Thermal Properties contain significantly more equations than the other sections. To aid the student, the most important equations are denoted with ▌ to the left of the equation. (Many of the equations not so designated can be derived from the marked equations.)

Since the equipartition theory only considers the case where all the degrees of freedom are fully excited, this is the high-temperature limit of the heat capacity of gaseous H_2O. In practice, at moderate temperatures, $C_{V,m}$ is much less. For example, $C_{V,m}$ (H_2O gas, $T = 298.15$ K) $= 3.038\ R$. Since translational and rotational energies are most easily excited, and $3\ R$ is their summed contribution, the remaining $0.038\ R$ reflects the fact that the vibrations of gaseous water molecules are only slightly active at room temperature because the vibrational energy spacings are greater than the available thermal energy, kT.

We will see later (Sections 6.4 and 6.5) how the equipartition theory can be used to assess heat capacity contributions in solids and liquids.

6.3 Real Heat Capacities and Heat Content of Real Gases

The heat content, or *enthalpy* (H), of a gas can be assessed through the relationships between H, T, and C_p *(heat capacity at constant pressure)*,* defined as

$$C_p = \left(\frac{\partial H}{\partial T}\right)_p \tag{6.3}$$

and

$$\Delta H = \int_{T_1}^{T_2} dH = \int_{T_1}^{T_2} C_p dT. \tag{6.4}$$

For polyatomic gases, we can get high-temperature values of C_V from equipartition considerations. We will show later (Section 6.8) that, for ideal gases, $C_{V,m}$ and $C_{p,m}$ are related as $C_{p,m} = C_{V,m} + R$.

At very low temperatures, not all degrees of freedom of a gas will be excited. The first modes to be excited will be those with the most closely spaced energy levels, the translational degrees of freedom; in the limit of $T \to 0$ K, a gas will have $C_{V,m} = 1.5\ R$ due to translation only. As the temperature is increased, the contribution of the two rotational modes of the diatomic gas to $C_{V,m}$ will gradually turn on, and then $C_{V,m} = 2.5\ R$. As the temperature is increased further, the vibrational mode will contribute to the heat capacity, and then $C_{V,m} = 3.5\ R$. This situation is shown schematically in Figure 6.1, where $C_{p,m}$ is shown as R higher than $C_{V,m}$ (i.e., the ideal gas approximation).

Returning to the theme of heat storage, and taking into account that the heat content over a temperature range is the integrated heat capacity (Equation 6.4), the higher the heat capacity, the greater the heat storage ability. This is

---

* Just as U and volume go together, H and pressure often appear together in thermodynamic equations.

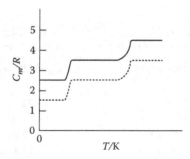

FIGURE 6.1
Stylized temperature dependence of the molar heat capacity of an ideal diatomic gas at constant pressure ($C_{p,m}$: ——) and at constant volume ($C_{V,m}$: ----).

true for all phases of matter, and we will return to this later when considering thermal energy storage applications.

Now let us consider what happens when a gas is cooled to very low temperatures. All gases liquefy when the temperature is sufficiently low. This may require a very low temperature; for example, helium liquefies at $T = 4.2$ K.

When a gas liquefies, it releases heat. Since enthalpy is a state function, the complete cycle from gas → liquid → gas would have $\Delta H = 0$.[*] Therefore, the enthalpy change on liquefaction ($\Delta_{liq}H = \Delta_{gas \to liquid}H$) is equal but of an opposite sign to the enthalpy change on vaporization ($\Delta_{vap}H$):

$$\Delta_{liq}H = -\Delta_{vap}H. \tag{6.5}$$

The enthalpy change on vaporization, $\Delta_{vap}H$, is a very useful quantity, and some typical enthalpies of vaporization are given in Table 6.2.

Since the Gibbs[†] energy change, ΔG, for a fixed quantity of material can be written at constant temperature as

$$\Delta G = \Delta H - T\Delta S \tag{6.6}$$

---

[*] In much of the discussion that follows, the thermodynamic quantities are implicitly expressed as molar quantities.

[†] Josiah Willard Gibbs (1839–1903) was an American theoretical physicist. Gibbs was a recipient of one of the first American PhD degrees, from Yale in 1863 (his thesis was titled "On the Form of the Teeth of Wheels in Spur Gearing"). After his PhD, Gibbs spent three years in Europe, then returned to New Haven, and rarely traveled anywhere thereafter. This rather reclusive scientist made many great contributions, mostly in the areas of thermodynamics. Gibbs' first scientific paper, published in 1873, set the record straight on the (then-confused) concept of entropy. Within five years, he published a 300-page memoir providing the basis of much of thermodynamics as it is taught to this day. Since most of the thermodynamic research at that time was carried out in Europe, especially Germany, it took some time (and Ostwald's 1892 translation of Gibbs' work to German) for Gibbs' work to make its lasting impression.

TABLE 6.2

Enthalpy Changes on Vaporization, $\Delta_{vap}H$, and Entropy Changes on Vaporization, $\Delta_{vap}S$, for a Variety of Materials at their Normal Boiling Points, T_b

Material	$\Delta_{vap}H/(J\ mol^{-1})$	$\Delta_{vap}S/(J\ K^{-1}\ mol^{-1})$	T_b/K
He	8.37×10^2	19.7	4.2
N_2	5.577×10^3	72.13	77.3
CH_4	8.180×10^3	73.26	111.7
NH_3	2.335×10^4	97.40	239.7
CH_3OH	3.527×10^4	104.4	337.9
CCl_4	3.000×10^4	85.8	349.9
C_6H_6	3.076×10^4	87.07	353.3
H_2O	4.0657×10^4	108.95	373.2
Hg	5.812×10^4	92.30	629.7
Zn	1.148×10^5	97.24	1180.0

where ΔS is the change in *entropy*, and the Gibbs energies of any two phases in equilibrium are identical (so $\Delta_{trs}G$, the change in Gibbs energy due to transition, equals zero), then

$$0 = \Delta_{trs}H - T\Delta_{trs}S \tag{6.7}$$

that is, $\Delta_{vap}H = T\Delta_{vap}S$ for the case of the vaporization transition. It is an experimental finding that many gases have values of $\Delta_{vap}S$ of about 90 J K^{-1} mol^{-1} (See Table 6.2). This finding was generalized by Trouton,* and is usually called *Trouton's rule*. It holds because the increase in disorder on going from the liquid phase (where molecules are translating and undergoing hindered rotations and perhaps muted vibrations) to the gas phase (where the molecules translate and rotate more freely than in the liquid) is approximately independent of the type of material involved. For materials with unknown values of $\Delta_{vap}H$, the value can be estimated from the boiling point and Trouton's value of $\Delta_{vap}S$.

There are exceptions to Trouton's rule, and one is H_2O, with $\Delta_{vap}S$ of 108.95 J K^{-1} mol^{-1}. The main reason for the exceptionally large increase in disorder on vaporization of water is that its liquid phase has a relatively low entropy due to the order associated with hydrogen-bond networks, which hinder molecular motion. This same network gives rise to the color of water through vibrational excitation, as seen in Chapter 2.

While we are looking at heat capacity and heat content, we should consider what happens to the heat capacity of a material at its boiling point. When a liquid is heated toward its boiling point, the material absorbs heat and the temperature rises until the boiling point is achieved. At that point,

---

* Frederick Thomas Trouton (1863–1922) was an Irish physicist who, in 1902, took up a professorship at University College London. He discovered what we now call Trouton's rule when he was an undergraduate student at Trinity College, Dublin.

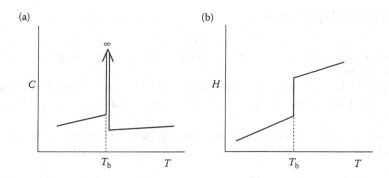

FIGURE 6.2
Boiling of a liquid at T_b, at a fixed pressure. (a) Schematic of the heat capacity as a function of temperature of a material in the vicinity of the boiling point. (b) Schematic of the corresponding enthalpy as a function of temperature in the vicinity of the boiling point.

heat is absorbed without further increase in temperature until all the material has been converted to vapor. From either C_V or C_p considerations, with q_V as heat taken up at constant volume and q_p as heat taken up at constant pressure,

$$C_V = \frac{q_V}{\Delta T} = \left(\frac{\partial U}{\partial T} \right)_V \tag{6.8}$$

and

$$C_p = \frac{q_p}{\Delta T} = \left(\frac{\partial H}{\partial T} \right)_p. \tag{6.9}$$

Since, on boiling, $\Delta T = 0$ (i.e., heat is input without temperature rise at the boiling point), $C_V = C_p = \infty$ at the boiling point, T_b, as depicted in Figure 6.2.

Because ΔH is the integration of C_p over temperature (Equation 6.4), H increases slightly with temperature over the liquid region and then has a step (integrating the infinite heat capacity) at the boiling point, followed by a further gentle increase as the temperature increases in the gas phase, as shown in Figure 6.2. An infinite heat capacity at the transition is not limited to the case of a liquid → gas phase transition, as we shall see later.

Before turning to the heat content of a nonideal gas, it is worth stating a few basic concepts concerning ideal gases. An *ideal gas* is a theoretical concept of a gas for which $pV = nRT$ is the equation of state for n moles under all circumstances. The equation of state for an ideal gas can be derived from the kinetic theory of gases, with two main assumptions: (1) the molecules of the gas have negligible volume in comparison with the total volume of the container, and (2) the molecules exert no attraction or repulsion on each other. All gases approach ideality as $p \to 0$ because these two conditions are most closely met at low pressures.

6.3.1 Joule's Experiment

Joule* carried out an experiment concerning ideal gases that led to an important discovery: The internal energy of an ideal gas depends only on temperature.

In Joule's experiment, two connected gas bulbs were placed in a water bath. One bulb was evacuated, and the other contained gas, as shown in Figure 6.3.

Joule opened the stopcock between the two bulbs and measured the resulting temperature change in the water bath, $\Delta_{bath}T$. He found that

$$\lim p_1 \to 0, \ \Delta_{bath}T = 0, \tag{6.10}$$

where p_1 is the pressure of the gas which expands into the vessel at $p_2 = 0$.

It can be shown, as follows, that this result leads to $\Delta U = 0$ for the gas in the limit of $p \to 0$ (i.e., approaching ideality). By the *first law of thermodynamics*, the change in internal energy, ΔU, is

$$\Delta U = q + w \tag{6.11}$$

and work, w, is given by

$$w = -\int p_{ext} dV \tag{6.12}$$

where p_{ext} is the external pressure. In this experiment, the gas expands within a fixed volume while doing no external work, so $w = 0$ here. This shows that expansion into a vacuum takes no work. Consideration of the heat, q,

$$q = C\Delta T \tag{6.13}$$

FIGURE 6.3
Joule's experiment. The gas at pressure p_1 was allowed to expand to the other bulb, where $p_2 = 0$, while the temperature of the water bath was recorded.

---

* James Prescott Joule (1818–1889) was an English scientist noted for the establishment of the mechanical theory of heat, and honored by the SI energy unit carrying his name. He carried out his work on heat while he was in his early 20s and then made other contributions to the field of thermodynamics, including the first estimate of the speed of gas molecules (1848). Lord Kelvin wrote of Joule, "His boldness in making such large conclusions from such very small observational effects is almost as noteworthy and admirable as his skill in extorting accuracy from them."

and Joule's finding that $\Delta_{\text{bath}}T = 0$ leads to $q = 0$. Since $w = 0$ and $q = 0$, it follows from Equation 6.11 that $\Delta U = 0$.

In other words, when a low-pressure (ideal) gas is expanded, it does so without a change in internal energy, U. Since U is most simply expressed as a function of T and V (just as H is a function of T and p), an infinitesimal change in U, written as dU, can be expressed in terms of an infinitesimal change in T, written as dT, and an infinitesimal change in V, written as dV:

$$dU = \left(\frac{\partial U}{\partial T}\right)_V dT + \left(\frac{\partial U}{\partial V}\right)_T dV. \tag{6.14}$$

From Joule's experiment, $dU = 0$ and $dT = 0$, so $(\partial U/\partial V)_T\, dV = 0$, but $dV \neq 0$ so $(\partial U/\partial V)_T = 0$ for an ideal gas. This shows that U is independent of volume for an ideal gas at constant temperature because the molecules are neither attracted nor repelled in an ideal gas. Therefore, for an ideal gas, U is a function of temperature only, in accordance with the equipartition theorem (i.e., U depends only on T, not on V or p).

6.3.2 Joule–Thomson Experiment

Joule teamed up with Thomson* to carry out an important experiment that has allowed quantification of the nonideality of a gas. This experimental apparatus was basically an insulated double piston, as shown schematically in Figure 6.4.

Initially the gas was all in the left chamber. When the left piston was pushed in, the gas moved through a porous plug (said to have been a linen handkerchief) into the right chamber. The gas state functions were transformed from initial values (on the left $p = p_L$, $T = T_L$, $V = V_L$; on the right $p = p_R$, $T = T_R$, $V = V_R = 0$) to final values (on the left $p = p'_L$, $T = T'_L$, $V = V'_L = 0$; on the right $p = p'_R$, $T = T'_R$, $V = V'_R$).

Since the piston was made of insulating material, there was no extraneous heat exchange with the surroundings (i.e., the experiment was carried out under an *adiabatic* condition: $q = 0$). This information can be used to show that the overall process was isenthalpic (i.e., at constant enthalpy, $\Delta H = 0$) as follows. Since H is defined as

$$H \equiv U + pV \tag{6.15}$$

---

* William Thomson (1824–1907) was a Belfast-born physicist later made Baron Kelvin of Largs and known as Lord Kelvin; the temperature unit is named after him. He graduated from Glasgow University at age 10 and was made professor of natural philosophy there at age 22. Thomson made major contributions to many areas: electrodynamics; thermodynamics including the absolute temperature scale and the second law (on his own and in collaboration with Joule); electromagnetism (with Faraday). He was knighted for his contribution to the laying of the first Atlantic cable, and made many other practical advances including the invention of a tide predictor and an improved mariner's compass. He was the acknowledged leader in physical sciences in the British Isles during his lifetime.

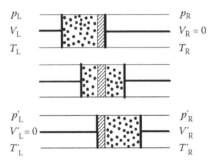

FIGURE 6.4
The Joule–Thompson double-piston experiment, shown as a function of time with the initial conditions at the top. In this experiment, gas was pushed through a porous plug in the center of a piston and taken from an initial set of pressure, volume, and temperature conditions on the left side of the piston (denoted by subscript L) to a final set of conditions (denoted by a prime) on the right side of the plug (denoted by subscript R).

it follows that

$$\Delta H = \Delta U + \Delta(pV). \tag{6.16}$$

Making use of the first law of thermodynamics (Equation 6.11), the change in pV for this process leads to

$$\Delta H = q + w + p'_R V'_R - p_L V_L \tag{6.17}$$

and here $q = 0$. From Equation 6.12,

$$w = -\int_{V_L}^{0} p_L dV - \int_{0}^{V'_R} p'_R dV = p_L V_L - p'_R V'_R \tag{6.18}$$

and substitution of this expression for w and $q = 0$ into Equation 6.17 leads to

$$\Delta H = p_L V_L - p'_R V'_R + p'_R V'_R - p_L V_L = 0, \tag{6.19}$$

which shows that the process takes place at constant enthalpy.

The Joule–Thompson experiment was carried out to see the change in temperature of the gas with respect to pressure changes, at constant enthalpy. The Joule–Thompson coefficient, μ_{JT}, so determined, is defined as

$$\mu_{JT} = \left(\frac{\partial T}{\partial p}\right)_H. \tag{6.20}$$

Experimentally, it is found that μ_{JT} for a given gas is positive at relatively low temperatures, negative at relatively high temperatures, and zero at

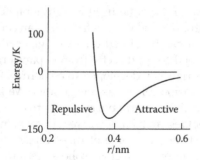

FIGURE 6.5
Intermolecular interaction energy as a function of intermolecular separation, r. The curve shown is for the interaction between two Ar atoms, and energy is expressed in units equivalent to thermal energy $(= kT$ at temperature T).

some intermediate temperature (the Joule–Thomson inversion temperature; the exact value of this temperature depends on the gas). This temperature dependence can be understood as follows.

Unlike an ideal gas in which the forces between molecules can be ignored, the molecules in real gases do interact each other. At infinitely large separations, the force between molecules is vanishingly small, but as the intermolecular separation, r, diminishes, the molecules begin to feel attraction for each other. It is this attraction that eventually leads to liquefaction of gases when they are compressed. This attraction lowers the energy of the system. At extremely short intermolecular distances, the intermolecular interaction becomes repulsive, increasing the energy of the system. This situation is shown schematically in Figure 6.5.

At relatively low temperatures, where $\mu_{JT} > 0$, the definition of μ_{JT} implies that as the pressure decreases, so does the temperature, and the gas cools on expansion. This cooling is because of intermolecular forces. The gas expansion (p is decreased) requires some work to overcome the attractive intermolecular forces. This work comes at the expense of the energy (or temperature) of the gas and causes the gas to cool on expansion.

At very high temperatures, where $\mu_{JT} < 0$, the opposite situation exists. At these very high temperatures, the temperature increases when the pressure decreases (gas is expanded). This warming is because high temperatures allow for many repulsive intermolecular interactions, so when the gas is expanded, energy is released, and manifested in increased temperature of the gas.

At the Joule–Thomson inversion temperature, there is an accidental cancellation of the attractive and repulsive forces, and there is no change in temperature on expansion of the gas; $\mu_{JT} = 0$ at this temperature.

An ideal gas, with its negligible intermolecular interactions, also has $\mu_{JT} = 0$. Therefore, μ_{JT} is a direct way to quantify the nonideality of a gas.

Most gases (exceptions noted below) are still below their Joule–Thomson inversion temperature at room temperature. Because these gases have $\mu_{JT} > 0$, they cool on expansion and can be used as refrigerants. Examples are NH_3, freons (chlorofluorocarbons), and CO_2. In fact, CO_2 can cool so much on expansion that an attachment can be placed on a CO_2 gas cylinder to produce dry ice, that is CO_2 in its solid form, directly from expansion of the gas from the cylinder. Freons were originally developed to be nonreactive gases that would efficiently cool on expansion for use as refrigerants. We know now that some freons are rather unreactive on earth but quite reactive with ozone in the upper atmosphere,* and recent research has led to replacement materials with suitable refrigeration capacity. However, some of these replacement gases contribute significantly to the greenhouse effect, so solid-state refrigerants (e.g., thermoelectric materials [see Chapter 12] and magnetocaloric materials [see Chapter 13]) are active areas of materials research.

Low-boiling gases are already above their Joule–Thomson inversion temperature at room temperature. Examples are H_2, He, and Ne. To use these gases as refrigerants that expand on cooling, they must first be pre-cooled below their Joule–Thomson inversion temperatures.

6.4 Heat Capacities of Solids

6.4.1 Dulong–Petit Law

In 1819, Dulong[†] and Petit[‡] found experimentally that, for many nonmetallic solids at room temperature, the molar heat capacity at constant volume $C_{V,m} \approx 3\ R \approx 25\ \mathrm{J\ K^{-1}\ mol^{-1}}$. Originally known as the *law of atomic heats*, this is now known as the *Dulong–Petit law*.

The Dulong–Petit law can be understood for monatomic solids by use of the equipartition theory, similar to the ideal monatomic gas presented in Section 6.2, except that atoms in solids vibrate and therefore have both kinetic and potential energy. For each atom in the monatomic solid, there are

---

* The 1995 Nobel Prize in Chemistry was awarded to Paul Crutzen (1933–, Max Planck Institute for Chemistry, Mainz, Germany), Mario Molina (1943–, MIT), and F. Sherwood Rowland (1927–2012, University of California–Irvine) for their work in atmospheric chemistry, particularly concerning the formation and decomposition of ozone.

† Pierre Louis Dulong (1785–1838) was a French physicist who, with Petit, discovered the law of atomic heats, now known more commonly as the Dulong–Petit law. This work was particularly important in establishing atomic masses. Dulong's work in his later life was hampered by inadequate funding (unlike most of his contemporaries, he had no funding from industrial connections) and the loss of an eye from his 1811 discovery of nitrogen chloride.

‡ Alexis Thérèse Petit (1791–1820) was a French physicist who, with Dulong, discovered the law of atomic heats (Dulong–Petit law). In 1810, at the age of 19, he was appointed a professor of Lycée Bonaparte in Paris. With Dulong, Petit was one of the first physical scientists to reject the caloric theory of heat (which treated heat as a fluid) and espouse the atomic theory of matter in 1819; this stand made little headway with his peers in France at that time.

three degrees of freedom, and each atom's three degrees of freedom can be expressed in terms of the vibrational motion of the atoms at a point of the lattice. From equipartition this gives, per atom,

$$U = 3 \times 2 \times \frac{kT}{2} = 3\,kT \tag{6.21}$$

so $C_V = 3\,k$ and, on a molar basis, with $kN_A = R$, this gives $C_{V,m} = 3\,R$ as found experimentally.

What the Dulong–Petit law does not address is the further experimental fact that all heat capacities decrease as the temperature is lowered (i.e., $C_{V,m} \to 0$ as $T \to 0$ K). In fact, this is not explainable in terms of classical theories and requires quantum mechanics.

6.4.2 Einstein Model

The *Einstein* model of heat capacity of a solid* (published by Einstein in 1906) was, historically, one of the first successes of quantum mechanics. In this model, thermal depopulation of vibrational energy levels is used to explain why $C_{V,m} \to 0$ as $T \to 0$ K.

Einstein considered each atom in the solid to be sitting on a lattice site, vibrating at a frequency v. For N atoms, this led to the following expression for the heat capacity:

$$C_V = 3Nk\left(\frac{h\nu}{kT}\right)^2 \frac{e^{h\nu/kT}}{\left(e^{h\nu/kT} - 1\right)^2} \tag{6.22}$$

where h is Planck's constant. The repeating factor $(h\nu/kT)$, which must be unitless because it appears as an exponent, leads to units of s^{-1} (also called Hertz[†]) for v. Often vibrational excitation information is given in the unit cm^{-1} ($\tilde{\nu}$, wavenumber), that is,

---

[*] Albert Einstein (1879–1955) was a German-born theoretical physicist. Einstein's early life did not predict great success; his entry to university was delayed because of inadequacy in mathematics, and when he completed his studies in 1901 he was unable to get a teaching post, so he took a junior position at a patent office in Berne. While working there he managed to publish three important papers: one on Brownian motion that led to direct evidence for the existence of molecules; the second connecting quantum mechanics and thermodynamics, proving that radiation consisted of particles (photons) each carrying a discrete amount of energy, which gave rise to the photoelectric effect; and the third concerning the special theory of relativity. In 1932 Einstein left Germany for a tour of America, but decided, as a Jew, that it was not safe to return to Germany, so he remained in the United States. Einstein received the Nobel Prize in Physics in 1921 for his work on the photoelectric effect.

[†] Heinrich Rudolf Hertz (1857–1894) was a German physicist. Hertz was an experimentalist who studied electrical waves and made important discoveries that led to significant advances in understanding of electricity. Hertz also discovered radio waves.

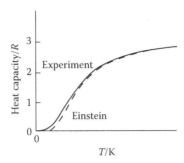

FIGURE 6.6
The Einstein heat capacity for a monatomic solid, compared with the experiment, for a monatomic solid.

$$E = h\nu = \frac{hc}{\lambda} = hc\tilde{\nu}, \tag{6.23}$$

where c is the speed of light.

Alternatively, $h\nu/k$ can be expressed directly in terms of the *Einstein characteristic temperature*, θ_E, defined as

$$\theta_E = \frac{h\nu}{k} = \frac{hc\tilde{\nu}}{k}. \tag{6.24}$$

As the term $h\nu/kT$ is the ratio of vibrational energy ($h\nu$) to thermal energy (kT), the use of θ_E makes the comparison even more direct. Written in terms of θ_E, the heat capacity for N atoms in the Einstein model can be expressed as

$$C_V = 3Nk \left(\frac{\theta_E}{T} \right)^2 \frac{e^{\theta_E/T}}{\left(e^{\theta_E/T} - 1 \right)^2}. \tag{6.25}$$

This model gives a temperature-dependent molar heat capacity of a solid (Figure 6.6), approaching $3R$ as $T \to \infty$ (provided the solid does not melt!) in agreement with the Dulong–Petit finding, and approaching 0 as $T \to 0$ K in agreement with the experiment. However, the Einstein model falls below the experimental heat capacity at low temperatures (Figure 6.6).

6.4.3 Debye Model

The 1912 *Debye* model* of heat capacity of a solid considers the atoms on lattice sites to be vibrating with a distribution of frequencies. Debye's model

---

* Peter Joseph Wilhelm Debye (1884–1966) was a Dutch-born chemical physicist who, at age 27, succeeded Einstein as professor of theoretical physics at the University of Zurich. He moved to the United States in 1940. Debye's contributions included fields as diverse as theory of specific heat, theory of dielectric constants, light scattering, x-ray powder (Debye–Scherrer) analysis, and Debye–Hückel theory of electrolytes. In 1936 he was awarded the Nobel Prize in Chemistry for his contribution to knowledge of molecular structure.

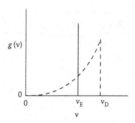

FIGURE 6.7
Comparison of Debye (---) and Einstein (————) distributions of frequencies. The correspond-ing frequencies, v_D ($=\theta_D k/h$) and v_E ($=\theta_E k/h$) are also shown.

assumed a continuum of frequencies in the distribution up to a maximum (cutoff) frequency, v_D, the Debye frequency. The dependence of the distribu-tion of frequencies on the frequency is shown in Figure 6.7, where the (single-frequency) Einstein model also is shown for comparison.

Mathematically, the two models can be compared in terms of the frequency distribution function for a single particle, $g(v)$, where $g(v)$ is the number of modes with frequency v:

$$\text{Einstein model}: \; g(v) = v_E \tag{6.26}$$

$$\text{Debye model}: \; g(v) = av^2 \qquad 0 \le v \le v_D \tag{6.27}$$

$$g(v) = 0 \qquad v > v_D \tag{6.28}$$

where a in Equation 6.27 is a constant, dependent on the material. This model leads to the following expression for the Debye heat capacity for N particles in three dimensions:

$$C_V = 9Nk \left(\frac{kT}{hv_D}\right)^3 \int_0^{\frac{hv_D}{kT}} \frac{\left(\frac{hv}{kT}\right)^4 e^{hv/kT}}{\left(e^{hv/kT}-1\right)^2} \, d\left(\frac{hv}{kT}\right) \tag{6.29}$$

and again, introducing a characteristic temperature, this time θ_D, the *Debye characteristic temperature*, where $\theta_D = hv_D/k$, leads to a simpler way of writing Equation 6.29:

$$C_V = 9Nk \left(\frac{T}{\theta_D}\right)^3 \int_0^{\frac{\theta_D}{T}} \frac{x^4 e^x}{\left(e^x - 1\right)^2} \, dx \tag{6.30}$$

where $x = hv/kT$.

The Debye heat capacity equation (Equation 6.30) shows that, theoreti-cally, $C_{V,m}$ is a universal function of θ_D/T, with θ_D as the scaling function

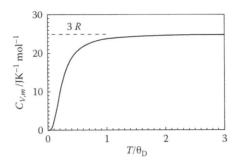

FIGURE 6.8
Heat capacity of a monatomic crystalline solid as a function of temperature scaled to the Debye temperature for the material. A material with a higher value of θ_D will have a given heat capacity at a higher temperature than another material with a lower value of θ_D. Note that the heat capacity of a monatomic solid approaches 3 R at high temperature.

for different materials. This is borne out very well in experimental data. Figure 6.8, which illustrates $C_{V,m}$ as a function of temperature expressed in units of θ_D, fits the data for many monatomic materials. As a scaling function, θ_D reflects the strength of the interatomic interactions; that is, θ_D is a measure of the force constant between molecules or atoms in the solid. Some typical values of θ_D are given in Table 6.3, where it can readily be seen that materials that are harder to deform have higher values of θ_D. The higher the value of θ_D, the less the vibrational modes are excited at a given temperature, and the lower the heat capacity (in comparison with materials with lower values of θ_D).

Like the Einstein model, the Debye model of heat capacity correctly describes the experimental situation in the temperature extrema ($C_{V,m} \to 3\,R$ as $T \to \infty$ and $C_{V,m} \to 0$ as $T \to 0$ K). The other major success of the Debye theory is that it correctly predicts the way in which $C_V \to 0$ as $T \to 0$ K for a monatomic solid. In the limit of $T \to 0$, the Debye heat capacity expression becomes, for N atoms,

TABLE 6.3

Debye Temperatures for Selected Materials[a]

Material	θ_D/K
Diamond	2230
Gold	225
Neon	75
Mercury (solid)	72

Source: C. Kittel, 2004. *Introduction to Solid State Physics*, 8th ed. John Wiley & Sons, Hoboken, NJ.

[a] Note that, at a given temperature, a material with a lower value of θ_D will have a higher heat capacity than a material with a higher value of θ_D.

$$\lim T \to 0 \quad C_V = \frac{12}{5}\pi^4 Nk\left(\frac{T}{\theta_D}\right)^3 \tag{6.31}$$

that is, for $T < \theta_D/100$, $C_V \propto T^3$, as observed experimentally for nonmetals. Equation 6.31 is referred to as the *Debye-T^3 law*.

For a monatomic solid, such as solid argon, the Debye model gives a more accurate description than the Einstein model because the frequency distribution (e.g., as determined by inelastic neutron scattering) is more Debye-like. For a polyatomic molecular solid, such as C_{60}, the rigid-molecule contribution to the heat capacity will be Debye-like, but the contributions of internal vibrations (which are localized at particular frequencies and can be observed in infrared and/or Raman spectroscopy experiments) are better treated by the Einstein model. This matter is considered further in Problem 6.10.

COMMENT: DEBYE TEMPERATURE AND AUDIO SPEAKERS

The Debye temperature, θ_D, and the speed of sound, v, are related as

$$\theta_D = \frac{vh}{2\pi k}\left(\frac{6\pi^2 N}{V}\right)^{1/3} \tag{6.32}$$

where N is the number of particles and V is their volume. The high Debye temperature of diamond, and consequently the high speed of sound of diamond, has led to the use of diamond-coating in diaphragms in tweeter speakers to reduce high-frequency sound distortion.

6.4.4 Heat Capacities of Metals

For metals, there is an additional factor to consider beyond the lattice contribution to the heat capacity. The extra degrees of freedom of the conducting electrons, that is, those with energy above the Fermi energy, must be taken into account. That this is a very small fraction of the valence electrons can be seen as follows.

The experimental molar heat capacity of a monatomic solid metal at high temperatures is usually a little more than $3\,R$. We can use equipartition theory to determine the contribution of free electrons to heat capacity. On the basis of vibrations alone, the internal energy of a monatomic solid is $3\,kT$ (see Section 6.4.1). If the atoms each had one free-valence electron, the total internal energy per atom would be

$$U = 3 \times 2 \times \frac{kT}{2} + 3 \times \frac{kT}{2} = 4.5\,kT \tag{6.33}$$

where the first term is from atomic vibrations and the second term is from the translation of one free electron per atom. This consideration would lead to $C_{V,m} = 4.5\,R$, which is far in excess of the experimental observation that the molar heat capacity of a monatomic metal is just slightly in excess of $3\,R$. Since $3\,R$ is the contribution from the vibrational degrees of freedom, only a small fraction of the valence electrons can be free to translate. Historically, this finding was the first indication that Fermi statistics (Equation 3.1) holds.

It can be shown from $P(E)$, the probability distribution function for electrons (Chapter 3), that the electronic contribution to the molar heat capacity at low temperatures is given by

$$C_{V,m}^{\text{elec}} = \gamma T \tag{6.34}$$

where γ depends on the particular metal. Therefore, the total molar heat capacity of a metal at very low temperatures is

$$\lim T \to 0 \quad C_{V,m}^{\text{metal}} = \gamma T + A T^3, \tag{6.35}$$

where the first term is the electronic contribution and the second term is the lattice contribution.

6.5 Heat Capacities of Liquids

As for many liquid properties, a description of a liquid in terms of its degrees of freedom falls between solids, where low-frequency vibrations are dominant, and gases, where translation, rotation, and perhaps vibrational degrees of freedom are active. Rigid crystals and liquids can be further distinguished by the excitation of configurational degrees of freedom in the latter. As is usual with other phases of matter, the heat capacity of a liquid usually increases as the temperature increases, due to the increased numbers of excited degrees of freedom at elevated temperatures, requiring more energy to invoke the same temperature rise.

The heat capacity of liquid water, as many other physical properties of water, illustrates a special case. The heat capacity of water near room temperature historically was used to define the calorie.* For convenience of comparison, the heat capacity of water can be converted to units of R, as $1\text{ cal K}^{-1}\text{ g}^{-1} \approx 18$ cal

---

* The calorie used to be defined as the amount of energy required to increase the temperature of exactly 1 g of water from 14.5 °C to 15.5 °C. Since this is a relative measurement and the joule is based on SI concepts, the calorie has been redefined in terms of the joule: $1\text{ cal} \equiv 4.184\text{ J}$.

K^{-1} $mol^{-1} \approx 9$ R. This value can be used to illustrate a point concerning degrees of freedom in water. As a triatomic molecule, water has a total of $3 \times 3 = 9$ degrees of freedom per molecule. If these were partitioned as 3 translational, 3 rotational, and 3 vibrational degrees of freedom, then the internal energy per molecule would be given by $U_{trans} + U_{rot} + U_{vib} = 3\,kT/2 + 3\,kT/2 + 3\,kT = 6\,kT$, and the molar heat capacity of water would be $C_{V,m} = 6\,R$. However, this value is insufficient to account for the known molar heat capacity of water ($9\,R$). The only way to increase $C_{V,m}$ within the equipartition model is to increase the twice-weighted vibrational degrees of freedom. Consider the extreme case of all 9 degrees of freedom being vibrational. This model leads to the internal energy per molecule of $9\,kT$ and a molar heat capacity, $C_{V,m}$, of $9\,R$, as observed experimentally. This simplified view of water is rather like that of a solid: All the degrees of freedom would be vibrational. This situation could arise because of the extended hydrogen-bonding network in water, which prevents free translation and rotation and allows water to look lattice-like in terms of its heat capacity. (It also gave water its color; see Chapter 2.) Although this view of water is oversimplified, detailed thermodynamic calculations of water are consistent with the existence of hydrogen-bonded clusters of water molecules in the liquid state.

6.6 Heat Capacities of Glasses

A *glass* can be defined as a rigid supercooled liquid formed by a liquid that has been cooled below its normal freezing point such that it is rigid but not crystalline. A glass is also said to be *amorphous*, which means without shape, showing its lack of periodicity on a molecular scale. A *supercooled liquid* (i.e., a liquid cooled below its normal freezing point) is *metastable* with respect to its corresponding crystalline solid. Metastability is defined as a local Gibbs energy minimum, whereas the global energy minimum is stable (see Figure 6.9). While a supercooled liquid is in an equilibrium state (the

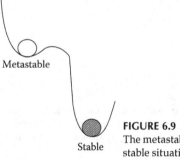

Metastable

Stable

FIGURE 6.9
The metastable situation is at a local energy minimum, whereas the stable situation is at a global energy minimum.

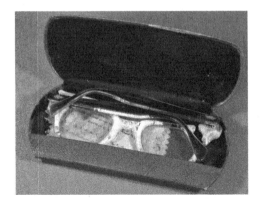

FIGURE 6.10
The frames of these eyeglasses are made of polymers that are predominantly in a glassy form. Over the course of about 40 years, they have begun to convert to a more stable crystalline form, causing destruction of the structure. The conversion is especially prominent at stress points, near the hinges and in the bridge.

molecules obey the Boltzmann distribution), a glass, which is obtained by cooling a supercooled liquid, is not in equilibrium with itself. This is because, in glasses, molecular configurations (i.e., the relative orientations and packing of the molecules) change slowly (sometimes hardly at all) so that Boltzmann distributions cannot be achieved. Given sufficient time, a glass would eventually convert to a crystalline form because it is more stable than the glass; an example of such a conversion is shown in Figure 6.10.

When a glass is heated, it does not have a well-defined melting point, but instead gradually becomes less viscous. Due to the irregularity of the molecular packing in a glass (see Figure 6.11 for a two-dimensional comparison of

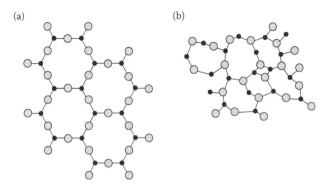

FIGURE 6.11
A two-dimensional schematic comparison of (a) crystalline SiO_2, which shows periodic atomic arrangement, and (b) glassy SiO_2, in which the atoms are not periodically arranged. Si atoms are shown as small black circles and O atoms are shown as larger gray circles. Note that this 2D presentation does not show the full bonding.

the arrangement of atoms in crystalline and glassy SiO_2), a glass can flow when a stress is applied.*

Common glass ($Na_2O \cdot CaO \cdot 6SiO_2$) also is called *soda lime glass*. This glass can be made by the reaction at elevated temperatures of soda with lime and sand. It accounts for about 90% of all glass manufactured today, and has been known for about 4500 years. Although this glass is very useful (and has better strength characteristics than many give it credit for), it has a major disadvantage, in that it cracks when either heated or cooled rapidly.

The addition of 12% B_2O_3 to soda glass gives a glass with better thermal shock fracture resistance. This glass is commonly known by the brand name *Pyrex®* and used in applications as common as lab glassware and home cookware, and as rare as Lindbergh's *Spirit of St. Louis* and the *Mercury* spacecraft.

When PbO replaces some of the CaO of soda glass, this gives a denser softer glass with a high refractive index. This glass is commonly known as *lead crystal* (although this is not technically a correct description of its structure, as it is still glass, not crystalline). The high refractive index of this glass allows its facets to "catch" light and cause its dispersion.

See the video "Amorphous Metal" under Student Resources at PhysicalPropertiesOfMaterials.com

The addition of 5% Al_2O_3 to soda glass gives a strong acid-resistant glass. When fibers of this glass are reinforced with an organic plastic, this is a composite material known as *Fiberglass®*.

The heat capacity of a typical glass is shown in Figure 6.12 in comparison with that of a crystalline solid. This figure shows that the heat capacity of a typical liquid is higher than that of the corresponding crystalline solid due to the increased number of degrees of freedom in the liquid. Decreasing in temperature from the liquid, the heat capacity of the supercooled liquid follows the trend of the liquid because no change in state occurred in passing below the melting point to the supercooled liquid state.

As the temperature of the supercooled liquid is lowered further, considerable entropy is removed according to

$$\Delta S = \int \frac{C_p}{T} dT. \tag{6.36}$$

In fact the rate of entropy removal is so high in comparison with the rate of removal in the crystalline solid, because the heat capacity of a supercooled liquid is more than that of the crystalline solid, that if the heat capacity were

---

* It has been proposed that window glass in ancient cathedrals is thicker at the bottom than at the top due to the flow of glass over the centuries. However, this effect has been shown to be insufficient to account for the observed difference in thickness. It is possible that it is due to mounting of panels of glass of irregular thickness: for stability, glaziers could have mounted the glass with the thickest edge to the bottom.

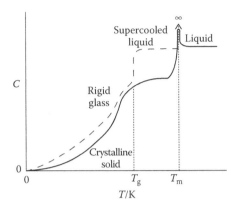

FIGURE 6.12
Heat capacity as a function of temperature showing typical crystalline solid, liquid, and rigid glass phases. The indicated values T_m and T_g are the melting and glass transition temperatures, respectively.

to remain so high as the temperature was lowered, this would eventually lead to a negative entropy for the supercooled liquid. This unphysical situation is avoided in that the heat capacity of the supercooled liquid drops down a step at a temperature characteristic of the material where concurrently rigidity sets in. Below this temperature, known as the *glass transition temperature*, T_g, the material is a *rigid glass*. The drop in heat capacity below T_g arises from the change in degrees of freedom from the supercooled liquid (an equilibrium state in which configurational degrees of freedom are active) to the rigid glass (a nonequilibrium state in which the heat capacity is essentially vibrational because the configurational part is frozen). The value of T_g is always approximate and less than the normal melting point of the pure crystalline material. For example, for SiO_2, $T_g \approx 1200\,°C$, and $T_m = 1610\,°C$.

The heat capacity of the rigid glass is still in excess of that of the crystalline solid, and this can be understood in terms of Debye's heat capacity theory applied to each phase. The fact that the heat capacity of the glass is higher than that of the crystal indicates that the effective Debye characteristic temperature of the glass is lower than that of the crystal. With reference to Figure 6.8, the temperature at which a given heat capacity is achieved in the glass will be lower than the temperature required for the crystal to have this same heat capacity, which implies that $\theta_D^{glass} < \theta_D^{crystal}$. This also means that atomic motion is easier in the glass than in the crystal. This conclusion is in line with the results of Table 6.3, which show that materials with lower Debye temperatures are softer. The relative ease of atomic motion in the glass reflects the lack of periodicity in its structure in comparison with the crystalline solid, as can be anticipated from the two-dimensional schematic views of the structures of glassy and crystalline SiO_2 in Figure 6.11.

Many chemical systems, both simple and complex, can exhibit glassy behavior. An appropriate thermal treatment is required to trick the material into forming the metastable supercooled liquid rather than the stable crystalline material. Examples of materials that form glasses include many biological materials, polymers (including plastics and paints), and foodstuffs. The glass transition temperature, T_g, can be very important in the characterization of materials from milk solids to paints, as is illustrated in the Tutorial on Thermal Analysis.

COMMENT: THERMODYNAMICS OF PIZZA

At one time or other, most people have burned the roof of their mouths by eating pizza. Although the pizza crust may have seemed to be at the correct temperature for eating, there is something about the pizza toppings (tomato sauce, cheese, etc.) that makes it a potential hazard. We might as well learn something from the experience: this situation illustrates the difference between temperature and heat.

If a piece of bread and a piece of pizza are in the same oven at the same temperature, and both are pulled out at the same time, we know from experience that the toast will be ready for safe consumption sooner than the pizza will be. Although they are at the same initial (oven) temperature, the toast cools faster because it has a lower heat capacity. On the other hand, the toppings on the pizza are very dense, and especially because of their high water content, have high heat capacities. We can say that the pizza has a higher thermal mass than the toast. For this reason, it takes much longer to dissipate the heat of the pizza. In other words, the toast and the pizza were at the same temperature when they were removed from the oven, but the toast had a much lower heat content (enthalpy) than the pizza. The higher enthalpy of the pizza means that it takes longer for it to cool off.

It's all thermodynamics!

6.7 Phase Stability and Phase Transitions, Including Their Order

A number of useful thermodynamic results that can be readily applied to phase stability, and phase transition information can be derived from some basic thermodynamic definitions. These will be used here to investigate the relationships among derivatives of Gibbs energy and to consider phase stability. Phase stability is very important in materials science because all the properties of a material—optical, thermal, electronic, magnetic, mechanical—depend critically on its phase.

Starting from the definitions of enthalpy, H (Equation 6.15), and Gibbs energy, G, for a fixed quantity of material:

$$G \equiv H - TS = U + pV - TS, \tag{6.37}$$

and Helmholtz* energy, A:

$$A \equiv G - pV = H - TS - pV = U - TS, \tag{6.38}$$

so for a closed system, infinitesimal changes in H, G, and A can, respectively, be written as

$$dH = dU + pdV + Vdp, \tag{6.39}$$

$$dG = dU + pdV + Vdp - TdS - SdT, \tag{6.40}$$

and

$$dA = dU - TdS - SdT. \tag{6.41}$$

If the work is pressure-volume work, and it is carried out reversibly, then

$$dU = \delta q + \delta w = \delta q_{rev} - pdV = TdS - pdV \tag{6.42}$$

where the second law of thermodynamics,

$$dS = \frac{\delta q_{rev}}{T} \tag{6.43}$$

has been used. Equations 6.39–6.42 can be used to give

$$dH = TdS + Vdp, \tag{6.44}$$

$$dA = -pdV - SdT, \tag{6.45}$$

and

$$dG = Vdp - SdT. \tag{6.46}$$

---

* Hermann Ludwig Ferdinand von Helmholtz (1821–1894) was a German physiologist and natural scientist. Helmholtz's thesis on connections between nerve fibers and nerve cells led to his studies of heat in animals, which, in turn, led to his theories of conservation of energy. Helmholtz made contributions to a wide variety of subjects: thermodynamics, invention of the ophthalmoscope, structure and mechanism of the human eye, the role of the bones in the middle ear, electricity and magnetism, and music theory.

These expressions for dU, dH, dG, and dA (Equations 6.42, 6.44, 6.45, and 6.46) are known collectively as the *fundamental equations*.

The fundamental equations can be used to derive important thermodynamic relations, known as the *Maxwell* relations*. Consider the fundamental equation for U, Equation 6.42. Since U is a state function of S and V, dU also can be written as

$$dU = \left(\frac{\partial U}{\partial S}\right)_V dS + \left(\frac{\partial U}{\partial V}\right)_S dV \qquad (6.47)$$

and the coefficients of the two forms of dU (Equations 6.42 and 6.47) can be equated, so

$$\left(\frac{\partial U}{\partial S}\right)_V = T \qquad (6.48)$$

and

$$\left(\frac{\partial U}{\partial V}\right)_S = -p. \qquad (6.49)$$

Since U is a *state function*, i.e., depends only on the state of the system and not the path to reach that state, the order of differentiation for its second derivative does not matter, and

$$\frac{\partial^2 U}{\partial V \partial S} = \frac{\partial^2 U}{\partial S \partial V}, \qquad (6.50)$$

so

$$\left(\frac{\partial T}{\partial V}\right)_S = -\left(\frac{\partial p}{\partial S}\right)_V \qquad (6.51)$$

which is the Maxwell relation from U.

Similarly, from the fundamental equation for G (i.e., Equation 6.46), since G is a state function of p and T, dG also can be written as

$$dG = \left(\frac{\partial G}{\partial p}\right)_T dp + \left(\frac{\partial G}{\partial T}\right)_p dT \qquad (6.52)$$

and the coefficients of the two forms of dG (i.e., Equations 6.46 and 6.52) can be equated, so

---

* James Clerk Maxwell (1831–1879) was a Scottish-born physicist who made many important contributions to electricity, magnetism, color vision, and color photography. His studies of Saturn's rings led to his seminal work on the kinetic theory of gases.

$$\left(\frac{\partial G}{\partial p}\right)_T = V \tag{6.53}$$

and

$$\left(\frac{\partial G}{\partial T}\right)_p = -S. \tag{6.54}$$

Since G is a state function, the order of differentiation for its second derivative does not matter, and

$$\frac{\partial^2 G}{\partial T \, \partial p} = \frac{\partial^2 G}{\partial p \, \partial T} \tag{6.55}$$

so

$$\left(\frac{\partial V}{\partial T}\right)_p = -\left(\frac{\partial S}{\partial p}\right)_T. \tag{6.56}$$

Equation 6.56 is the Maxwell relation from G.

Two further Maxwell relations can be derived, one from each of the other fundamental equations. It is left as an exercise for the reader to show that

$$\left(\frac{\partial T}{\partial p}\right)_S = \left(\frac{\partial V}{\partial S}\right)_p \tag{6.57}$$

and

$$\left(\frac{\partial p}{\partial T}\right)_V = \left(\frac{\partial S}{\partial V}\right)_T. \tag{6.58}$$

To consider the stability of a given phase, it is first useful to summarize the pressure-temperature stability of a "typical" material, as given in Figure 6.13.

As the pressure is held constant and the temperature is increased along the dashed line in Figure 6.13, the stable phase goes from solid to liquid to gas. This can be shown by consideration of G as a function of T at constant p (Figure 6.14).* This figure shows the driving force for phase changes; at a given temperature and pressure, the phase with the lowest G for a given amount of matter is the most stable. At the equilibrium between two phases (such as the solid–liquid equilibrium at the melting point or the liquid–gas equilibrium at the boiling point), the Gibbs energies of the equilibrium phases are equal ($G_{\text{solid}} = G_{\text{liquid}}$ at the melting point, $G_{\text{liquid}} = G_{\text{gas}}$ at the boiling point).

---

* The quantity of material must remain constant. For example, in the discussion of Figures 6.14–6.16, all the quantities can be considered to be molar values (G_m, V_m, S_m, H_m, $C_{p,m}$).

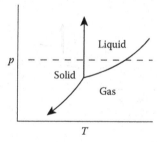

FIGURE 6.13
The pressure–temperature phase diagram of a typical pure material. The solid lines denote the phase boundaries, and the dashed line indicates an isobaric change in temperature, as discussed in the text.

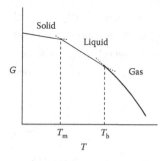

FIGURE 6.14
Gibbs energy, G, as a function of temperature for a typical material. T_m and T_b are the melting and boiling temperatures, respectively. Solid lines indicate the stable phase in a given region. The quantity of material must be constant.

The slope of the curve in the $G(T)$ plot, $(\partial G/\partial T)_p$, is identical to $-S$ (Equation 6.54). Since $S > 0$ (always), all the slopes are negative in Figure 6.14. While the entropy increases on going from solid to liquid to gas, the slope becomes more negative from solid → liquid → gas. Even within a given phase, the entropy increases as the temperature increases, so the curves are concave down in Figure 6.14.

Within the constraints of model calculations, one can calculate G_m for various phases under specific temperature and pressure conditions, and the predicted most stable phase is the one with the lowest G_m. Comparison with experimental phase stabilities can indicate the validity of the assumed interatomic interactions.

The *Ehrenfest* classification of phase transitions* hinges on the behavior of G near the phase transformation. In this classification, phase transitions can be *first order* (first derivatives of G are discontinuous) or *second order* (first derivatives of G are continuous but second derivatives of G are discontinuous).

---

* Paul Ehrenfest (1880–1933) was an Austrian-born theoretical physicist who studied for his PhD under Boltzmann's supervision. As a teacher, Einstein described Ehrenfest as "peerless" and "the best teacher in our profession I have ever known." His students nicknamed him "Uncle Socrates," for his probing but personable style. Ehrenfest's contributions to thermodynamics and quantum mechanics stemmed from his ability to ask critical, probing questions. Depression due to the plight of his Jewish colleagues and personal difficulties led Ehrenfest to take his own life.

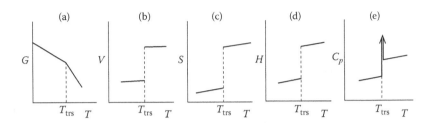

FIGURE 6.15
Schematic view of (a) $G(T)$, (b) $V(T)$, (c) $S(T)$, (d) $H(T)$, and (e) $C_p(T)$ for a first-order phase transition. T_{trs} designates the transition temperature. The quantity of material must remain constant.

In a first-order phase transition, the first derivatives of $G((\partial G/\partial p)_T = V$ and $(\partial G/\partial T)_p = -S)$ are discontinuous. As shown in Figure 6.15, since $G^\alpha = G^\beta$ at the equilibrium of phase α with phase β, and given the preceding discussion, we find that $\Delta_{trs}V \neq 0$ and $\Delta_{trs}S \neq 0$ at a first-order phase transition. From $\Delta_{trs}G = 0$ and Equation 6.6, it follows that

$$\Delta_{trs}S = \frac{\Delta_{trs}H}{T}. \tag{6.59}$$

Since there is a change in the amount of disorder at a first-order transition,

$$\Delta_{trs}S \neq 0 \tag{6.60}$$

and therefore

$$\Delta_{trs}H \neq 0 \tag{6.61}$$

for a first-order phase transition. From Equation 6.4, with the integration to give $\Delta_{trs}H$ over the fixed temperature T_{trs}, in order for $\Delta_{trs}H$ be nonzero, C_p must be infinite at the transition. This is illustrated in Figure 6.15. Examples of first-order transitions (i.e., transitions with the signatures of $\Delta_{trs}V \neq 0$, $\Delta_{trs}S \neq 0$, $\Delta_{trs}H \neq 0$, $C_p = \infty$) include melting, boiling, sublimation, and some solid-solid transitions (e.g., graphite-to-diamond).*

A second-order phase transition does not have a discontinuity in $V (= (\partial G/\partial p)_T)$ or $S (= -(\partial G/\partial p)_p)$, but the second derivatives of $G ((\partial^2 G/\partial T\partial p)_T)$ $= (\partial V/\partial T)_p$ and $(\partial^2 G/\partial p\partial T) = -(\partial S/\partial p)_T)$ are discontinuous. This situation is illustrated in Figure 6.16; $\Delta_{trs}V = 0$, $\Delta_{trs}S = 0$, and $\Delta_{trs}H = 0$ at the temperature of the second-order transition, although V, S, and H each have anomalous behavior in the region of the transition. The fact that $\Delta_{trs}H = 0$ for a second-order transition leads to its finite heat capacity at the transition, as shown

---

* As in the previous discussion, the quantities must remain constant. For example, Equations 6.59 to 6.61 can be written on a molar basis.

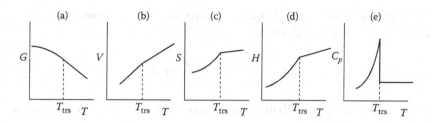

FIGURE 6.16
Schematic view of (a) $G(T)$, (b) $V(T)$, (c) $S(T)$, (d) $H(T)$, and (e) $C_p(T)$ for a second-order phase transition. T_{trs} is the transition temperature. The quantity of matter remains constant throughout the temperature changes.

in Figure 6.16. The λ-shape of the heat capacity at this transition leads this to be called a λ-*transition* (lambda transition). An example of such a transition (which is much rarer than a first-order transition) is observed in liquid helium (see Chapter 9).

6.8 $(C_p - C_V)$: An Exercise in Thermodynamic Manipulations

The aim here is to derive a useful thermodynamic relationship between C_p and C_V and also to look at one special case of its use.

From the definitions of C_p (Equation 6.3) and C_V (Equation 6.1):

$$C_p - C_V = \left(\frac{\partial H}{\partial T}\right)_p - \left(\frac{\partial U}{\partial T}\right)_V \tag{6.62}$$

where we see that $(C_p - C_V)$ is a measure of the difference in energy required to increase temperature at constant pressure (which allows for $p - V$ work of expansion on heating) relative to increasing the temperature at constant volume (where there is no expansion work).

The definition of H (Equation 6.15) leads to an expression for dH in terms of dU (Equation 6.39), that can be differentiated with respect to temperature at constant pressure to give

$$\left(\frac{\partial H}{\partial T}\right)_p = \left(\frac{\partial U}{\partial T}\right)_V + p\left(\frac{\partial V}{\partial T}\right)_p \tag{6.63}$$

and substitution of Equation 6.63 into Equation 6.62 leads to

$$C_p - C_V = \left(\frac{\partial U}{\partial T}\right)_p + p\left(\frac{\partial V}{\partial T}\right)_p - \left(\frac{\partial U}{\partial T}\right)_V. \tag{6.64}$$

This expression can be simplified by finding another expression for $(\partial U/\partial T)_p$. Since U is a function of V and T,

$$dU = \left(\frac{\partial U}{\partial V}\right)_T dV + \left(\frac{\partial U}{\partial T}\right)_V dT \tag{6.65}$$

which leads, on differentiation with respect to temperature at constant pressure, to

$$\left(\frac{\partial U}{\partial T}\right)_p = \left(\frac{\partial U}{\partial V}\right)_T \left(\frac{\partial V}{\partial T}\right)_p + \left(\frac{\partial U}{\partial T}\right)_V. \tag{6.66}$$

Substitution of Equation 6.66 into Equation 6.64 leads to

$$C_p - C_V = \left(\frac{\partial U}{\partial V}\right)_T \left(\frac{\partial V}{\partial T}\right)_p + p\left(\frac{\partial V}{\partial T}\right)_p. \tag{6.67}$$

Equation 6.67 shows that the difference between C_p and C_V is the total work done by thermal expansion (expressed per K of temperature rise). This can be seen further by noting that the common factor of $(\partial V/\partial T)_p$ is the thermal coefficient of the volume. Therefore, the second term in Equation 6.67 is the quantity of $p - V$ work done against the external pressure, p, per K of temperature rise. In the first term, $(\partial U/\partial V)_T$ is the effective internal "pressure" (p_{int}) that results from the chemical bonds that hold the material together. Therefore, the first term also represents $p - V$ work, now done against internal pressure. For typical solids, $p_{int} > 100\,\text{MPa}$. Therefore, for solids at atmospheric pressure ($p \sim 0.1\,\text{MPa}$), the first term in Equation 6.67 dominates. To relate Equation 6.67 to more experimentally accessible parameters, thermodynamic relationships can be used. From the fundamental equation for U (Equation 6.42), differentiation with respect to volume at constant temperature leads to

$$\left(\frac{\partial U}{\partial V}\right)_T = T\left(\frac{\partial S}{\partial V}\right)_T - p. \tag{6.68}$$

Substitution of Equation 6.68 into Equation 6.67 gives

$$C_p - C_V = T\left(\frac{\partial V}{\partial T}\right)_p \left(\frac{\partial S}{\partial V}\right)_T. \tag{6.69}$$

Although $(\partial S/\partial V)_T$ is still awkward on its own, a Maxwell relation (Equation 6.58) allows

$$C_p - C_V = T\left(\frac{\partial V}{\partial T}\right)_p \left(\frac{\partial p}{\partial T}\right)_V \tag{6.70}$$

for all systems, with no approximations. However, Equation 6.70 is more usually written in terms of the *coefficient of thermal expansion* (α) and the *isothermal compressibility* (β_T), defined as follows:

$$\alpha = \frac{1}{V}\left(\frac{\partial V}{\partial T}\right)_p \tag{6.71}$$

and

$$\beta_T = \frac{-1}{V}\left(\frac{\partial V}{\partial p}\right)_T. \tag{6.72}$$

From these definitions, the units of α are K^{-1} and units of β_T are reciprocal pressure (e.g., Pa^{-1}). The advantage of using α and β_T over $(\partial V/\partial T)_p$ and $(\partial V/\partial p)_T$ is that α and β_T are *intensive properties* (i.e., independent of the size of the system), whereas $(\partial V/\partial T)_p$ and $(\partial V/\partial p)_T$ require knowledge of the size of the system (i.e., they are *extensive properties*, dependent upon the size of the system). From the definition of α (Equation 6.71), it follows that

$$C_p - C_V = TV\alpha\left(\frac{\partial p}{\partial T}\right)_V. \tag{6.73}$$

To obtain $(\partial p/\partial T)_V$, consider that V is a function of p and T so that

$$dV = \left(\frac{\partial V}{\partial p}\right)_T dp + \left(\frac{\partial V}{\partial T}\right)_p dT \tag{6.74}$$

which leads to

$$\left(\frac{\partial p}{\partial T}\right)_V = -\frac{\left(\frac{\partial V}{\partial T}\right)_p}{\left(\frac{\partial V}{\partial p}\right)_T} \tag{6.75}$$

and

$$\left(\frac{\partial p}{\partial T}\right)_V = \frac{-V\alpha}{-V\beta_T} = \frac{\alpha}{\beta_T}. \tag{6.76}$$

Substitution of Equation 6.76 into Equation 6.73 leads to the following general expression for $(C_p - C_V)$:

$$C_p - C_V = \frac{TV\alpha^2}{\beta_T} \tag{6.77}$$

for isotropic materials. If the material is anisotropic, α and β_T will depend on the direction, and a tensor form of Equation 6.77 specific to the system's symmetry must be employed.

In general, $(C_p - C_V)$ is small compared with C_p for most solids; it is especially small at low temperatures (since $\alpha \to 0$ as $T \to 0$; see Chapter 7) and may be < 1% of C_p at $T < \theta_D$. $(C_p - C_V)$ typically is a few percent of C_p for a solid at room temperature. (See problem 6.6.)

For the special case of ideal gases, $(C_p - C_V)$ has an especially simple form. An ideal gas is described by the equation of state $pV = nRT$, so α for an ideal gas can be written as

$$\alpha = \frac{1}{V}\left(\frac{\partial V}{\partial T}\right)_p = \frac{p}{nRT}\left(\frac{\partial}{\partial T}\left(\frac{nRT}{p}\right)\right) = \frac{p}{nRT}\left(\frac{nR}{p}\right) = \frac{1}{T} \tag{6.78}$$

and β_T for an ideal gas can be written as

$$\beta_T = \frac{-1}{V}\left(\frac{\partial V}{\partial p}\right)_T = -\frac{p}{nRT}\left(\frac{\partial}{\partial p}\left(\frac{nRT}{p}\right)\right) = \frac{p}{nRT}nRT\,p^{-2} = \frac{1}{p} \tag{6.79}$$

and substitution of these results into Equation 6.77 leads to

$$C_{p,m} - C_{V,m} = \frac{TV\left(T^{-2}\right)}{p^{-1}} = \frac{Vp}{T} = R \tag{6.80}$$

for an ideal gas, where the result is written in terms of the molar heat capacities $C_{p,m}$ and $C_{V,m}$, i.e., $n = 1$.

6.9 Thermal Energy Storage Materials: A Tutorial

Thermal energy storage materials have many applications, from scavenging heat from hot industrial exhaust to keeping electrical components from overheating during soldering, to storing the energy from solar thermal collectors, to making use of off-peak electricity.

a. Thermal energy storage materials such as rock, brick, water, and concrete absorb energy when heated through what is called *sensible heat storage* (i.e., heat storage due to their heat capacities). These materials are used as "night heaters" for electrical thermal storage, taking advantage of off-peak electricity rates. The room temperature *specific heat capacities,** C_s, of some common materials are given in Table 6.4.

---

* The term "specific" (e.g., specific heat capacity or specific gravity) is used to denote the value of a physical property per unit mass.

TABLE 6.4

Specific Heat Capacities of Some Common
Materials at Room Temperature

Material	$C_s/(\text{J K}^{-1}\,\text{g}^{-1})$
Common brick	0.920
Balsa wood	2.93
Marble	0.879

Which of these materials can store the most energy per unit mass at room temperature? What additional information is needed to make this decision per unit volume of material? Do you expect this information to influence which is the best heat storage material?

b. Some other materials such as Glauber salts (hydrates of $CaSO_4$ and Na_2SO_4), paraffins (long-chain saturated hydrocarbons), and fatty acids can store energy via melting. These are known as phase change materials (PCMs), pioneered by Mária Telkes.* Sketch a graph of the following as a function of temperature for a material that stores energy through fusion (i.e., PCM via melting): (i) the heat capacity through the region from solid to liquid, and (ii) the enthalpy through the region from solid to liquid.

c. What do you expect to be the most important criteria for a successful thermal energy storage material that uses fusion (i.e., melting)? Think of thermodynamics but also think of other matters.

d. Would there be any advantages or disadvantages to heat storage materials that make use of solid–solid phase transitions? Discuss.

e. A commercial product, Re-Heater™, contains sodium acetate trihydrate. This salt melts at 58 °C with an enthalpy change of 17 kJ mol⁻¹. Sketch the enthalpy versus temperature diagram for this salt from 0 °C to 100 °C.

f. Sodium acetate trihydrate can be supercooled quite easily to room temperature. Using a dashed line, draw the enthalpy curve for the supercooling process on your diagram from (e).

g. A supercooled liquid is an example of a metastable state. In terms of the Gibbs energy, G, this state is at a local minimum, and the stable state is at a global minimum. Sketch a Gibbs energy curve that shows both these states.

h. When a supercooled liquid is perturbed, for example when a seed crystal is added, the barrier between the stable and the metastable

---

* Mária Telkes (1900–1995) was an Hungarian-American scientist who pioneered the use of phase change materials for thermal energy storage. She also is known for creating the first thermoelectric power generator and for her designs of solar heating systems.

states is lowered and the stable state is achieved. This is *a physical change*. What do we usually call this situation for *chemical* changes?

i. How does a seed crystal lower the Gibbs energy path between the metastable supercooled liquid state and the stable crystalline state?

j. If you were able to lower the Gibbs energy path between the super-cooled liquid state of sodium acetate trihydrate and its solid state, from the enthalpy dia-gram of part (e) and (f), what would happen energetically?

See the video "Heat Pack" under Student Resources at PhysicalPropertiesOfMaterials.com

k. In a closed system, it is not easy to add a seed crystal. How else could you induce crystallization in supercooled liquid sodium acetate trihydrate?

l. This application gives off heat when activated. Does the spontaneous crystallization process always give off heat? Can you think of a similar principle that takes up heat when activated?

6.10 Thermal Analysis: A Tutorial

There are three main techniques in thermal analysis, and they each are known by their initials: TGA (thermogravimetric analysis), DTA (differential thermal analysis), and DSC (differential scanning calorimetry).

a. *Thermogravimetric analysis* (TGA) is an analytical tool to carry out "gravimetry" (i.e., mass determination) of a sample as the temperature changes. This tool is straightforward: the sample is heated and its mass is determined as a function of temperature. A typical plot is mass versus temperature. Suppose that a commercial sample of mass 10.0 mg is heated from room temperature and experiences a 3.0% mass loss at 80 °C. Sketch the resulting thermogravimetric curve.

b. Thermogravimetry can be used to characterize pure samples and blends, and it is a very useful technique. An even more useful thermoanalytical tool is *differential thermal analysis*. In this technique, a sample and a reference (e.g., an empty sample cell) are both monitored as a function of temperature. The experiment can be carried out either in the heating or cooling mode, but the rate of sample heating (or cooling) is usually kept constant. In this experiment, the same power is input to the sample and the reference, and the temperature difference between the sample and the reference is measured as a function of temperature. If the heat capacities of the sample and reference are exactly the same throughout the temperature range of

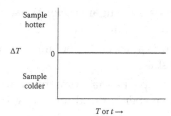

FIGURE 6.17
A flat signal from DTA when the sample and reference are perfectly matched in heat capacity.

the measurement, the signal is "flat" as shown in Figure 6.17. What would the signal look like if the sample container held a solid that melted during the temperature range of the scan?

c. Why do you suppose this technique is called *differential* thermal analysis?

d. What would the signal look like if a rigid glass were heated to well into the region where it is a supercooled liquid?

e. *Differential scanning calorimetry* is similar to DTA in several respects. Again, it is a differential technique, with a sample and a reference, and the setup is heated (or cooled) at a constant rate. DSC is distinguished from DTA by what is monitored: in DTA the temperature difference between the sample and the reference is measured, and in DSC the difference in power to the sample and reference to keep heating (or cooling) them both at the same rate is measured. Sketch and label the DSC curve for the melting of a crystalline solid.

f. It would be very useful in a technique such as DTA or DSC if the area "under" peaks could be converted to enthalpy change of the corresponding process. Only one of the DTA and DSC has the area scaling to enthalpy change. Which one? *Hint*: Consider what the peak area units are in each case.

g. How could one scale the area under the peak to determine the enthalpy change?

h. As an example of an amorphous solid, consider the following. Sugar can be crystalline (e.g., big sugar crystals that one can grow out of sugar/water solutions; also white sugar—its crystallinity is particularly obvious under a microscope). It also can be amorphous (or glassy, or *vitreous*; all mean the same thing). If sugar is heated to a high temperature (above the melting point) and then cooled very quickly—for example, by dropping it into cold water—it is transformed to a rigid glassy state. What is the confectionery name for this state of sugar?

i. Why should you not waste your time trying to make this confectionery unless you have a candy thermometer?

j. Is the rigid glassy state of sugar stable? What happens to this state of sugar if left for a long time?

k. If the melted sugar is allowed to cool more slowly, will it form the same rigid glassy state?

l. What are the white specks that sometimes form on the surface of chocolates that have been around for a long time? Should you eat these chocolates?

m. Why would a chocolate factory own a differential scanning calorimeter?

n. Although DSC can be used to determine the heat capacity of a sample, the precision of the results can be considerably greater than their accuracy. (The most accurate technique to determine heat capacities near and below room temperature is adiabatic calorimetry, in which the heat capacity is determined by the energy input and temperature rise [Equation 6.9] under constant pressure, adiabatic conditions.) Suggest some possible systematic errors in DSC determination of heat capacity.

o. The setting of cement takes place in two exothermic stages. The first stage starts at 143 °C, with an enthalpy change of 11 J g^{-1}, and the second stage starts at 196 °C with an enthalpy change of 5.8 J g^{-1}. Sketch the corresponding DSC curve.

6.11 Learning Goals

- Equipartition of energy
- Heat capacity of a gas: monatomic gas, nonlinear triatomic gas, real gas
- Trouton's rule
- Joule's experiment (U is independent of volume for ideal gas)
- Joule–Thomson coefficient
- Heat capacities of condensed phases: Dulong-Petit law; Einstein model; Debye model; metals; liquids; glasses
- Glass properties, glass transition temperature
- Fundamental equations
- Maxwell relations
- Phase transitions (first order, second order)
- $C_p - C_V$
- Thermal energy storage materials (sensible heat storage and PCMs)
- Thermal analysis (TGA, DTA, DSC)

6.12 Problems

6.1 The ideal gas law is a limiting law.

    a. Derive the relationship between pressure and density (ρ) for an ideal gas.

    b. Confirm the relationship (as a limiting law) on the basis of the following data for dimethyl ether at $T = -13.6\,°C$.

p/Pa	101325	70928	50662	30398	20265	10133
$\rho/\text{g L}^{-1}$	2.4110	1.4684	1.0615	0.6635	0.4501	0.2247

Use a graph to answer this question. It may prove useful to know that the molecular formula of dimethyl ether is CH_3OCH_3.

6.2 The Einstein function for heat capacity is given by Equation 6.22.

    a. Calculate the value of C_V from the Einstein function for 1 mol of a monatomic material with an Einstein characteristic frequency, expressed in wavenumbers as 100 cm^{-1}, for the following temperatures: 5.00, 10.00, 25.00, 50.00, 75.00, 100.0, 150.0, 200.0, 250.0, and 300.0 K. Give your answer for $C_{V,m}$ in terms of R, the gas constant.

    b. What is the percentage difference between the calculated value of $C_{V,m}$ at $T = 300.0$ K and the value from the Dulong–Petit law?

    c. On the same graph, show $C_{V,m}$ both for your results from (a) and for a frequency of 200 cm^{-1}, calculated for the same temperatures as in (a).

6.3 Derive the Maxwell relations from the fundamental equations for

    a. H;

    b. A.

6.4 Derive a plot of G as a function of pressure (analogous to $G(T)$ given in Figure 6.14)) for a material that goes from gas to liquid to solid as the pressure is increased. Show your reasoning as fully as possible.

6.5 A new molecular species was investigated with a view to using it as a commercial refrigerant. The temperature drops for a series of initial pressures, p_i, expanding into 101325 Pa were measured at 0 °C, with the following results:

p_i/Pa	3.2×10^6	2.4×10^6	1.8×10^6	1.1×10^6	8×10^5	5×10^5
$-\Delta T/\text{K}$	22	18	15	10	7.4	4.6

    a. What is the value of the Joule–Thomson coefficient for this gas at these conditions? Hint: Consider the definition of the Joule–Thomson coefficient and also the definition of the partial derivative function.

b. Would another gas that would be a better refrigerant have a lower or higher Joule–Thomson coefficient? Explain.

6.6 Calculate $(C_p - C_V)$ in J K^{-1} mol^{-1} for: (i) liquid water and (ii) solid Zn, given the following information for 25 °C:

	α/K^{-1}	β_T/Pa^{-1}	V_m/cm^3 mol^{-1}
H_2O(l)	4.8×10^{-4}	4.5×10^{-10}	18
Zn(s)	8.93×10^{-5}	1.5×10^{-11}	7.1

and calculate the percentage difference from the C_p values: 75 J K^{-1} mol^{-1} for water and 23 J K^{-1} mol^{-1} for Zn.

6.7 Sketch the differential scanning calorimeter trace for water on cooling from 15 °C to −10 °C. Label the temperature axis (including values and units), but just include the label and the units (i.e., no values) on the other axis. Label the exothermic and endothermic directions on the y-axis.

6.8 A new gaseous species has been made in the laboratory and it is found to be the triatomic molecule ABC. It is not known if ABC is linear or nonlinear. Explain how heat capacity measurements could differentiate between these possibilities. Be as explicit as possible, including stating expected values of the heat capacities and the conditions under which you would carry out the measurements.

6.9 Very careful calorimetry can be used to carry out quality control tests on pacemakers to determine the reliability of their batteries. Suggest the physical basis for these tests.

6.10 Although the Debye model is better able to describe the heat capacity of a monatomic solid, the Einstein model is useful in molecular solids with more than one atom per molecule. In this case, there are vibrational motions of the whole molecule (*acoustic phonons*) that are well described by the Debye model and motions due to internal vibrations (*optic phonons*) that are described well by the Einstein model. In a unit cell with N atoms, there are 3 acoustic (Debye) vibrational modes and $3N-3$ optic (Einstein) vibrational modes. Calculate $C_{V,m}$ for a diatomic molecular solid with 2 atoms per unit cell and a Debye characteristic temperature of 300 K and all Einstein modes at 200 cm^{-1}. Plot your result as a function of temperature from $T = 0$ to 500 K.

6.11 An advertisement claims that a novel product could allow planting of outdoor crops earlier than would otherwise be possible in cold climates. It might even be possible, using this device, to plant before all danger of frost has passed. This product is a plastic container composed of vertical columns filled with water, which then surrounds the seedling with a cylinder of water. A top view is shown schematically in Figure 6.18.

FIGURE 6.18
A seedling is surrounded by cylinders of water. This is a top view.

    a. How does this device work? Explain in thermodynamic terms.

    b. Would this device be as effective if it was filled with another liquid, e.g., oil? Explain.

    c. Sketch the temperature as a function of time at the plant for both cases (a) and (b) when the external temperature drops from 15 °C to –10 °C.

6.12 A gas at room temperature and at a constant pressure of 101 kPa is going to be used to heat a material that is at a cooler temperature. You are offered the option of using 1 mol of argon (Ar) gas or 1 mol of nitrogen (N_2) gas for this purpose. Which will be more efficient? Explain your reasoning.

6.13 a. Which form of carbon do you expect to have the higher value of θ_D: diamond or graphite? Justify your answer.

    b. Sketch the molar heat capacity from $T = 0$ K to a temperature high enough for equipartition to hold, for graphite and diamond. Put both on the same graph and put a numeric scale on the heat capacity axis, in terms of R.

6.14 A consumer product made of a synthetic material (which looks very much like leather) is supple at room temperature but found to become brittle in the outdoors in the Canadian winter. It is thought that this is because the material is not very crystalline (i.e., it is highly amorphous), and it undergoes a transformation to a rigid glass at a temperature of about –10 °C. Sketch the expected differential scanning calorimetry trace for such a material, from –30 °C to 20 °C.

6.15 When a natural material that is mostly amorphous but partially crystalline ages over tens of years, there is an increase in the ratio of the crystalline-to-amorphous portion of the material. Sketch the

DSC curves of two pieces of this material, one newly produced and one well aged.

6.16 A bowl of frozen vegetables is to be cooked in a microwave oven. When the bowl is covered during the cooking process, two minutes is sufficient cooking time to produce steaming hot vegetables. If the bowl is not covered, after two minutes the vegetables are still cold. Why?

6.17 When a bicycle tire is inflated with a hand pump, the valve stem is noticeably warmed (temperature increase of about 10 K). Why?

6.18 For silver, $\theta_D = 225$ K and $\gamma = 6.46 \times 10^{-4}$ J K^{-2} mol^{-1}. Calculate $C_{V,m}$ of silver at 0.1 K and also at 1 K. Comment on the relative contributions of the lattice (Debye) heat capacity and the electronic heat capacity at each temperature.

6.19 Air in a container is at a pressure of about 100 kPa and at room temperature. Inside the container there is a sensitive thermometer, and when the air is pumped out of the container with a vacuum pump, the temperature is seen to fall. Explain why the temperature falls and what this indicates about the gas.

6.20 The fuels on the space shuttle Challenger leaked out during the flight, causing the shuttle to explode. The fuels were meant to be held in place by O-rings, i.e., soft rubbery materials that could be compressed to make tight seals. However, at the cold launch temperature (near 0 °C), the soft rubbery material had become a rigid glass. On this basis, sketch the DSC (differential scanning calorimetric) curve for this rubber from 0 °C to 25 °C and indicate any features.

6.21 Polymers have been proposed as the basis of new electrooptic devices, but one of the problems is that the required molecular alignment of amorphous polymers can decay over time. In order for a polymer to be useful in an electrooptic device, should its T_g be very high or very low? Explain.

6.22 The speed of sound varies considerably with depth in the earth's crust. In the upper mantle, it is 8 km s^{-1}, and it increases rapidly with depth to about 12 km s^{-1} in the lower mantle.

   a. How does the Debye characteristic temperature of the earth's crust vary with depth?

   b. What does this variation of the speed of sound indicate about the interactions within the crustal materials as a function of depth? Does this make sense?

   c. Suggest an application of the variation of speed of sound with depth in the earth's crust.

6.23 Fast-cooled slag from smelters is glassy and therefore less reactive with water from the environment than slow-cooled slag (which is more crystalline). Therefore, fast-cooled slag is preferred because it leads to less contamination of the environment from rainwater, in the form of runoff. Explain how DSC could be used for quality control of the slag-cooling process.

6.24 a. Sketch the DSC curve of a material that exhibits both a glass transition temperature at $T = 40\,°C$ and then melting at $T = 100\,°C$.

b. If a given material shows both a glass transition and a melting point in a DSC curve, what can we deduce about its crystallinity? Explain.

6.25 Sketch the DSC curve from 300 K to 400 K for an amorphous material with a glass transition temperature of 320 K; this material undergoes exothermic chemical decomposition starting at 395 K. Mark all features.

6.26 The *Neumann–Kopp law* is a generalized experimental finding that in many cases the heat capacity of a solid can be well approximated by the sum of the heat capacities of the constituent elements, weighted by their contribution to the total composition. For example, a generic salt represented by M_2X_5 would have a molar heat capacity close to $2C_{V,m}(M) + 5C_{V,m}(X)$, where M and X are both solids. The Neumann–Kopp law can work well over a wide range of temperatures when M, X, and M_2X_5 are all solids. Provide a reasonable theoretical explanation for the Neumann–Kopp law.

6.27 *Plasticizers* are added to many polymers to increase their flexibility. The most common plasticizers are phthalate derivatives. Plasticizers are absorbed into the polymer interstices, acting as a lubricant that enhances polymer chain mobility. The addition of a plasticizer to a polymer reduces the glass transition temperature. Sketch and label the DSC curves for the same amorphous polymer with and without plasticizer, from below T_g to above T_g. Indicate the exothermic and endothermic directions.

6.28 The molar heat capacity at constant volume, $C_{V,m}$, for liquid water decreases as the temperature increases throughout the liquid range, whereas for most condensed materials (solids and liquids), $C_{V,m}$ increases as the temperature increases. Explain why water is anomalous, and what this indicates about the "structure" of water as a function of temperature.

6.29 A useful roof covering for some large buildings is a polymer composed of polyvinyl chloride, known by the trade name Sarnafil™. A study of Sarnafil roof coverings has shown their glass transition temperatures to be in the range −54 °C to −11 °C, with a trend to increased T_g as the roof material ages. Briefly discuss the implications of increased T_g of Sarnafil with age of the material with respect to utility as a roof material in a wide range of climates.

6.30 Materials such as $Ge_2Sb_2Te_5$ undergo thermally induced reversible changes between crystalline phases and amorphous phases. The two phases differ in physical properties such as electrical resistivity and reflectivity of light. The change in reflectivity on change of phase makes $Ge_2Sb_2Te_5$ useful for optical data storage such as CDs and DVDs. The difference in electrical conductivity can be sufficient to represent two logic states, leading to phase-change random access memory (PRAM). Although PRAM technology can have significant advantages, such as fast read/write speeds and simple fabrication, there are some challenges. Discuss your expectations of PRAM use with regard to power consumption and thermal interference (i.e., spread of heat from one bit to the next) and discuss ways to overcome these limitations.

6.31 Two potential thermal energy storage materials are available for use in storing solar thermal energy. Both melt at an appropriate temperature (42 °C) and they have similar enthalpy changes ($\Delta_{fus}H \sim 120 J\ g^{-1}$) and therefore both should capture solar thermal energy by melting. One material crystallizes at 41 °C, whereas the other material supercools to 10 °C before crystallizing. Discuss potential applications of each of these PCMs for thermal energy storage.

6.32 A pyrometric cone is a cone-shaped piece of ceramic developed to indicate the temperature of a ceramic kiln. The ceramic softens and droops at a specific temperature, according to its composition. Consider two cones, one that softens at a lower temperature than the other. Sketch the DSC curves for materials samples taken from these two cones.

> See the video "Drooping Cones" under Student Resources at PhysicalPropertiesOfMaterials.com

6.33 Phase change materials can be used to store thermal energy for later use or to buffer temperature rises. Sketch the interior temperature as a function of time (7 am to the following day at 7 am) for a building that receives a lot of thermal energy from the sun. This building has no air conditioning or heat source (other than the sun) and little insulation. On the same graph, sketch the interior temperature of the same building in which the walls contain PCM that melts at 27 °C.

6.34 A large portion of household energy consumption is used for heating. It can be as high as 60% in cold climates. Even in hot climates, considerable home energy is used to heat water. Conversion of solar energy to electrical energy with a photovoltaic device is about 20% efficient or a little higher. Conversion of this electrical energy to heat can be nearly 100% efficient. Collection of solar thermal energy can be as high as 90% efficient. Based on energy needs alone, present an argument for which is better to add to a home in your geographic area: solar thermal collectors or photovoltaic cells.

6.35 Dodecanoic acid is potentially useful as a phase change material for thermal energy storage. It melts at 43 °C with a high enthalpy of fusion, 180 J g^{-1}. Furthermore, it is inexpensive, and quite safe, as it is also a foodstuff. It can be derived from the kernel of the oil palm. Its embodied energy is 12 GJ t^{-1} (Noël et al., 2015. *Int. J. Life Cycle Assessment* 20, 367). If 200 kg of dodecanoic acid is required for thermal energy storage for a domestic hot water heating in a home, and the energy requirement to heat hot water for this household is 15 GJ per year, what is the energy payback time to recoup the energy of the dodecanoic acid in a solar thermal system?

6.36 Lithium, which has atomic number 3, has the highest specific heat capacity at room temperature of any solid element, where "specific" refers to the quantity on a mass basis.

a. At room temperature, what would the value of the molar heat capacity of lithium expected to be, expressed in terms of the gas constant, R? Explain your reasoning briefly.

b. Why is lithium's specific heat capacity so high? Explain.

Further Reading

General References

Many introductory physical chemistry and thermodynamics textbooks discuss the concepts presented here (heat capacity, Maxwell Relations, phase transitions, etc.). In addition, the following items address many of the general points raised in this chapter.

A. Barton, 1997. *States of Matter: States of Mind.* Institute of Physics Publishing, Bristol.
P. W. R. Bessonette and M. A. White, 1999. Realistic thermodynamic curves describing a second-order phase transition. *Journal of Chemical Education,* 76, 220.
A. Bondi, 1968. *Physical Properties of Molecular Crystal, Liquids and Glasses.* John Wiley & Sons, Hoboken, NJ.
W. D. Callister, Jr. and D. G. Rethwisch, 2013. *Materials Science and Engineering: An Introduction,* 9th ed. John Wiley & Sons, Hoboken, NJ.
R. DeHoff, 2006. *Thermodynamics in Materials Science,* 2nd ed. CRC Press, Boca Raton, FL.

M. de Podesta, 2002. *Understanding the Properties of Matter*, 2nd ed. Taylor & Francis, Washington, DC.

D. R. Gaskell, 2017. *Introduction to Thermodynamics of Materials*, 6th ed. CRC Press, Boca Raton, FL.

M.B. Johnson and M.A. White, 2014. Thermal methods. In *Inorganic Materials: Multi Length-Scale Characterisation*, D.W. Bruce, D. O'Hare and R. I. Walton, Eds. Wiley, West Sussex, pp. 63–119.

T. Matsuo, A. Inaba, O. Yamamuro, M. Hashimoto, and N. Sotani, 2002. Rubber elasticity in the introductory thermodynamics course. *Journal of Thermal Analysis and Calorimetry*, 69, 1015.

J. B. Ott and J. Boerio-Goates, 2000. *Chemical Thermodynamics. Vol. I: Principles and Applications; Vol. II: Advanced Applications*. Academic Press, Oxford.

L. Qiu and M. A. White, 2001. The constituent additivity method to estimate heat capacities of complex inorganic solids. *Journal of Chemical Education*, 78, 1076.

R. L. Scott, 2006. The heat capacity of ideal gases. *Journal of Chemical Education*, 83, 1071.

S. Stølen and T. Grande, 2004. *Chemical Thermodynamics of Materials: Macroscopic and Microscopic Aspects*. John Wiley & Sons, Hoboken, NJ.

E. J. Windhab, 2006. What makes for smooth, creamy chocolate? *Physics Today*, June, 82.

Glass and Amorphous Matter

S. G. Benka, 2015. History matters for glass hardness. *Physics Today*, December, 24.

L. Berthier and M. D. Ediger, 2016. Facets of glass physics. *Physics Today*, January, 40.

K. Binder and W. Kob, 2005. *Glassy Materials and Disordered Solids*. World Scientific, Singapore.

R. H. Brill, 1963. Ancient glass. *Scientific American*, November, 120.

R. J. Charles, 1967. The nature of glasses. *Scientific American*, September, 126.

Y.-T. Cheng and W. L. Johnson, 1987. Disordered materials: A survey of amorphous solids. *Science* 235, 977.

E. Donth, 2001. *The Glass Transition: Relaxation Dynamics in Liquids and Disordered Materials*. Springer-Verlag, New York.

W. S. Ellis, 1998. *Glass*. Avon Books, New York.

W. S. Ellis, 1993. Glass: Capturing the dance of light. *National Geographic*, 184, 37.

C. H. Greene, 1961. Glass. *Scientific American*, January, 92.

S. J. Hawkes, 2000. Glass doesn't flow and doesn't crystallize and it isn't a liquid. *Journal of Chemical Education*, 77, 846.

M. Jacoby, 2017. What's that stuff? Glass. *Chemical and Engineering News*, November 27, 28.

W. B. Jensen, 2006. The origin of Pyrex. *Journal of Chemical Education*, 83, 692.

K. E. Kolb and D. K. Kolb, 2000. Glass – sand + imagination. *Journal of Chemical Education*, 77, 812.

R. C. Plumb, 1989. Antique windowpanes and the flow of supercooled liquids. *Journal of Chemical Education*, 66, 994.

K. E. Spear, T. M. Besmann, and E. C. Beahm, 1999. Thermochemical modeling of glass: Application to high-level nuclear waste glass. *MRS Bulletin*, April, 37.

Y. M. Stokes, 1999. Flowing windowpanes: Fact or fiction? *Proceedings of the Royal Society of London A*, 455, 2751.

S. Szabó, K. Mazák, D. Knausz, and M. Rózsahegyi, 2001. Enchanted glass. *Journal of Chemical Education*, 78, 329.

E. D. Zanotto, 1998. Do cathedral glasses flow? *American Journal of Physics*, 66, 392.

Liquid Crystals

A. J. Leadbetter, 1990. Solid liquids and liquid crystals. *Proceedings of the Royal Institute of Great Britain*, 62, 61.

Thermal Analysis

R. Blaine, 1995. Determination of calcium sulfate hydrates in building materials using thermal analysis. *American Laboratory*, September, 24.

M. A. White, 1996. Chapter 4: Thermal analysis and calorimetry methods. In *Volume 8 of Comprehensive Supramolecular Chemistry*, J.-M. Lehn, J. L. Atwood, D. D. MacNicol, J. E. D. Davies, and F. Vögtle, Eds. Pergamon Press, Oxford.

Thermal Energy Storage

J. A. Noël, S. Kahwaji, L. Desgrosseilliers, D. Groulx and M. A. White, 2016. Phase change materials. In *Storing Energy: With Special Reference to Renewable Energy Sources*, T. M. Letcher, Ed. Elsevier, Amsterdam, pp. 249–272.

J. A. Noël, P. M. Allred and M. A. White, 2015. Life cycle assessment of two biologically produced phase change materials and their related products. *International Journal of Life Cycle Assessment*, 20, 367.

M. A. White, 2001. Heat storage systems. In *2002 McGraw-Hill Yearbook of Science and Technology*. McGraw-Hill, New York, p. 151.

Thermodynamics of Water and Ice

S. W. Benson and E. D. Siebert, 1992. A simple two-structure model for liquid water. *Journal of the American Chemical Society*, 114, 4269.

E.-M. Choi, Y.-H. Yoon, S. Lee, and H. Kang, 2005. Freezing transition of interfacial water at room temperature under electric fields. *Physical Review Letters*, 95, 085701.

P. G. Debenedetti and H. E. Stanley, 2003. Supercooled and glassy water. *Physics Today*, June, 40.

B. J. Murray and A. K. Bertram, 2006. Formation and stability of cubic ice in water droplets. *Physical Chemical Chemical Physics*, 8, 186.

V. Petrenko and R. W. Whitworth, 2002. *Physics of Ice*. Oxford University Press, Oxford.

R. Rosenberg, 2005. Why is ice slippery? *Physics Today*, December, 50.

H. E. Stanley, 1999. Unsolved mysteries of water in its liquid and glass states. *MRS Bulletin*, May, 22.

Websites

For links to relevant websites, see PhysicalPropertiesOfMaterials.com

7

Thermal Expansion

7.1 Introduction

Most materials expand when they are heated, regardless of the phase of matter. This *thermal expansion* (quantified as the *coefficient of thermal expansion*, or CTE) can be related directly to the forces between the atoms. Materials with different thermal expansion values, when used together in applications, can lead to mechanical problems (loose fits or thermal stress). We begin here with consideration of the basis of thermal expansion in gases (ideal and nonideal) and then move on to the solid state.

7.2 Compressibility and Thermal Expansion of Gases

The defining equation (also called equation of state) for an ideal gas is $pV_m = RT$, where V_m is the molar volume ($= V/n$). From the equation of state, it can be readily seen that as the temperature is increased at constant pressure for an ideal gas, the volume increases. In other words, gases expand on isobaric warming (i.e., they undergo thermal expansion).

For a nonideal gas, corrections must be introduced to the equation of state to account for the shortcomings of the ideal gas model. The *compressibility factor*, Z, is defined as

$$Z = \frac{pV_m}{RT} \tag{7.1}$$

and clearly, $Z = 1$ for an ideal gas and $Z \neq 1$ for a nonideal gas. All gases approach ideality as $p \to 0$, that is, $Z \to 1$ in the limit $p \to 0$.

For a nonideal gas in which $Z < 1$, at a given temperature and pressure, the molar volume is less than the ideal molar volume. This finding means that the gas is more compressed than if it had been ideal, which implies that the intermolecular forces are dominantly attractive under these conditions (e.g., very low temperature).

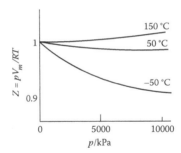

FIGURE 7.1
Variation of compressibility factor, Z, for N_2, with temperature and pressure.

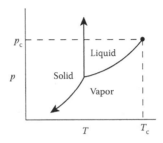

FIGURE 7.2
Stability of phases of matter for a typical pure material as functions of temperature and pressure, showing the critical pressure, p_c, and critical temperature, T_c, beyond which gas and liquid are no longer distinguishable.

The opposite would be true if Z > 1: At a given temperature and pressure, the molar volume would be greater than the ideal molar volume, reflecting the dominance of repulsive intermolecular interactions under these conditions (e.g., very high temperature).

For a real gas, Z varies both with temperature and with pressure and with the type of gas. An example is shown in Figure 7.1 for N_2.

It is worth considering how the value of Z might be qualitatively related to the Joule–Thompson coefficient, as both can be used to quantify nonideality in a gas. This is left as an exercise for the reader.

As it turns out, many gases can be placed together on a series of plots similar to that shown in Figure 7.1 by using appropriate scaling factors. The scaling factors are related to the exact conditions under which that particular liquid is no longer distinguishable from its gas phase; this point is called the *critical point* (see Figure 7.2), and the temperature, pressure, and volume at this point are referred to as the *critical temperature* (T_c), *critical pressure* (p_c), and *critical volume* (V_c), respectively. T_c, p_c, and V_c depend on the gas, and some typical values are listed in Table 7.1.

When the temperature is considered in units of T_c, this defines the *reduced temperature*, $T_r = T/T_c$; the *reduced pressure* is $p_r = p/p_c$, and *reduced volume* is

TABLE 7.1

Critical Temperature, Pressures, and Volumes for Some Gases

Gas	Boiling Point/K	T_c/K	p_c/kPa	V_c/(cm³ mol⁻¹)
He	4.2	5.2	228	61.55
Ne	27.3	44.75	2721	44.30
H_2	20.7	33.2	1297	69.69
O_2	90.2	154.28	5037.2	74.42
N_2	77.4	125.97	3393	90.03
CO_2	194.7	304.16	7379	94.23
H_2O	373.2	647.3	22140	55.44
NH_3	239.8	405.5	11370	72.02
CH_4	109.2	190.25	4620	98.77

Source: A. W. Adamson, 1973. *A Textbook of Physical Chemistry.* Academic Press, Oxford.

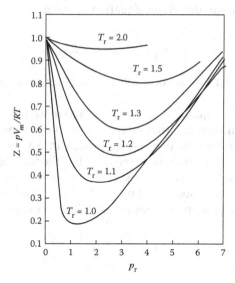

FIGURE 7.3
Compressibility factor, Z, as a function of reduced pressure and reduced temperature, for a number of gases. (Data from O.A. Hougen et al., 1959. *Chemical Process Principles II Thermodynamics*, 2nd ed. John Wiley & Sons, Hoboken, NJ.)

$V_r = V/V_c$. A plot of compressibility, Z, as a function of p_r for various values of T_c (see Figure 7.3) is very nearly universal for all gases. This so-called *Hougen–Watson plot* is especially useful as it can be used to derive p, V, and T relationships for a gas in conditions in which the ideal equation of state does not hold.

The near-universality of the Hougen–Watson plot suggests that there could be a relatively simple expression relating p, V, and T at high pressures where $pV_m \neq RT$. There are many high-pressure equations of state, and we now consider one: the *van der Waals* equation of state*. This equation of state accounts simply for the major shortcomings in the ideal gas model.

First, real molecules do take up space. The ideal gas model assumed the volume of the gas molecules themselves to be negligible with respect to the overall volume, which is why it breaks down at higher pressures. Since the real volume is greater than the ideal volume, the van der Waals equation of state replaces V_m^{ideal} with $\left(V_m^{real} - b\right) = \left(V_m - b\right)$, where b is the effective volume occupied by the molecules in 1 mol of gas; it follows that b depends on the type of gas.

Secondly, the molecules in real gases experience intermolecular forces, whereas the ideal gas model assumed there were no intermolecular forces. At long range, the forces are dominantly attractive, so the observed pressure is less than the ideal pressure. On the basis of experiments, van der Waals proposed that p^{ideal} could be replaced with $p + a/V_m^2$ where p is the observed pressure and a depends on the gas.

With these two substitutions, the ideal gas equation of state can be modified to produce the van der Waals equation of state:

$$\left(p + \frac{a}{V_m^2}\right)\left(V_m - b\right) = RT. \tag{7.2}$$

Values of the van der Waals constants (a and b) for some gases are given in Table 7.2.

The van der Waals equation of state is still too inaccurate to describe every gas in every circumstance. Other more advanced equations of state (including those taking anisotropy of molecular shape into account) are also available, and the interested reader is referred to the "Further Reading" section at the end of this chapter for references. However, it is worth mentioning here that Kamerlingh Onnes[†] suggested in 1901 that the equation of state for a gas could be expressed as a power series (also called *virial equation*) such as

$$Z = \frac{pV_m}{RT} = 1 + \frac{B(T)}{V_m} + \frac{C(T)}{V_m^2} + \frac{D(T)}{V_m^3} + \cdots \tag{7.3}$$

---

[*] Johannes Diderik van der Waals (1837–1923) was a Dutch physicist. Although his career got off to a slow start (his PhD thesis [the subject was the equation of state that now carries his name] was not published until he was 35 years old), van der Waals' accomplishments concerning equations of state for real gases were widely acclaimed. He was awarded the 1910 Nobel Prize in Chemistry for his work on states of matter.

[†] Heike Kamerlingh Onnes (1853–1926) was a Dutch physicist. In his inaugural lecture as professor of experimental physics at Leiden University, Kamerlingh Onnes stated, "In my opinion it is necessary that in the experimental study of physics the striving for quantitative research, which means for the tracing of measure relations in the phenomena, must be in the foreground. I should like to write 'Door meten tot weten' ['Through measuring to knowing'] as a motto above each physics laboratory." This outlook led him to build a laboratory for the first low-temperature research, including the first liquefaction of H_2 (20.4 K) in 1906 and He (4.2 K) in 1908. Kamerlingh Onnes was awarded the 1913 Nobel Prize in Physics for low-temperature research.

TABLE 7.2

van der Waals Constants for Selected Gases

Gas	$a/L^2 \text{ kPa mol}^{-2}$	$b/L \text{ mol}^{-1}$
He	3.457	0.02370
Ne	21.35	0.01709
H_2	24.76	0.02661
O_2	137.8	0.03183
N_2	140.8	0.03913
CO_2	364.0	0.04267
H_2O	553.6	0.03049
NH_3	422.5	0.03707
CH_4	228.3	0.04278

Source: A. W. Adamson, 1973. *A Textbook of Physical Chemistry*. Academic Press.

where the coefficients $B(T)$, $C(T)$, and $D(T)$ are called the second, third, and fourth *virial coefficients*, respectively. Virial coefficients can be directly related to intermolecular forces.

In these considerations of nonideality, it may be interesting to note that, although gases deviate from nonideality at pressures as low as a few atmospheres (i.e., a few hundred kPa), the properties of the gas in the sun (at pressures of billions of atmospheres and extremely high temperatures) can be expressed well by the ideal gas law,[*] presumably due to accidental cancellation of attractive and repulsive forces at these extreme conditions.

The thermal expansion, α, of a gas can be calculated for any equation of state, ideal, or nonideal. When it is not possible to manipulate the equation of state to derive thermal expansion analytically, α can be evaluated numerically by calculation of the volume at two close temperatures and use of the definition of α from Chapter 6, Equation 6.71. See Problem 7.2 at the end of this chapter.

7.3 Thermal Expansion of Solids

To a first approximation, a solid can be described as atoms (or molecules) interacting with their neighbors as if they were connected with springs. For two atoms separated by a distance x away from their equilibrium distance, the restoring force between them, F, can be expressed, to a first (harmonic) approximation by *Hooke's*[†] *law*:[‡]

---

[*] D. B. Clark, 1989. *Journal of Chemical Education*, 65, 826.

[†] Robert Hooke (1635–1703) was an English scientist and inventor. Although during his lifetime his scientific reputation suffered due to conflicts with his contemporaries over scientific priority, Hooke is now considered second only to Newton as the 17th century's most brilliant English scientist. Hooke's law was published in 1676 as *Uttensio sic vis*, Latin for "as the tension, so the force."

[‡] The most important equations in this chapter are designated with ▮ to the left of the equation.

$$F = -k'x \tag{7.4}$$

where k' is the Hooke's law force constant.

The potential energy, $V(x)$, can be derived from the force, $F(x)$, by the general relation

$$F(x) = -\frac{dV(x)}{dx} \tag{7.5}$$

that is,

$$V(x) = -\int F(x)dx \tag{7.6}$$

where Equation 7.5 is a form of Equation 5.2. For a force given by Equation 7.4, the potential energy is

$$V(x) = \frac{k'}{2}x^2 = cx^2 \tag{7.7}$$

where $V = 0$ at $x = 0$ (the equilibrium distance). Since the potential function is parabolic in shape (see Figure 7.4 for a one-dimensional representation), the average separation of the atoms will be independent of temperature in the *harmonic approximation*. Although the energy of the atoms will be nearer the bottom of the well at low temperatures and higher up the well at higher temperatures, the average value of x, represented by the symbol $<x>$ (where $<>$ denotes the average over statistical fluctuations), would be independent of temperature due to the symmetric shape of the parabolic potential. Of course, we know that most solids do expand as the temperature increases, and we must look further to see how this comes about.

A true intermolecular potential is not quite parabolic in shape; as shown in Figure 7.5, it is only parabolic at the very bottom of the well. A consequence of this deviation from the harmonic potential of Equation 7.7 is the fact that bonds can be dissociated (broken) at large separation.

Higher up the well, the two sides are no longer symmetric. A better representation of the intermolecular potential in terms of x, the displacement from minimum-energy separation, is given by the *anharmonic potential*:

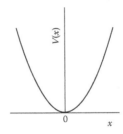

FIGURE 7.4
A harmonic potential.

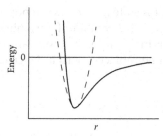

FIGURE 7.5
A comparison between a true intermolecular potential (——) and a harmonic potential (- - -), as a function of intermolecular separation, r.

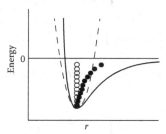

FIGURE 7.6
Thermal expansion for a harmonic (- - -) and an anharmonic (——) potential. The asymmetry in the anharmonic intermolecular potential leads to an increase in $\langle x \rangle$ (displacement from the minimum-energy separation) as the temperature is increased; o, with a harmonic (parabolic) intermolecular potential, $\langle x \rangle$ is independent of the energy level occupied; •, with a true (anharmonic) intermolecular potential, $\langle x \rangle$ depends on the energy of the system, and at higher energies (i.e., higher temperatures), $\langle x \rangle$ is increased.

$$V(x) = cx^2 - gx^3 - fx^4. \tag{7.8}$$

The first term in Equation 7.8 represents the harmonic potential (as in Equation 7.7), the second and third terms (anharmonic terms) represent the asymmetry of the mutual repulsion and the softening of the vibrations at large amplitude. The anharmonic terms lead to $\langle x \rangle$ increasing as the temperature increases because at higher temperatures the asymmetry of the potential leads to a larger value of the intermolecular separation distance. This situation is shown schematically in Figure 7.6. This increase in $\langle x \rangle$ as temperature increases is the source of thermal expansion in solids, as will now be quantified.

For any property generalized as L, where the displacement of the atoms from their equilibrium position is one example of L, the value of L averaged over statistical fluctuations, represented by $\langle L \rangle$, will be given by

$$\langle L \rangle = \frac{\sum_i L_i n_i}{\sum_i n_i} \tag{7.9}$$

where L_i is the value of L for state i, and n_i is the population of state i, and the sum is over all states.* Since the number of possible states for the position of an atom in a solid is extremely large, the sum can be replaced with an integral. Furthermore, n_i can be replaced with a *Boltzmann factor* to weight each state

$$n_i = ne^{-V_i/kT} \tag{7.10}$$

where V_i is the energy of state i, and n is a normalization factor. Therefore, Equation 7.9 can be replaced with

$$\langle L \rangle = \frac{\int_{-\infty}^{\infty} L\, e^{-V(L)/kT}\, dL}{\int_{-\infty}^{\infty} e^{-V(L)/kT}\, dL}. \tag{7.11}$$

For the average value of the displacement from minimum-energy separation, $<x>$, becomes

$$\langle x \rangle = \frac{\int_{-\infty}^{\infty} x\, e^{-V(x)/kT}\, dx}{\int_{-\infty}^{\infty} e^{-V(x)/kT}\, dx}. \tag{7.12}$$

From the potential energy (Equation 7.8), the exponentials in Equation 7.12 can be written as

$$e^{-V(x)/kT} = e^{-\left(cx^2 - gx^3 - fx^4\right)/kT} = e^{-cx^2/kT}\, e^{\left(gx^2 + fx^4\right)/kT} \tag{7.13}$$

and since, for small values of x,

$$e^{\left(gx^3 + fx^4\right)/kT} = 1 + \frac{gx^3}{kT} + \frac{fx^4}{kT} + \cdots \tag{7.14}$$

where the higher-order terms can be neglected, Equation 7.12 can be written as

$$\langle x \rangle = \frac{\int_{-\infty}^{\infty} e^{-cx^2/kT}\left(x + \frac{gx^4}{kT} + \frac{fx^5}{kT}\right) dx}{\int_{-\infty}^{\infty} e^{-cx^2/kT}\left(1 + \frac{gx^3}{kT} + \frac{fx^4}{kT}\right) dx}. \tag{7.15}$$

---

* It might help to consider an example: Consider $<L>$ to be the average grade on a test where n_i is the number of students with grade L_i, and the sum is over all possible grades.

Since the second and third terms in the denominator are $\ll 1$, Equation 7.15 can be approximated as

$$\langle x \rangle = \frac{\int_{-\infty}^{\infty} e^{-cx^2/kT}\left(x + \frac{gx^4}{kT} + \frac{fx^5}{kT}\right)dx}{\int_{-\infty}^{\infty} e^{-cx^2/kT}dx}. \tag{7.16}$$

Both the numerator and the denominator are standard integrals, and they can be solved analytically to give

$$\langle x \rangle = \frac{\dfrac{3\sqrt{\pi}}{4}\dfrac{g}{c^{5/2}}(kT)^{3/2}}{\left(\dfrac{\pi kT}{c}\right)^{1/2}}, \tag{7.17}$$

which simplifies to

$$\langle x \rangle = \frac{3gkT}{4c^2}. \tag{7.18}$$

Equation 7.18 predicts that the dimensions of a material (represented by the $T = 0$ K dimension plus <x>, all multiplied by a factor to account for the total number of atoms in the system) will increase by a factor that is linear in temperature. Experimental results show that the dimensions of a material increase nearly linearly in temperature above a certain temperature. An example is shown in Figure 7.7 for the lattice constant of solid Ar. Note the near-linear behavior at high temperatures in keeping with Equation 7.18 and the nearly constant lattice parameter at very low temperatures, as the Ar interactions become more harmonic and the CTE approaches zero.

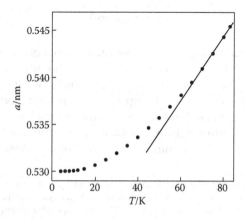

FIGURE 7.7
The lattice constant, a, as a function of temperature for solid argon. (Data from O. G. Peterson et al., 1966. *Physical Review*, 150, 703.)

Thermal expansion, α, was defined previously (Chapter 6, Equation 6.71) in terms of volume as

$$\alpha = \alpha_V = \frac{1}{V}\frac{dV}{dT}, \tag{7.19}$$

but α also can be defined in a particular direction in a solid. For a noncubic solid, the thermal expansion will usually be different in the different directions, and it will be important to take account of this in considering the overall thermal expansion. If the unit cell of a solid has one side of dimension a, the thermal expansion in this direction, α_a, is defined as

$$\alpha_a = \frac{1}{a}\frac{da}{dT} \tag{7.20}$$

and in the case of Ar, α_a at any temperature would correspond to the $1/a$ times the slope of the curve shown in Figure 7.7. For all materials, the dimensions approach a constant value as the temperature approaches absolute zero, so $\alpha \rightarrow 0$ as $T \rightarrow 0$. This finding is a consequence of the Maxwell relation (Chapter 6, Equation 6.56) and the fact that $S \rightarrow 0$ as $T \rightarrow 0$.

Typical values of linear coefficients of thermal expansion, α_a, for a variety of materials at different temperatures are given in Table 7.3. Note that for an isotropic material, the volumetric CTE value is three times the linear CTE, i.e., $\alpha_V = 3\alpha_a$.

Thermal expansion of a crystalline solid can be measured by scattering x-rays from a material and determining the lattice spacing as a function of temperature with *Bragg's law**:

See the video "Thermal Expansion of Plastic" under Student Resources at PhysicalPropertiesOfMaterials.com

$$n\lambda = 2d\sin\theta \tag{7.21}$$

which is the condition for constructive interference of the x-ray beam from successive layers of the solid; n is an integer; λ is the wavelength of the x-ray beam (of the order of the lattice spacing); d is the lattice spacing; and θ is the angle of incidence, as defined in Figure 7.8. In some instances, it is preferable to scatter neutrons rather than x-rays; the principles are much the same although x-rays are more readily available at laboratory sources, while neutrons require a specialized source. However, since x-rays scatter from the

---

* Sir William Henry Bragg (1862–1942), English scientist, was the founder of the science of crystal structure determination by x-ray diffraction. With his son (William Lawrence Bragg, 1890–1971) he was awarded the Nobel Prize in Physics in 1915 for seminal work showing the positions of atoms in simple materials such as diamond, copper, and potassium chloride. Bragg's law was first presented by Lawrence Bragg, in 1912.

TABLE 7.3

Values of Linear Coefficients of Thermal Expansion, α_a, of Selected Materials at Various Temperatures

Material	T/K	α_a/K^{-1}
Al	50	3.8×10^{-6}
Al	300	2.32×10^{-5}
Diamond	50	4×10^{-9}
Diamond	300	1.0×10^{-6}
Graphite (*c*-axis)	300	29×10^{-6}
Graphite (*a*-axis)	300	-1×10^{-6}
Cu	50	3.8×10^{-6}
Cu	300	1.68×10^{-5}
Ag	50	8×10^{-6}
Ag	300	1.93×10^{-5}
Quartz(cr)	50	-3.3×10^{-7}
Quartz(cr)	300	7.6×10^{-6}
Ice	100	1.27×10^{-5}
Ice	200	3.76×10^{-5}
Polycarbonate	300	7.0×10^{-5}
Polyethylene	300	1.0×10^{-4}
Pyrex	50	5.62×10^{-5}
Pyrex	300	-2.3×10^{-6}
ZrW_2O_8	0.3–693	-8.7×10^{-6}

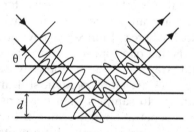

FIGURE 7.8

The Bragg diffraction condition for x-rays or neutrons scattered from successive layers of a solid. Only some angles and wavelengths allow the scattered beam to be enhanced by constructive interference; if the scattered beams are not completely in phase, then less intense scattering is observed. This diagram shows the diffracted waves from successive layers to be in phase, resulting in constructive interference.

electrons of the atoms involved and neutrons scatter from the nuclei (and also interact magnetically, especially with unpaired electrons), the two techniques can be quite complementary.

Another way to determine thermal expansion is to use the sample to prop apart the plates of a capacitor and measure the capacitance as a function of temperature. The capacitance depends directly on the distance between the plates, so this method can lead to a very accurate assessment of the dimensions of the sample as a function of temperature and, hence the thermal expansion in the *capacitance dilatometer.*

A related way to determine thermal expansion is to use a *push-rod dilatometer.* In this instrument, a push-rod connected to the ferromagnetic core of a transformer pushes against the sample so that changes in the sample length are recorded as changes in the voltage of the transformer, due to the movement of the ferromagnetic core.

As can be seen from Equation 7.18, thermal expansion reflects both the parameter c (which characterizes the strength of the interaction potential) and g (which represents the asymmetry of the intermolecular potential). More asymmetric potentials (larger g values) lead to larger thermal expansion, which makes sense considering that a harmonic potential, which is perfectly symmetric, has no thermal expansion, that is, $\alpha^{harmonic} = 0$. Larger values of c, which imply stiffer interactions and steeper walls to the potential, lead to lower coefficients of thermal expansion. Therefore, measurement of thermal expansion of a solid can directly provide information concerning the intermolecular potential.

COMMENT: X-RAY VERSUS NEUTRON SCATTERING

Neutrons have several advantages over x-rays for the study of the structure of matter. For example, neutrons of wavelength 1 nm have an energy of 78 J mol^{-1}, whereas 1-nm x-rays have an energy of 1.2×10^6 J mol^{-1}, which exceeds that of a typical chemical bond, $\approx 2 \times 10^5$ J mol^{-1} Furthermore, typical x-ray wavelengths for x-ray diffraction are even shorter than 1 nm, so they are more energetic still, and therefore x-rays can cause more damage to samples than neutrons used for diffraction. Another factor is penetration in the sample: 1-nm x-rays penetrate less than 1 mm into the sample, whereas 1-nm neutrons can penetrate 100 cm, probing the bulk properties of matter. Furthermore, since x-rays scatter from electrons and neutrons scatter from nuclei, neutrons can be used to "see" light atoms such as hydrogen much more accurately than x-rays can. Neutrons have an additional advantage because their intrinsic spin allows interaction with magnetic centers in the structure, helping delineate magnetic structures (ferromagnetic, antiferromagnetic, etc.; see Chapter 13). However, x-rays can be produced by laboratory-based sources, whereas neutron sources for scattering investigations require a nuclear reactor or an accelerator.

Thermal expansion can be an important property of a material as it can have some influence over other properties such as mechanics of fitting.* Thermal expansion can influence the color of a liquid crystal as in a mood ring (a piece of costume jewelry that changes color with temperature and could be used to indicate mood). Thermal expansion also can lead to stress in a material, and this can greatly influence mechanical stability. For a uniform change in temperature of a material, the *thermal shock fracture resistance*, R_S, is given by

$$R_S = \frac{\kappa \sigma}{|\alpha| E}, \tag{7.22}$$

where κ is the thermal conductivity (see Chapter 8), σ is the strength of the material (see Chapter 14), α is the volumetric thermal expansion coefficient, and E is the elastic (or Young's) modulus (see Chapter 14), and higher values of R_S are less susceptible to thermal shock fracture. It is apparent from Equation 7.22 that materials with high thermal expansion coefficients can easily fracture thermally because large temperature gradients lead to higher

---

COMMENT: MATERIALS WITH NO THERMAL EXPANSION

Although most materials increase in dimension as the temperature is increased, one rather curious material, an alloy of 35 atomic% nickel and 65 atomic% iron, has virtually zero thermal expansion throughout the temperature range from −50 °C to 150 °C. The discovery of this material was the subject of the 1920 Nobel Prize in Physics to Charles E. Guillaume of France. This material, commonly called *Invar* (to represent its *invariant* thermal expansion), has many practical uses: It is used in laser optics labs where dimensional stabilities of the order of the wavelength of light are required; it was used to determine the diurnal (day-night) height variations of the Eiffel Tower, and hulls of supertankers can be made of Invar in order to avoid strains due to thermal expansion associated with welding processes. The unusual absence of thermal expansion in Invar is related to its magnetic properties, but it is not yet fully understood. A few other materials have near-zero thermal expansion, including Zerodur® (a lithium-aluminosilicate glass ceramic from Schott AG, $\alpha_a = 0.05 \pm 0.10 \times 10^{-6}$ K^{-1}, 273 K < T < 573 K) and ZrMgMo$_3$O$_{12}$ ($\alpha_a = 0.16 \pm 0.02 \times 10^{-6}$ K^{-1}, 298 K < T < 723 K; see Romao et al., 2015, *Chemistry of Materials* 27, 2633).

---

* Woodwind musical instruments only have the correct pitch when they are at the correct temperature; if they are cold, they sound flat. This is not due to contraction of the cold instrument, as the shorter cold instrument would sound sharp. The important factor here is the contraction of the cold gas. The velocity of sound in the gas decreases as the gas density increases (i.e., as the temperature decreases), and this lower velocity of sound decreases the frequency of the sound as temperature decreases, making a cold woodwind instrument sound flat. Although thermal expansion of the instrument also occurs as it warms up, the thermal expansion of the instrument is a minor factor for pitch relative to the thermal expansion of the air.

thermal expansion in some spots than in others, leading to stress that can exceed the strength of the material.

COMMENT: THERMIOTIC MATERIALS CONTRACT WHEN HEATED

Most materials expand when heated (see Table 7.3), that is, $\alpha > 0$ for most materials at most temperatures in accord with Equation 7.18, but, in some instances, $\alpha < 0$. A few materials have $\alpha < 0$ in one dimension as they are heated, but generally they expand in another dimension such that the overall thermal expansion is positive. Graphite at $T = 300$ K is one such example. However, when $\alpha < 0$ in all directions, these materials are said to be *thermomiotic*, i.e., they experience *negative thermal expansion* (NTE) over the temperature range for which $\alpha < 0$. Examples of materials that contract when heated include quartz at very low temperature and Pyrex® at room temperature. A familiar material with anomalous thermal expansion is H_2O: When liquid water is heated from 0 °C to 4 °C, it contracts. Ice also shows NTE at very low temperatures. Zirconium tungstate, ZrW_2O_8, has the unusual property of shrinking uniformly in all three dimensions as the temperature is increased from 0.3 to 1050 K! When the linked polyhedral units within this material are heated, their thermal agitation leads to a more closely packed structure, and NTE results. A few related ceramics and other families of compounds also show negative thermal expansion. Such thermomiotic materials are of interest because they open the possibility of materials that can change temperature with high resistance to thermal stress fracture (see Equation 7.22).

7.4 Examples of Thermal Expansion: A Tutorial

Typically thermal expansion of a solid at room temperature is of the order 10^{-5} K^{-1}.

a. How much does the length of a 1 m long rod of a typical material change for 3 K increase in temperature?

b. In some applications, it is important not to have changes in material dimensions as the temperature changes. One such example is the pendulum of a clock where the time accuracy depends on the stability of the length of the pendulum. Many of the decorative metals have large values of thermal expansion near room temperature. Given that there will be variations in the temperature from time to time in the vicinity of the clock, it has been suggested that a bimetallic arrangement could be used for the pendulum, with a decorative metal (large thermal expansion) on the outside face, securely attached to a less attractive metal (smaller thermal expansion) on the inside face. In the case of extreme temperature fluctuations, will the pendulum remain straight?

See the video "Bimetallic Strip" under Student Resources at PhysicalPropertiesOfMaterials.com

c. Can the stress induced by differential thermal expansion in a bimetallic strip be used to amplify thermal expansion effects? Consider the change in the pointer position in the thermostat shown in Figure 7.9, in comparison with the change in dimension found in (a).

d. Suggest an application taking advantage of the thermal expansion of a material.

e. Explain why, in certain circumstances, thermal expansion can lead to thermal stress fracture of a material. (An example could be crazing of pottery for which the glaze and green body have mismatched thermal expansion coefficients, exacerbated by the high temperature of the firing furnace.)

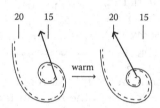

FIGURE 7.9
The bimetallic strip inside a thermostat. As the temperature warms to 20 °C, the coil tightens because the thermal expansion coefficient of the outer material (solid line) is greater than that of the inner material (dashed line).

7.5 Learning Goals

- Coefficient of thermal expansion
- Compressibility and thermal expansion of gases; Hougen–Watson plot; van der Waals equation of state
- Thermal expansion of solids; anharmonicity; temperature-dependence
- Bragg's law, x-ray and neutron diffraction
- Other dilatometry techniques
- Materials with no thermal expansion
- Materials with negative thermal expansion
- Thermal shock fracture resistance

7.6 Problems

7.1 44.0 g of CO_2 are contained in a 1.00 L vessel at 31 °C.

   a. What is the pressure of the gas (in kPa) if it behaves ideally?

   b. What is the pressure of the gas if it obeys the van der Waals equation of state? See Table 7.2 for data. Give your answer in kPa.

   c. Estimate the pressure of the gas (in kPa) based on the Hougen–Watson plot (Figure 7.3) and the values of the critical constants for CO_2 given in Table 7.1.

7.2 Based on the data given in Problem 7.1, calculate the volume coefficient of thermal expansion, α, for CO_2 at 31 °C and P = 101325 Pa, for the following cases:

   a. the gas is ideal;

   b. the gas obeys the van der Waals equation of state.

7.3 An apparatus consisting of a rod made of material A, placed inside a tube made of material B as shown in Figure 7.10, is being constructed. The aim is to

> See the video "Thermal Expansion" under Student Resources at PhysicalPropertiesOfMaterials.com

have the rod fit tightly in the tube at room temperature. One way to achieve this is to cool both the rod and the tube (e.g., in liquid nitrogen) and put them together when they are cold and the fit is loose. As they warm to room temperature, the fit will be tight. This procedure is called *cold welding*. It relies on different thermal expansions of the two materials for the fit to be tight at room temperature and loose when much colder. Which material needs to have greater thermal expansion for this to work: A (rod) or B (tube)? Explain.

7.4 Boiling water is poured into two glasses; the only difference between the glasses is that one is made of thick glass and the other is made of thin glass. Which glass is more likely to break? Why?

7.5 Consider the curves for intermolecular potential as a function of molecular separation, r, as shown in Figure 7.11.

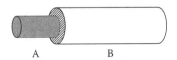

A   B

FIGURE 7.10
Rod A is placed inside tube B.

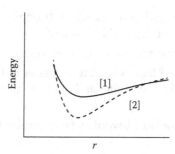

FIGURE 7.11
Two intermolecular potentials.

    a. Which case, [1] or [2], has the stronger intermolecular potential? Explain briefly.

    b. Which case, [1] or [2], is more anharmonic? Explain briefly.

    c. Which case, [1] or [2], has greater thermal expansion? Explain briefly.

7.6 Owens–Corning Fiberglas Corp. has introduced a form of glass fiber for insulation. The fiber is composed of two types of glass fused together. At the cooling stage in the production, the components separate so that one side of the fiber is made of one component and the other side has the other composition. The key to the applications of this material is that the fibers are coiled, not straight as for regular glass fibers: The coils in the new fibers allow them to spring into place, giving them excellent insulating properties. What causes the fiber to be coiled?

7.7 When a polymer is cooled below its glass transition temperature, T_g, there is an abrupt drop in its coefficient of thermal expansion. Given that the configurational degrees of freedom are frozen below T_g and become active above T_g, what does this imply about the harmonicity of the dominant intermolecular interactions in the rigid glass compared with the more flexible state? Explain.

7.8 The linear thermal expansion coefficient for a borosilicate crown glass is zero at $T = 0$ K, gradually rises to 80×10^{-5} K^{-1} at $T = 600$ K, and then rapidly rises to 500×10^{-5} K^{-1} at $T = 800$ K. How will this increase in thermal expansion coefficient influence the high-temperature applications of this material?

7.9 On the basis of a tree's structure, would you expect thermal expansion of wood to be isotropic or anisotropic? Explain how this could influence the design of a wooden musical instrument.

7.10 A gas-filled piston is central to the operation of a pneumatic window opener that opens and closes according to the temperature.

The piston is attached to the window frame such that its extension determines the opening of the window.

a. Explain the principle on which this works.

b. If the gas were ideal, would it be better (all other factors being equal) to have it operate at low pressure or high pressure? Explain.

c. If the gas were not ideal, would it be preferable to have $Z > 1$ or $Z < 1$? Explain.

7.11 Invar was used on a space flight to determine the outside temperature of a space capsule during its voyage. Given that Invar does not change dimensions in the expected temperature range ($-50\,°C$ to $150\,°C$), why was it chosen?

7.12 For a particular application, it is necessary to introduce a piece of wire into a glass tube. This will be done by heating the glass and inserting the wire. Given a choice of platinum (linear expansion 8.6×10^{-6} K^{-1}) or copper (linear expansion coefficient 1.7×10^{-5} K^{-1}) to insert in glass with a linear thermal expansion coefficient of 8.6×10^{-6} K^{-1}, which metal should you choose? Explain.

7.13 At some sinks, when the hot water tap is opened only slightly, the water first flows, and then slows until only a trickle, finally shutting off completely. A second opening of the hot water tap does not usually lead to a repeat of this behavior, and the cold water tap does not behave this way. Give an explanation for this phenomenon.

7.14 The coefficient of thermal expansion, α, of a sample could be measured by placing the sample between parallel circular copper plates to form a capacitor. Since the capacitance of the capacitor is directly proportional to the area of one of the plates and inversely proportional to the distance between the plates, α can be determined by monitoring the capacitance as a function of temperature. However, the sample is usually placed outside the capacitor plates (which are in vacuum) in such a way that it moves one of the capacitor plates when the temperature is changed. Why would it be better to determine the capacitance of the cell in this geometry than with the sample between the plates?

7.15 The speed of sound in the ocean varies with the depth, depending mostly on the temperature of the water (i.e., salinity is of secondary importance). The speed is highest in warmer water and can vary by a few percent from its average value of $1500\,$m s^{-1}, depending on the depth.

a. Explain why the speed of sound varies as it does with temperature. Relate this explanation to how the thermal expansion coefficient of the water varies with temperature.

b. Under certain conditions, such as shallow ocean water in the summer, the water temperature first decreases with increasing depth, reaches a minimum at a certain depth (approximately 50 m), and then increases again to a nearly constant value at depths greater than 150 m.

   i. Sketch the sound speed (x-axis) as it changes with depth (y-axis) under these conditions.

   ii. Like light, sound waves bend (refract) toward the lower velocity. Use this information to show how a sound wave can be guided by the ocean to travel great distances without loss (similar to fiber optics acting as a *wave guide*; see Chapter 4).

7.16 a. Graphite is an anisotropic material due to its layered structure. Explain why the coefficient of thermal expansion is lower along the layer directions (a-axis) than perpendicular to the layers (c-axis). See Table 7.3 for the data.

   b. From the data in Table 7.3, explain the relative thermal expansion coefficients for diamond and graphite and why they are not the same given that both are forms of carbon.

7.17 Near room temperature, polymers have higher thermal expansion coefficients than metals, which have higher thermal expansion coefficients than ceramics. Explain this order. Consider intermolecular forces.

7.18 Rubies can exhibit thermochromism. If a ruby of composition 10% Cr_2O_3, 90% Al_2O_3 is heated from room temperature to 500 °C, it will turn from red to green. The color is due to absorption in both cases. Suggest a reason for the change in color.

7.19 Consider the bimetallic strip in Figure 7.9 and devise a thermometer that can indicate the minimum and maximum temperature reached during a given time period.

7.20 Is the thermal stress associated with heating a material higher or lower if the material is a poor conductor of heat? Explain why.

7.21 From Equation 7.22, determine the SI units for thermal shock fracture resistance. Explain briefly the meaning of R_S, in consideration of these units.

7.22 What difference does a 2 K increase in the temperature of room temperature water make to a column of water of depth 500 m? The linear thermal expansion coefficient for water at this temperature is 7×10^{-5} K^{-1}. On this basis, do you expect thermal expansion of ocean water due to climate change to have a significant influence on ocean levels, given that the average ocean depth is about 3 km?

7.23 Equation 7.22 describes the thermal shock fracture resistance. The thermal expansion coefficient appears in this equation as its absolute value, $|\alpha|$. Some materials exhibit negative thermal expansion coefficients; explain the physical processes that occur in thermal shock of such materials.

7.24 *Glass ceramics* can have coefficients of thermal expansion less than 3×10^{-7} K^{-1}, making them useful for cooktop panels and precision optical applications. Glass ceramics, for example with composition $Li_2O \cdot Al_2O_3 \cdot nSiO_2$, are produced under carefully controlled crystallization conditions, leading to 30%–60% crystalline content in a glass matrix. The crystallites are very small, on the order of microns (10^{-6} m), and they are thermomiotic, while the glass matrix exhibits normal positive thermal expansion, yielding an overall near-zero thermal expansion coefficient.

a. Why is low magnitude of thermal expansion coefficient required for precision optical applications? Give an example.

b. In cooktop use, the glass ceramic surface is above a series of electrical elements that provide the heat. Why is the low thermal expansion coefficient of the glass ceramic important? What other properties of the glass ceramic would be important for this application?

Further Reading

General References

New frontiers in the application of neutron scattering to materials science. Special issue. D. Richter and J. M. Rowe, Eds. *MRS Bulletin*, December 2003, 903–928.

A. W. Adamson, 1973. *A Textbook of Physical Chemistry*. Academic Press, Oxford.

T. H. K. Barron and G. K. White, 1999. *Heat Capacity and Thermal Expansion at Low Temperatures*. Springer, New York.

R. S. Berry, S. A. Rice, and J. Ross, 2000. *Physical Chemistry*, 2nd ed. Oxford University Press, New York.

W. D. Callister, Jr. and D. G. Rethwisch, 2013. *Materials Science and Engineering: An Introduction*, 9th ed. John Wiley & Sons, Hoboken, NJ.

M. de Podesta, 2002. *Understanding the Properties of Matter*, 2nd ed. CRC Press, Boca Raton, FL.

R. E. Hummel, 2004. *Understanding Materials Science*, 2nd ed. Springer, New York.

C. Kittel, 2004. *Introduction to Solid State Physics*, 8th ed. John Wiley & Sons, Hoboken, NJ.

H. McGee, J. McInerney, and A. Harms, 1999. The virtual cook: Modeling heat transfer in the kitchen. *Physics Today*, November, 30.

Experimental Methods

G. K. White and P. J. Meeson, 2002. *Experimental Techniques in Low-Temperature Physics,* 4th ed. Oxford University Press, New York.

Materials with Unusual Thermal Expansion

A. N. Aleksandrovskii, V. B. Esel'son, V. G. Manzhelii, G. B. Udovidchenko, A. V. Soldatov, and B. Sundqvist, 1997. Negative thermal expansion of fullerite C_{60} at liquid helium temperatures. *Low Temperature Physics,* 23, 943.

G. D. Barron, J. A. O. Bruno, T. H. K. Barron, and A. L. Allan, 2005. Negative Thermal Expansion. *Journal of Physics – Condensed Matter,* 17, R217.

S. W. Benson and E. D. Siebert, 1992. A simple two-structure model for liquid water. *Journal of the American Chemical Society,* 114, 4269.

M. Freemantle, 2003. Zero-expansion conductor. *Chemical and Engineering News,* October 20, 6.

M. B. Jakubinek, C. A. Whitman, and M. A. White, 2010. Negative thermal expansion materials: Thermal properties and implications for composite materials. *Journal of Thermal Analysis and Calorimetry,* 99, 165.

D. Lindley, 2004. Bake, shake and shrink. *Physical Review Focus,* November, story 21, 14.

T. A. Mary, J. S. O. Evans, T. Vogt, and A. W. Sleight, 1996. Negative thermal expansion from 0.3 to 1050 Kelvin in ZrW_2O_8. *Science,* 272, 90.

J. Pellicer, J. A. Manzanares, J. Zúñiga, P. Utrillas, and J. Fernández, 2001. Thermodynamics of rubber elasticity. *Journal of Chemical Education,* 78, 263.

D. G. Rancourt, 1989. The Invar problem. *Physics in Canada,* 45, 3.

C.P. Romao, K.J. Miller, C.A. Whitman, M.A. White, and B.A. Marinkovic, 2013. Negative thermal expansion (thermomiotic) materials. In *Comprehensive Inorganic Chemistry II,* J. Reedijk and K. Poeppelmeier, Eds. Elsevier, Oxford, Vol. 4, pp. 128–151.

C. P. Romao, F. A. Perras, U. Werner-Zwanziger, J. A. Lussier, K. J. Miller, C. M. Calahoo, J. W. Zwanziger, M. Bieringer, B. A. Marinkovic, D. L. Bryce, and M. A. White, 2015. Zero Thermal expansion in $ZrMgMo_3O_{12}$: NMR crystallography reveals origins of thermoelastic properties. *Chemistry of Materials,* 27, 2633.

Websites

For links to relevant websites, see PhysicalPropertiesOfMaterials.com.

8

Thermal Conductivity

8.1 Introduction

Thermal conductivity determines whether a material feels hot or cold to the touch, and why some materials are good thermal insulators while others efficiently conduct heat. In this chapter, we consider thermal conductivities of gases and then of solids.

The symbol used to represent thermal conductivity is κ. (Some sources use λ, but we prefer to reserve λ for wavelength [already introduced] and mean free path [introduced later in this chapter].) The higher the value of κ, the better a material conducts heat. As for electrical conductivity, the values of thermal conductivity can vary by orders of magnitude from one material to another and even within a given material when the temperature is changed.

8.2 Thermal Conductivity of Gases

The theoretical thermal conductivity of gases, as derived by Peter Debye, is based on the kinetic theory of gases.

The model of a gas for kinetic theory is a swarm of molecules, each of mass m, all in continual random motion. The molecules are considered to have negligible size with respect to the average distance they travel between collisions. The average distance traveled by a gas molecule between collisions is called the *mean free path* and is denoted λ. In this model, the molecules within the gas do not interact except when they collide; in other words, long-range attractive forces are neglected.

The kinetic theory of gases has many important successes. For example, it leads to the equation of state for an ideal gas, $pV_m = RT$. The kinetic theory of gases also leads to the distribution of molecular speeds within a gas, as shown in Figure 8.1. That figure shows that the distribution shifts to higher

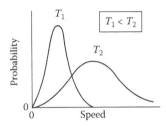

FIGURE 8.1
Distributions of molecular speeds in a gas at two different temperatures.

speeds as the temperature is increased. In fact, it leads to a relationship between the weighted average speed, $\bar{v}$, and the temperature of the gas:

$$\bar{v}^2 = \text{const} \times T \tag{8.1}$$

which follows from consideration of the translational kinetic energy, KE:

$$KE = \frac{1}{2}m\bar{v}^2 = \text{const} \times kT \tag{8.2}$$

where the latter proportionality comes from the equipartition theory (Chapter 6).

Thermal conductivity is a *transport property*, involving the motion of heat (energy) from hot to cold. Other transport properties are *effusion* (mass transported out of a small orifice), *diffusion* (general mass transport), *viscosity* (momentum transfer), and *electrical conductivity* (charge transport). Much of the formalism of the derivation of various transport properties follows similar lines, so the derivation of the relationship between thermal conductivity of a gas and other physical parameters will be presented here in some detail.

The *flux*, $J(i)$, of a physical property, i, can be defined as the amount of i flowing per unit time per unit area. For example, in a box with a temperature gradient (one end hot, one end cold, as shown in Figure 8.2), J(energy) is the energy flux or the amount of energy flowing per unit area per unit time. SI units of J(energy) are J m^{-2} s^{-1} = W m^{-2}. The way the box of Figure 8.2 has been drawn, the temperature gradient is parallel to the z-axis, and thus we can consider this situation to be a one-dimensional problem in which the fluxes can be represented by the flux in the z-direction, J_z.

The energy flux along the z-direction can be written as

$$J_z(\text{energy}) = -\kappa\left(\frac{dT}{dz}\right) \tag{8.3}$$

where κ is the *coefficient of thermal conductivity* or, often, simply *thermal conductivity*; SI units of κ are W m^{-1} K^{-1}. Since $(dT/dz) < 0$ here, J_z (energy) > 0, and

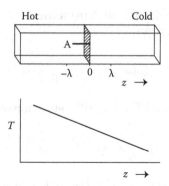

FIGURE 8.2
A box of gas with a temperature gradient, used to consider the thermal conductivity of a gas.
T is the temperature of the gas as a function of position z.

energy flows from low to high z values (Figure 8.2), that is, from hot to cold.
Equation 8.3 is Fourier's first law of heat flux.*

The purpose now is to express κ in terms of physically meaningful
parameters. To do so, it is useful to start by expressing the energy flux in
terms of the matter flux in the gas. If each molecule moving under the temperature gradient carries energy $\varepsilon(z)$, where z is its position, we can calculate
the net flux of molecules per unit area per unit time. In the box in Figure 8.2,
consider the area A to be defined as exactly 1 unit area. To calculate J_z(energy),
we need to know the net number of molecules passing through A per second.

As discussed above, the mean free path, λ, is defined as the average distance
that a molecule in a gas can travel before colliding with another molecule.
On average, molecules that hit area A from the left have traveled one mean
free path (λ) since their last collision. Other molecules will be too far away
(and will bump into other molecules before reaching A), or too close (and
will pass through A). The number of molecules hitting A from the left per
unit time is then ($N\bar{v}/6$), where $\bar{v}$ is the average speed of the molecules and
N is the number density. The factor of 1/6 accounts for the fact that, of the six
directions (up, down, left, right, in, out), only one (to the right) leads to area
A. The expression ($N\bar{v}/6$) can be seen to lead to units of molecules $m^{-2}\ s^{-1}$
for the matter flux to A from the left, as one would expect. With an energy
designated as $\varepsilon(-\lambda)$ for each molecule at a distance $-\lambda$ from A (see Figure 8.2),
this leads to left-to-right energy flux, J_z(energy)$^{L\to R}$:

$$J_z\left(\text{energy}\right)^{L\to R} = \frac{N\bar{v}}{6}\varepsilon\left(-\lambda\right). \qquad (8.4)$$

---

* Jean Baptiste Joseph Fourier (1768–1830) was a French mathematician and scientist. His study
  of the conduction of heat introduced mathematical expansions as functions of trigonomet-
  ric series, not appreciated by his contemporaries, but now known as the important area of
  Fourier series.

There is also a flux of molecules to A from the right, of $(N\bar{v})/6$, giving right-to-left energy flux, $J_z(\text{energy})^{R\to L}$:

$$J_z\left(\text{energy}\right)^{R\to L} = \frac{N\bar{v}}{6}\varepsilon(\lambda). \tag{8.5}$$

The net energy flux at A will be the difference between these two fluxes:

$$J_z\left(\text{energy}\right) = \frac{1}{6}N\bar{v}\left(\varepsilon(-\lambda) - \varepsilon(\lambda)\right) = \frac{1}{6}N\bar{v}\,\Delta\varepsilon. \tag{8.6}$$

where $\Delta\varepsilon$ is the difference in energy per molecule at distance λ to the left of A (i.e., at $z = -\lambda$) compared to a distance λ to the right of A (i.e., at $z = \lambda$), that is, $\Delta\varepsilon = \varepsilon(-\lambda) - \varepsilon(\lambda)$. Furthermore, $\Delta\varepsilon$ can be written in terms of the energy gradient, $d\varepsilon/dz$:

$$\Delta\varepsilon = \left(\frac{d\varepsilon}{dz}\right)\Delta z = -\left(\frac{d\varepsilon}{dz}\right)2\lambda \tag{8.7}$$

where $\Delta z = -2\lambda$. From Equations 8.6 and 8.7, the energy flux can be written:

$$J_z\left(\text{energy}\right) = \frac{1}{6}N\bar{v}\left(\frac{d\varepsilon}{dz}\right)2\lambda = -\frac{1}{3}N\bar{v}\,\lambda\left(\frac{d\varepsilon}{dz}\right). \tag{8.8}$$

Letting E be the total energy per unit volume, it follows that $\varepsilon = E/N$ and

$$\frac{d\varepsilon}{dz} = \frac{1}{N}\frac{dE}{dz}. \tag{8.9}$$

Furthermore, dE is related to dT by the molar heat capacity at constant volume, $C_{V,m}$, the number of moles, n, and the volume, V:

$$dE = \frac{C_{V,m}\,n}{V}\,dT \tag{8.10}$$

so the energy gradient, dE/dz, can be expressed in terms of the temperature gradient, dT/dz:

$$\frac{dE}{dz} = \frac{C_{V,m}\,n}{V}\frac{dT}{dz}. \tag{8.11}$$

Substitution of Equations 8.11 and 8.9 into Equation 8.8 leads to

$$J_z\left(\text{energy}\right) = -\frac{1}{3}\bar{v}\,\lambda\,C\left(\frac{dT}{dz}\right), \tag{8.12}$$

where $C = C_{V,m}n/V$ is the heat capacity per unit volume (SI units of C are J K^{-1}m^{-3}). Comparison of Equations 8.3 and 8.12 leads to*

$$\kappa = \frac{1}{3} C \bar{v} \lambda. \qquad (8.13)$$

Equation 8.13 is known as the *Debye equation for thermal conductivity*, and it will prove to be very useful for both gases and solids.

Equation 8.13 shows that κ for a gas is governed by three factors: its heat capacity per unit volume, the mean free path of the molecules that carry the heat, and the average speed of these molecules. Each of these factors is temperature-dependent, and since their temperature dependences do not cancel, it follows that κ for a gas also is temperature-dependent.

What about the pressure dependence of κ? From the kinetic theory of gases, it can be shown that the mean free path, λ, can be written as

$$\lambda = \frac{V}{\sqrt{2}\pi \, d^2 n \, N_A} \qquad (8.14)$$

where V is the volume, n is the amount of gas in moles, d is the diameter of the gas molecule under consideration and N_A is Avogadro's number. From Equation 8.13, it follows that

$$\kappa = \frac{\bar{v} \, C_{V,m}}{3\sqrt{2} \, \pi \, d^2 \, N_A} \qquad (8.15)$$

which shows thermal conductivity, κ, to be independent of the pressure of the gas. This finding is true at moderate pressures, from about 1 to ~10^4 kPa. However, Equation 8.15 shows κ to depend on $C_{V,m}$ and d, both of which depend on the nature of the gas, so at a given temperature, the thermal conductivity of a gas will depend on its composition, but not pressure. This conclusion can be used to detect different gases and is used, for example, in thermal conductivity detectors in gas chromatography.

The physical interpretation of the pressure independence of thermal conductivity is as follows. As the pressure is increased, more molecules per unit volume are available to transport the heat, i.e., the heat capacity per unit volume increases, and this would increase κ. However, the presence of more molecules decreases the mean free path, which decreases κ. These equal and opposite effects lead to κ being independent of pressure.

However, if the pressure is extremely low, as in the case of a Thermos™ bottle, pressure can be important in determining κ. The mean free path, λ, increases as the pressure decreases, but eventually λ is limited by the

---

* The most important equations in this chapter are designated with ▌ to the left of the equation.

TABLE 8.1

Thermal Conductivities of Selected Materials at $T = 300$ K

Material	$\kappa/(\text{W m}^{-1}\,\text{K}^{-1})$
Aerogel	0.2
Al_2O_3	36.0
Argon (gas)	0.02
Boron	2.76
Copper	398
Diamond	2310
Graphite	2000 along a-axis
	9.5 along c-axis
Helium (gas)	0.15
Iodine	0.449
MgO	60.0
Nitrogen (gas)	0.025
Black phosphorus	12.1
White phosphorus	0.235
Sapphire	46
SiO_2 (crystalline)	10.4 along c-axis
	6.2 along a-axis
SiO_2 (amorphous)	1.38

dimensions of the container, and it stays at this constant value no matter how much further the pressure is reduced. As the number density of molecules has been able to fall faster than the mean free path has been able to grow at very low pressure, this situation leads to very poor thermal conduction at very low pressures, and the use of vacuum flasks for insulation becomes possible.

Typical values of thermal conductivities of some gases and solids are presented in Table 8.1.*

8.3 Thermal Conductivities of Insulating Solids

In a gas, the molecules carry the heat. What carries the heat in a nonmetallic solid? The answer to this question was considered by Debye: Vibrations of the atoms in the solid, known as *lattice waves* (also known as *phonons*), carry the heat. A phonon is a quantum of crystal wave energy, and it travels at the speed of sound in the medium. The presence of phonons in a solid is a

---

* Data for solids are taken from Y. S. Touloukian, R. W. Powell, C. Y. Ho, and P. G. Klemens, 1970. *Thermal Conductivity: Nonmetallic Solids*. Plenum, New York. Data for gases are from A. W. Adamson, 1973. *A Textbook of Physical Chemistry*. Academic Press, Oxford.

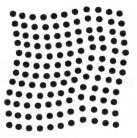

FIGURE 8.3
Depiction of a lattice wave (phonon) in a two-dimensional solid. At $T = 0$ K, the atoms would be arranged on a regular grid, but this instantaneous picture shows the atoms displaced from their equilibrium positions, although the displacements here are greatly exaggerated for illustrative purposes.

manifestation of the available thermal energy (kT); the more heat, the greater the number of phonons (i.e., there are more excited lattice waves). A depiction of a lattice wave is shown in Figure 8.3.

The thermal conductivity of a solid can be measured by heating the material (with power $\dot{q}$, this power being applied at one end and lost at the same rate at the other end, to achieve steady-state conditions; see Figure 8.4) while determining the temperature gradient, dT/dx. The thermal conductivity, κ, is given by

$$\kappa = \frac{\dot{q}}{A} \frac{dx}{dT}, \tag{8.16}$$

where A is the cross-sectional area of the material.

By analogy with the derivation of the thermal conductivity in gases (Equation 8.13), for nonmetallic solids, thermal conductivity is given by

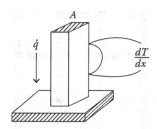

FIGURE 8.4
The thermal conductivity, κ, of a single crystal (or well-compacted block) of a material can be determined by direct measurement. For a crystal of cross-sectional area A, with a vertical temperature gradient of dT/dx, the thermal conductivity is given by $\kappa = \dot{q}(dx/dT)A^{-1}$, where $\dot{q}$ is the power supplied to the top of the crystal. The measurement would be carried out at steady-state conditions, such that the rate of power introduced to the top of the crystal equals the loss at the bottom, and the average temperature of the crystal remains constant. In addition, for high-accuracy measurements, care would need to be taken to reduce extraneous heat exchange with the surroundings; for example, the crystal and its platform would be in vacuum, and the crystal could be surrounded with a temperature-matched heat shield.

$$\kappa_{\text{solid}} = \frac{1}{3} C \, \bar{v} \, \lambda. \tag{8.17}$$

Equations 8.13 and 8.17 are identical, but some of the parameters have different meanings for gases (Equation 8.13) and insulating solids (Equation 8.17).

Explicitly, C is the heat capacity per unit volume (as before), but for insulating solids, $\bar{v}$ is the mean *phonon* speed, and λ is the mean free path of the *phonons*.

To understand the thermal conductivity of nonmetallic solids, we must first return to the simple harmonic oscillator. If all the atoms in a solid were connected by spring-like forces to each other (with all the force constants given by Hooke's law; Equation 7.4), then the potential in all directions is shaped like a three-dimensional parabola; a schematic view of such a solid is given in Figure 8.5.

As the lattice is heated, there would be an increased probability of exciting high-energy phonons, and more of the atoms would be vibrated away from their equilibrium lattice sites. (Incidentally, the following will indicate just how far off their lattice site they could be. Vibration increases as temperature increases, and Lindemann* showed in 1912 that, at the melting point, the vibrations result in about 10% increase in volume over the volume at $T = 0$ K. This corresponds to about 3% increase in the lattice parameters over the $T = 0$ K values.)

If the lattice were perfectly harmonic, the phonons would carry heat perfectly, with no resistance to heat flow, and the thermal conductivity would then be infinite. Clearly, nonmetallic solids must have some heat-flow resistance mechanism since observed thermal conductivities are not infinite.

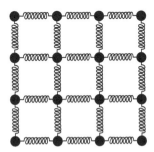

FIGURE 8.5
A solid can be considered to be a network of atoms, each connected to other atoms by "springs".

---

* Frederick Alexander Lindemann (1886–1957) was a German-born physicist who spent most of his career at the University of Oxford in England. In addition to his melting point theory and his theory of unimolecular reactions, Lindemann was personal scientific advisor to Winston Churchill during World War II.

This problem was solved by Peierls* when he showed that phonon–phonon collisions can lead to "turning back" ("Umklapp" in German) processes, as shown in Figure 8.6. This model requires the presence of anharmonicity in the intermolecular potential. Recall that anharmonicity also was responsible for thermal expansion and now we see that it also leads to thermal resistance (finite thermal conductivity).

The thermal conductivity for a "typical" simple crystalline nonmetallic solid is shown in Figure 8.7, and its shape as a function of temperature can be understood simply in terms of the Debye equation (Equation 8.17) as follows.

At high temperatures $(T > \theta_D)$, C is approximately constant and $\bar{v}$ (which depends on the material as it reflects the force constant of interactions

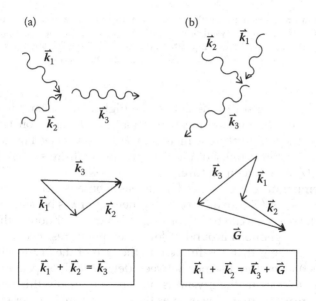

FIGURE 8.6

(a) Two incident phonons (each shown as a wave) can interact to give a resultant phonon with a motion in the same general direction as the incident phonons. This process is known as a *normal* or *N process*; it does not resist heat flow. The vector view shows $\vec{k}_1$ and $\vec{k}_2$ as incident momentum (wave) vectors and $\vec{k}_3$ as the resultant wavevector. (b) Two incident phonons (each shown as a wave) can interact to give a resultant phonon with one component of the motion in a direction opposite to the general motion of the incident phonons. This process, since it "turns back" heat flow, is known as an *Umklapp process*. It is these sorts of processes that lead to thermal resistance, thereby reducing thermal conductivity. The vector view shows $\vec{k}_1$ and $\vec{k}_2$ as incident wavevectors, and $\vec{k}_3$ is the resultant wavevector. $\vec{G}$ is a reciprocal lattice vector and represents a momentum transfer to the lattice, required for overall momentum conservation.

---

* Sir Rudolf Ernst Peierls (1907–1995) was a German-born theoretical physicist. His 1929 theory of heat conduction (predicting an exponential increase in thermal conductivity as a perfect crystal is cooled) was finally verified experimentally in 1951. Peierls also made important contributions to nuclear science (including work on the Manhattan Project), quantum mechanics, and other aspects of solid-state theory.

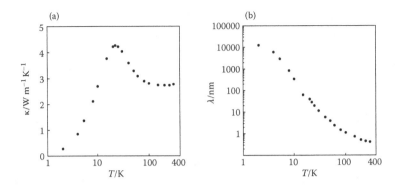

FIGURE 8.7
(a) The temperature dependence of the thermal conductivity of a "typical" simple insulating crystalline solid, a single crystal of very pure $Y_2Ti_2O_7$, and (b) the corresponding mean free phonon path as a function of temperature, determined using Equation 8.17. (From M. B. Johnson et al., 2009. *Journal of Solid State Chemistry* 182, 725. With permission.)

within the solid; the stiffer the lattice, the higher $\bar{v}$) is nearly independent of temperature. The phonon mean free path, λ, depends on temperature such that, as the temperature is increased, λ decreases (see Figure 8.7b) due to the increased probability of phonon–phonon collisions. Therefore, from Equation 8.17, at high temperatures, as T increases, κ decreases (Figure 8.7a).

As the temperature is decreased, κ increases because the mean free path, λ, is getting longer (Figure 8.7). This lengthening of λ is because at low temperatures the phonons can travel for longer distances without colliding since there are fewer phonons around at lower temperatures. Eventually, λ will have grown so long that it is limited by the size of the crystal (if the crystal is perfect) or by the distance between defects such as packing faults or even other isotopes (if the crystal is not perfect). Below this temperature, λ cannot increase further. However, as the temperature is lowered below the Debye characteristic temperature, θ_D, the heat capacity, C, falls. Given that the average phonon speed, $\bar{v}$, is approximately independent of temperature, the Debye equation (Equation 8.17) shows that κ will decrease with decreasing temperatures at low temperatures (Figure 8.7a).

8.4 Thermal Conductivities of Metals

In metals there are two mechanisms to carry heat, the phonons and the free electrons, which is why metals are good thermal conductors. For a metal, the total thermal conductivity, κ, can be written as the sum of the phononic contribution, κ_{phonon}, and the electronic contribution, κ_{elec}:

$$\kappa = \kappa_{phonon} + \kappa_{elec} \tag{8.18}$$

where κ_{phonon} is given by Equation 8.17 and, by analogy, κ_{elec} is given by

$$\kappa_{elec} = \frac{1}{3} C_{elec} \bar{v}_{elec} \lambda_{elec} \tag{8.19}$$

where C_{elec} is the electronic contribution to the heat capacity per unit volume, $\bar{v}_{elec}$ is the mean speed of the conducting electrons, and λ_{elec} is the mean free path of the conducting electrons.

Since, typically, κ for a metal is 10–100 times κ for a nonmetal, it is clear that most of the heat in a metal is carried by conducting electrons not phonons.

Nevertheless, for parallel reasons, the thermal conductivity of a metal has a temperature dependence similar to that of a nonmetal. The temperature dependence of the thermal conductivity of Cu is shown in Figure 8.8 as a typical example.

See the video "Ice Melting on Different Platforms" under Student Resources at PhysicalPropertiesOfMaterials.com

COMMENT: IS MIRACLE THAW™ A MIRACLE?

A commercial thawing tray (Figure 8.9) is advertised to thaw frozen food using a "superconducting" material. In reality, the thawing tray is a sheet of aluminum painted black. Aluminum is a metal and therefore a good conductor of heat. The black paint ensures efficient "capture" of ambient thermal energy, which is then transferred to the food.

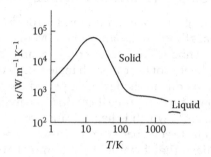

FIGURE 8.8
Temperature dependence of the thermal conductivity of a "typical" metal. The data shown are for Cu.

FIGURE 8.9
Ice melts rapidly on the food-thawing tray.

COMMENT: POPPING CORN

Popcorn "pops" when it is heated because the inner moisture is vaporized, and this breaks open the hull. Measurements of the thermal conductivities of various types of corn show that popcorn hull is two to three times better as a thermal conductor than nonpopping corn kernels. This indicates that popcorn husks have more crystalline (less amorphous) cellulose than other corn kernels, and this has been confirmed by spectroscopic experiments. The high thermal conductivity of popping corn ensures efficient heat transfer into the kernel, while the high mechanical strength associated with its more organized crystalline structure means that the kernel can sustain a high pressure before popping.

COMMENT: SPACE VEHICLE TILES

Thermally insulating tiles play an important role in protecting space vehicles as they pass through the atmosphere. Most of the tiles are made of amorphous silica fibers. The fact that they are amorphous means that they are poor conductors of heat, as such materials have very short phonon mean free paths due to the lack of a periodic structure. The fiber structure cuts down further on heat conduction and reduces the density to about 0.1 g cm^{-3}, which is helpful since vehicle mass is a major fuel consumption factor on space flights. In some areas of the exterior of the space vehicle, tiles have added alumina-borosilicate content, which adds mechanical strength. The tiles are prepared by making a water slurry of the fibers, casting it into soft blocks to which colloidal silica binder has been added, and then sintering at high temperature to form a rigid block that can be machined to exact dimensions.

Thermal expansion is an important factor in the placement of the tiles; if they buckled against one another during heating, this would cause the tiles to fall off. For this reason, gaps of about 0.1 mm are left between tiles.

8.5 Thermal Conductivities of Materials: A Tutorial

a. Refer to the Debye equation (Equation 8.17), which describes the thermal conductivity of an insulating solid. Consider the influence of the perfection of a crystal on the phonon mean free path. Show what the presence of lattice imperfections would do to the thermal conductivity by sketching $\lambda(T)$ and $\kappa(T)$, both for a perfect crystal and for the same crystal with a few percent of imperfections (e.g., impurities).

b. On the same $\lambda(T)$ graph produced in (a), sketch the mean free path for the same composition of material, now in an amorphous state.

c. Again considering the Debye equation and the microscopic picture of an amorphous material (compared with a perfect crystal), sketch how the thermal conductivity of an amorphous material would be expected to look as a function of temperature (see Figure 6.11 for a schematic structure of an amorphous material).

d. *Ceramics* are hard, brittle heat-resistant, inorganic, nonmetallic materials. The word "ceramics" comes from the Greek *keramikos*, which means "of pottery," and many traditional ceramics are clay products. Often they are prepared by shaping and then firing at high temperatures. Other traditional ceramics include materials such as silicate glass and cement. Advanced ceramics include carbides (SiC), pure oxides (e.g., Al_2O_3 and $BaTiO_3$), and nitrides (e.g., Si_3N_4). Why does coffee stay warm longer in a ceramic mug than in a metal cup?

e. Based on the Debye equation, how would you expect the thermal conductivity of a harder material to compare with a softer material? Does this explain why marble is a good heat conductor (and therefore a poor energy storage material [Tutorial in Chapter 6])? What can you predict about diamond's thermal conductivity? (See also Table 8.1.)

f. Metals also have high thermal conductivities. Why is diamond so special with regard to its thermal conductivity in consideration of its other properties?

g. How could one make diamond have an even higher thermal conductivity?

h. An article titled "Hot Rocks" in the business pages of a newspaper reported on the finding that diamonds made from isotopically purified ^{12}C are better conductors of heat than ordinary diamonds. What do you think of the title of the article? Would these diamonds feel hot to the touch?

8.6 Learning Goals

- Thermal conductivity
- Thermal conductivity of gases from kinetic theory (Debye model)
- Thermal conductivity of insulating solids (Debye model) including temperature dependence
- Phonons, anharmonicity (U-processes, N-processes)
- Determination of thermal conductivity of solids
- Thermal conductivity of metals
- Thermal conductivity of amorphous materials

8.7 Problems

8.1 a. Calculate the thermal conductivity for He at 25 °C and 101 kPa given the following information: at these conditions, $\bar{v}$ of He is 1360 m s^{-1} and the mean free path is 7.2×10^{-7} m.

b. Compare the results from (a) with the following thermal conductivities at the same conditions (all in W m^{-1} K^{-1}): air, 2.4×10^{-2}; crystalline SiO$_2$, 10; Cu, 400. Explain the relative thermal conductivities in terms of the Debye equation.

8.2 Several expressions for κ_{gas}, the thermal conductivity of a gas, have been presented in this chapter. One of the expressions was given by Equation 8.15. Use this equation and the temperature dependence of the variables in it to show how κ_{gas} changes as temperature is increased.

8.3 Figure 8.10 shows the variation of thermal conductivity of a semiconductor after doping with n- or p-type impurities. Explain the

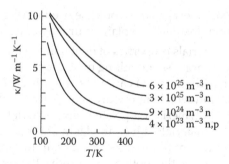

FIGURE 8.10
The temperature dependence of the thermal conductivity of a semiconductor with n- and p-type impurities at various concentrations, indicated as number of impurity sites per m^3.

origin of the change in thermal conductivity on introduction of these impurities. The numbers on the figure indicate the concentration of the impurities.

8.4 At a trade show exhibiting new building materials, one of the energy-efficient products was a window with good thermal insulation. The window had two panes of glass, and the space between them contained argon gas. The advertisement said that this window gave much better insulation than could be achieved when the space was filled with air (the more common case for double-paned windows). Explain why argon is a better thermal insulator than air. Assume both to be at the same pressure. It may be helpful to know that the molecular mass of argon is 39.95 g mol^{-1} and that of air (which is about 80% N_2 and 20% O_2) is about 29 g mol^{-1}, and to consider all factors in the Debye expression for thermal conductivity. State and justify any assumptions.

8.5 Ceramics are very useful inorganic materials composed of metallic and nonmetallic ions held together by partly ionic and partly covalent bonds. Ceramics are electrical insulators, characterized by their hardness, strength, and heat and chemical resistance. They can be amorphous or crystalline. Explain how the microstructure, i.e., structure on the micron length scale, changed by small differences in the composition of a ceramic, can influence its thermal conductivity. Take into account the fact that the porosity (and hence density) of a ceramic can be modified by the introduction of impurities.

8.6 Incandescent light bulbs usually have an argon atmosphere sealed inside them. The purpose of the argon is to prevent the filament from oxidizing (in air) or subliming (in vacuum). A portion of the energy that is used to light an incandescent bulb is "lost" as heat.

Suggest another inert gas that could be used in place of argon but is a better thermal insulator. Explain your reasoning.

8.7 Appropriate materials properties of dental filling materials are key to their success. A large "silver" filling (made of an amalgam, i.e., a mixture of silver with mercury) can make a tooth very sensitive to hot or cold food. However, temperature sensitivity is not so problematic when a white polymeric composite filling material is used. Explain why these two materials, silver amalgam and polymeric composite, lead to such different thermal sensitivities. In this case, consider the nerve endings inside the tooth to be sensitive thermometers.

8.8 Two countertop materials look very similar: One is natural marble and the other is a synthetic material. They can be distinguished easily by their feel: One is cold to the touch and the other is not. Explain which is which and the basis of the difference in their thermal conductivities.

8.9 *Stainless steel* is iron with added chromium and possibly also nickel. It has good corrosion resistance because Cr forms passivating Cr_2O_3. When stainless steel is used for pots for cooking, the pot bottoms are often made of copper. Why is stainless steel not used on the pot bottoms? Why is copper better for the pot bottoms? Explain.

8.10 Black phosphorus has a density of 2.7 g cm^{-3}, while that of white phosphorus is 1.8 g cm^{-3} (both values are at room temperature). Use this information to explain the difference in thermal conductivities of the two forms of phosphorus (Table 8.1).

8.11 *Brass* is formed from copper with zinc; a common brass formulation gives two phases, one rich in Cu and the other with more Zn. Copper is very difficult to weld (welding is a method of joining metals that involves heating them together, possibly with another material) because it has a very high thermal conductivity. On the basis of its thermal conductivity, would brass be expected to be easier or more difficult to weld than copper? Explain.

8.12 The baking temperatures for a cake mix, as listed on the box, are different depending on whether the pans are glass or metal. Which type of pan requires lower baking temperature? Why?

8.13 An aerogel is a very low-density material, typically made of silica with about 50%–99% air in its structure. The thermal conductivity of an aerogel is extremely low in comparison with more typical solids (see Table 8.1). Why? Consider the Debye equation (Equation 8.13 or 8.17.)

8.14 Some negative thermal expansion (NTE) materials have low-frequency phonons associated with the collective movement of the atomic groups. These phonons can interact with the heat-carrying phonons to reduce the phonon mean free path. On this basis, would you expect thermomiotic materials to have lower or higher thermal conductivities than other similar materials? Explain your reasoning.

8.15 Thermal conductivity, κ, is an important property of a material, but, for the function of a material, a key related property is *thermal diffusivity*, a, which is a measure of the speed of heat flow through a material, defined as

$$a = \frac{\kappa}{\rho C_s} \qquad (8.20)$$

where ρ is the density and C_s is the specific heat. Explain why a is proportional to κ and inversely proportional to ρ and C_s.

8.16 *Moissanite* is a form of SiC with many properties similar to diamond. For example, it is transparent and colorless and has a high refractive index, so it sparkles. Moissanite has a hardness of 9.25 on the Mohs scale, while diamond has a hardness of 10. (Harder materials have higher values of Mohs hardness, with significantly larger steps between higher values of hardness.) Jewelers often use an instrument to determine the thermal conductivity of a diamond to show that it is genuine. Do you expect that moissanite could fool this instrument, leading unsuspecting consumers to buy the cheaper moissanite? Explain.

8.17 A subclass of polymorphism (see Chapter 9) is known as *polytypism*. In a layers structure, different polytypes have different periodicity of the layers. For example, there could be three types of layers, A, B, and C, that have the same composition but are shifted laterally. One polytype could be ABCABC and another could be ABABCABABC and so forth. Examples of materials exhibiting polytypism are SiC, CdS_2, GaSe, and some micas and clays. Would two different polytypes of a material be expected to exhibit the same thermal conductivity perpendicular to the layers? What about parallel to the layers? Explain.

8.18 Consider the role of the thermal conductivity of a material in the value of the thermal shock fracture resistance, R_S (Equation 7.22). Does a higher value of the thermal conductivity make a material more or less susceptible to thermal fracture? Explain both in terms of the equation and in terms of physical processes involved.

8.19 Thermal conductance, K, is related to thermal conductivity, κ, as follows:

$$K = \frac{\kappa A}{L} \qquad (8.21)$$

where A is the material's cross-sectional area and L is its length.

  a. What are the units of thermal conductance?
  b. The reciprocal of thermal conductance is thermal resistance, R_{th}. What are the units of R_{th}?
  c. Only one of the two, K or κ, is an intensive property of the material (i.e., independent of dimensions) and the other is extensive (depends on material dimensions). Which is which? And which is more usefully tabulated for materials?

8.20 Thermal resistivity, ρ_{th}, is defined as the reciprocal of thermal conductivity. Thermal resistivity due to inter-particle contacts can be a difficulty for heat transfer in materials. Explain why this thermal resistance can be especially problematic for nanostructured materials.

8.21 If a glass jar is too tightly closed, it is common to run hot water over the lid to make it easier to open. How does this process work? Consider the thermal expansion of the metal and the glass and also the thermal conductivities of each material.

Further Reading

General References

Many physical chemistry textbooks contain derivations of the kinetic theory of gases. In addition, the following address many of the general points raised in this chapter.

High thermal conductivity materials. Special issue. K. Watari and S. L. Shinde, Eds. *MRS Bulletin*, June 2001, 440–455.

A. W. Adamson, 1973. *A Textbook of Physical Chemistry*. Academic Press, Oxford.

W. D. Callister, Jr. and D. G. Rethwisch, 2013. *Materials Science and Engineering: An Introduction*, 9th ed. John Wiley & Sons, Hoboken, NJ.

B. S. Chandrasekhar, 1998. *Why Things Are the Way They Are*. Cambridge University Press, Cambridge.

M. de Podesta, 2002. *Understanding the Properties of Matter*, 2nd ed. CRC Press, Boca Raton, FL.

A. B. Ellis, M. J. Geselbracht, B. J. Johnson, G. C. Lisensky, and W. R. Robinson, 1993. *Teaching General Chemistry: A Materials Science Companion*. American Chemical Society, Washington, DC.

R. E. Hummel, 2004. *Understanding Materials Science*, 2nd ed. Springer, New York.

C. Kittel, 2004. *Introduction to Solid State Physics*, 8th ed. John Wiley & Sons, Hoboken, NJ.

I. Maasilta and A. J. Minnich, 2014. Heat under the microscope. *Physics Today*, August, 27.

V. Murashov and M. A. White, 2004. Chapter 1.3: Thermal conductivity of insulators and glasses. In *Thermal Conductivity*, T. Tritt, Ed. Kluwer/Plenum, New York, pp. 93–104.

T. N. Narasimhan, 2010. Thermal conductivity through the 19th century. *Physics Today*, August, 36.

R. Sun and M. A. White, 2004. Chapter 3.1: Ceramics and glasses. In *Thermal Conductivity*, T. Tritt, Ed. Kluwer/Plenum, New York, pp. 23–254.

R. L. Sproull, 1962. Conduction of heat in solids. *Scientific American*, December, 92.

Experimental Techniques

M. E. Bacon, R. M. Wick, and P. Hecking, 1995. Heat, light and videotapes: Experiments in heat conduction using liquid crystal film. *American Journal of Physics*, 63, 359.

V. V. Murashov, P. W. R. Bessonette, and M. A. White, 1995. A first-principles approach to thermal conductivity measurements of solids. *Proceedings of the Nova Scotian Institute of Science*, 40, 71.

Implications of Thermal Conductivity

W. H. Corkern and L. H. Holmes Jr., 1991. Why there's frost on the pumpkin. *Journal of Chemical Education*, 68, 825.

W. J. González-Espada, L. A. Bryan, and N.H. Kang, 2001. The intriguing physics inside an igloo. *Physics Education*, 26, 290.

Thermal Conductivities of Specific Materials

D. R. Askeland, P. P. Fulay and W. J. Wright, 2010. *The Science and Engineering of Materials*, 6th ed. Cengage, Stamford, CT.

W. J. da Silva, B. C. Vidal, M. E. Q. Martins, H. Vargas, A. C. Pereira, M. Zerbetto, and L. C. M. Miranda, 1993. What makes popcorn pop. *Nature*, 362, 417.

H T. Gspann, S. M. Juckes, J. F. Niven, M. B. Johnson, J. A. Elliott, M. A. White, and A. H. Windle, 2017. High thermal conductivities of carbon nanotube films and micro-fibres and their dependence on morphology. *Carbon*, 114, 160.

C. A. Kennedy and M. A. White, 2005. Unusual thermal conductivity of the negative thermal expansion material, ZrW_2O_8. *Solid State Communications*, 134, 271.

J.-H. Pöhls, M. B. Johnson, and M. A. White, 2016. Origins of Ultralow Thermal Conductivity in Bulk [6,6]-Phenyl-C_{61}-Butyric Acid Methyl Ester (PCBM). *Physical Chemistry Chemical Physics*, 18, 1185.

Y. S. Touloukian, R. W. Powell, C. Y. Ho, and P. G. Klemens, 1970. *Thermal Conductivity: Nonmetallic Solids*. Plenum, New York.

M. A. White, V. Murashov, and P. Bessonette, 1996. Thermal conductivity of food-thawing trays. *Physics Teacher*, 34, 4.

K. M. Wong, 1990. Space shuttle thermal protection system. *California Engineer*, December, 12.

Websites

For links to relevant websites, see PhysicalPropertiesOfMaterials.com

9

Thermodynamic Aspects of Phase Stability

9.1 Introduction

The thermal stability of a material, with respect to phase change and/or chemical change, can be very important in determining uses of a material. In this chapter we examine such properties, first for pure materials, then for two-component materials, and finally for three-component materials.

9.2 Pure Gases

A typical phase diagram for a pure material as a function of temperature and pressure is shown schematically in Figure 9.1, where several isotherms near the critical point are indicated.

The thermodynamic stability of a gas depends on its molar Gibbs energy being less than a condensed phase (liquid or solid). As we have seen already, an ideal gas cannot condense (because it has no intermolecular forces), but we know that real gases do condense to give liquids and/or solids as we compress them, due to attractive intermolecular forces.

If we consider the pressure as a function of volume for the isotherms of Figure 9.1, this gives a series of curves as shown in Figure 9.2. At T_1, which is greater than the critical temperature, compression of the gas just increases the pressure; it does not lead to condensation. However, at a temperature below the critical temperature (e.g., T_3), a decrease in the volume of the gas leads first to an increase in pressure and then to a sudden drop in volume at constant pressure, corresponding to the drop in volume due to liquefaction of the gas. Further compression at T_3 leads to a rapid increase in pressure since liquids are not very compressible. At a lower temperature, T_4, liquefaction also occurs. At $T_2 = T_c$, the critical temperature, liquid and gas are indistinguishable throughout the whole volume range.

Although the ideal equation of state does not lead to the liquefaction of a gas, the van der Waals equation of state shows instabilities in the

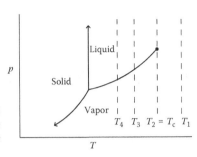

FIGURE 9.1
The generalized phase diagram for a pure material as a function of temperature (T) and pressure (p). The dashed lines correspond to constant–temperature lines (isotherms) considered further in Figure 9.2.

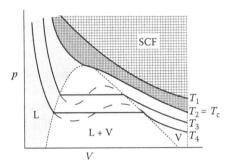

FIGURE 9.2
The pressure (p) as a function of volume (V) for various isotherms of a real gas. The temperatures correspond to those shown in Figure 9.1. The dashed lines correspond to the solutions to the van der Waals equation of state, and the solid lines correspond to experimental findings. The shading indicates distinctions among the regions, L: liquid; V: vapor; L + V: liquid and vapor; SCF: supercritical fluid. At the critical temperature, T_2 in Figure 9.1, compression of the gas leads to an inflection point in the p versus V curve shown here, corresponding to $(\partial p/\partial V)_T = 0$ and $(\partial^2 p/\partial V^2)_T = 0$. This inflection point is the apex of the curve joining the ends of the constant pressure volume liquefaction lines, and is another way to define the critical point of a gas.

pressure–volume curves below the critical temperature, T_c. The solutions of the van der Waals equation of state lead to the $p(V)$ dashed curves indicated in Figure 9.2; the portion of those curves that leads to decreases in pressure with decreasing volume indicates their unphysical nature in this region. The isobaric condensation of the gas exactly balances the areas of the peak and inverse peaks of the $p(V)$ plot from the van der Waals equation of state (dashed lines in Figure 9.2).

9.3 Phase Equilibria in Pure Materials: The Clapeyron Equation

Phase diagrams such as Figure 9.1 can be understood by examining the experimental pressure-temperature boundaries of the phases. In view of this comment, it would be useful to derive an expression for dp/dT along the phase equilibrium lines.

To do so, it is first useful to consider the fact that the molar Gibbs energies are identical for two single-component phases that are in equilibrium. This is true for liquid–solid equilibrium ($G^{\text{solid}} = G^{\text{liquid}}$ at the melting point), liquid–vapor equilibrium ($G^{\text{liquid}} = G^{\text{vapor}}$ at the boiling point), or any equilibrium of phase α with phase β ($G^{\alpha} = G^{\beta}$).*

If initial temperature and pressure conditions allow phase α to be in equilibrium with phase β,

$$G^{\alpha}_{\text{initial}} = G^{\beta}_{\text{initial}} \tag{9.1}$$

and the temperature and pressure are changed (changing G by dG) to final conditions such that the equilibrium of phases α and β is maintained, then

$$G^{\alpha}_{\text{final}} = G^{\beta}_{\text{final}}. \tag{9.2}$$

Since

$$G^{\alpha}_{\text{final}} = G^{\alpha}_{\text{initial}} + dG^{\alpha} \tag{9.3}$$

and

$$G^{\beta}_{\text{final}} = G^{\beta}_{\text{initial}} + dG^{\beta}, \tag{9.4}$$

it follows that

$$dG^{\alpha} = dG^{\beta}. \tag{9.5}$$

The fundamental equation for dG (Equation 6.46) can be written for phase α and also for phase β, in terms of the change in pressure (dp) and in temperature (dT):

$$dG^{\alpha} = V^{\alpha}dp - S^{\alpha}dT = dG^{\beta} = V^{\beta}dp - S^{\beta}dT. \tag{9.6}$$

This equation can be rearranged to give the quantity we were aiming for, dp/dT:[†]

$$\frac{dp}{dT} = \frac{\Delta_{\text{trs}}S}{\Delta_{\text{trs}}V} = \frac{\Delta_{\text{trs}}H}{T\Delta_{\text{trs}}V} \tag{9.7}$$

---

* This discussion assumes the amount of material to be constant. All thermodynamic quantities discussed in this section (G, H, S) can equally be expressed as their molar values (G_m, H_m, S_m).
† The most important equations in this chapter are designated with ▮ to the left of the equation.

where the transition changes are given by $\Delta_{trs}S = S^{\alpha} - S^{\beta}$ and $\Delta_{trs}V = V^{\alpha} - V^{\beta}$ and the second equality makes use of the generalization

$$\Delta G = \Delta H - T\Delta S \tag{9.8}$$

where, due to the equality of G for any two phases in equilibrium,

$$\Delta_{trs}G = 0 \tag{9.9}$$

and hence

$$\Delta_{trs}S = \frac{\Delta_{trs}H}{T_{trs}} \tag{9.10}$$

where $\Delta_{trs}H$ is the enthalpy change of the transition and T_{trs} is the temperature of the transition. Equation 9.7 is called the *Clapeyron* equation*. The Clapeyron equation can be applied to any first-order phase transition, and holds exactly.

To see how the Clapeyron equation can be used to determine the sign of the slope of phase boundaries, we consider two examples.

For the transition from solid to vapor, the Clapeyron equation gives

$$\left(\frac{dp}{dT}\right)_{sol-vap} = \frac{\Delta_{subl}H}{T\Delta_{subl}V} \tag{9.11}$$

where $\Delta_{subl}H$, the enthalpy change on sublimation, is positive (it always takes heat to convert a material from solid to vapor), and the volume change on sublimation, $\Delta_{subl}V = V_{vapor} - V_{solid}$, also is positive. Since T is always positive (on the kelvin scale), it follows that (dp/dT) on the solid–vapor equilibrium line (i.e., the slope of the solid–vapor equilibrium line) is always positive. For the transition from solid to liquid, the Clapeyron equation gives

$$\left(\frac{dp}{dT}\right)_{sol-liq} = \frac{\Delta_{fus}H}{T\Delta_{fus}V}. \tag{9.12}$$

Since $\Delta_{fus}H$, the enthalpy change on fusion (melting) is positive, and T is positive, and $\Delta_{fus}V = V_{liquid} - V_{solid}$ is positive for most materials, dp/dT along the solid–liquid line is positive for most materials. However, for some materials, $V_{liquid} < V_{solid}$; an example is water (we know this because ice floats) and other examples are Ga, Sb, Bi, Fe, Ge, and diamond. The larger volume for the solid leads to $dp/dT < 0$. We will see this negative slope in $p(T)$ in the experimentally determined water phase diagram later in this chapter.

---

* Benoit Pierre Émile Clapeyron (1799–1864) was a French engineer who specialized in bridges and locomotives. His only publication in pure science concerned the expression of vapor pressure as a function of temperature.

To generalize, given that it takes heat ($\Delta H > 0$) to proceed from a lower-temperature phase to a higher-temperature phase (another way to look at this is that the molar entropy of the higher-temperature phase exceeds that of the lower-temperature phase), the slope of p–T equilibrium lines in the phase diagram of a pure material can indicate, using the Clapeyron equation, the relative densities of the two phases.

Let us turn now to some specific phase diagrams for pure materials.

9.4 Phase Diagrams of Pure Materials

Figure 9.3 shows the pressure–temperature phase diagram of carbon dioxide. This diagram is an example of the simple p–T phase diagrams discussed earlier. One particularly interesting feature here is that, at atmospheric pressure (~100 kPa), the solid converts directly to vapor, at a temperature of 194.7 K; this can be seen in the sublimation of dry ice, $CO_2(s)$, at room temperature, one of the few examples of conversion from solid to vapor at ambient pressure without first passing through the liquid phase.

The relatively moderate critical point of CO_2 ($p_c = 7380$ kPa and $T_c = 304.2$ K) leads to supercritical CO_2 relatively easily. Fluids beyond the critical point are called *supercritical fluids*, and they are neither gas nor liquid. Supercritical fluids have unusual properties such as much lower viscosity than the liquid and much higher density (and therefore greater solvent power) than the gas. Supercritical CO_2 has such diverse uses as the carrier for some chromatographic separations and the extraction solvent for decaffeination of coffee.

Figure 9.4 shows the pressure–temperature phase diagram of sulfur. The solid lines represent the stable phase diagram. The main new feature in this phase diagram is the existence of two solid phases, one of monoclinic

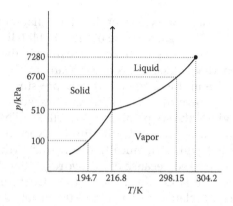

FIGURE 9.3
The pressure–temperature phase diagram for pure CO_2. The scales are not linear.

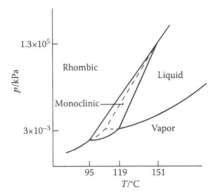

FIGURE 9.4
The pressure–temperature phase diagram for sulfur. The solid lines indicate the stable phases, and the dashed lines indicate metastability.

structure and the other of orthorhombic structure. The existence of more than one solid phase is called *polymorphism*, literally "many shape types." The two polymorphs of sulfur have different temperature and pressure regions of stability. (For elements, polymorphs are also called *allotropes*).

Figure 9.4 also shows, by the dashed lines, the metastable phase diagram for sulfur. By appropriate thermal treatment (rapid cooling from above the melting point), it is possible to trick sulfur into transforming directly from a supercooled liquid to the rhombic form, without first passing through the monoclinic form. In the region where the monoclinic form is most stable (i.e., the T and p range where monoclinic has the lowest G_m of all possible phases), the liquid or rhombic form can exist only metastably. That they can exist at all is testament to the high activation energy required for the conversion from the metastable form to the monoclinic form in this temperature–pressure region.

Figures 9.5a and b show two views of the phase diagram of water. At low pressure (Figure 9.5a), the negative slope of the p-T solid–liquid line is readily apparent. This line indicates, as we have seen earlier in the discussion of the Clapeyron equation, that $H_2O(s)$ is less dense than $H_2O(l)$ under these conditions. The phase of ice in this temperature and pressure region is generally referred to as ice Ih, where the "h" stands for "hexagonal." (The hexagonal structure is the origin of the six points in a snowflake.) The hexagonal structure of this form of ice (shown in Figure 9.6) is a very open structure (i.e., low-density structure) with dynamically disordered hydrogen bonds. The structure is not close packed because of the hydrogen-bonding requirements of the H_2O molecules. It is the openness of this structure that leads to the lower density of ice Ih relative to water, and the negative value of dp/dT in this region. It is worth considering that the negative value of dp/dT leads to a decrease in the melting point of ice when pressure is applied. Although this

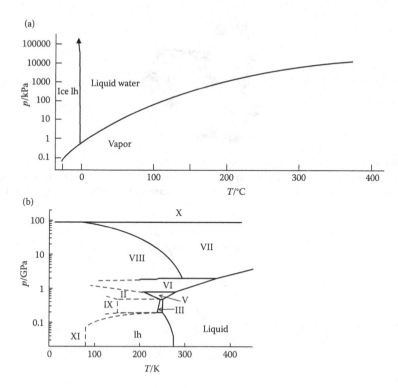

FIGURE 9.5
The pressure–temperature phase diagram of H_2O, (a) at low pressure (note the negative p-T slope) and (b) over a wider pressure range.

has been cited as the basis for ice skating, calculations[*] show that the temperature drop is insufficient to account for a liquid layer between the skate and ice.

At higher pressures, more polymorphs of ice are observed, as illustrated in Figure 9.5b. The large number of polymorphs for H_2O arises from the many ways in which H_2O molecules can be arranged to satisfy their hydrogen-bonds to their neighbors, with different structures having lowest G_m values in different pressure and temperature regions. Note that at high pressures (above about 0.2 GPa), the values of dp/dT along the melting line for ice (in phase ice III) have become positive, indicating that the higher-pressure polymorphs of ice are denser than water. This situation is because the higher-pressure forms are compressed relative to the low-density ice Ih.

Figure 9.7 shows the pressure–temperature phase diagram of CH_4 in the solid region. (The solid–liquid–vapor triple point of methane is at 90.69 K.) There are at least four solid phases for CH_4, and polymorphism in this case can be attributed to the shape of the molecule. Although we may consider methane to be tetrahedral, as in the ball-and-stick model of Figure 9.8a, the shape of any molecule

---

[*] L. F. Louks, 1986. *Journal of Chemical Education*, 63, 115.

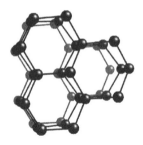

FIGURE 9.6
The oxygen positions in the hexagonal structure of ice Ih. The hydrogens are not shown because they are dynamically disordered. The very open structure makes ice Ih float in liquid water.

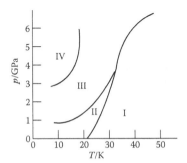

FIGURE 9.7
The low-temperature phase diagram of CH_4, showing only the solid phases.

is really more closely related to the electron density distribution which, for CH_4, is more nearly represented as a sphere with slight protuberances at tetrahedral locations (Figure 9.8b). The nearly spherical shape of the methane molecule means that it takes very little energy to rotate it even in the solid state. In fact, in Phase I of CH_4, the methane molecules are nearly freely rotating, although they are located on particular lattice sites (Figure 9.9). Another way to express this is to say that they are translationally ordered but rotationally, dynamically disordered. Phase I of CH_4 is an *orientationally disordered phase*.

As the temperature is lowered, CH_4 molecules become more ordered, and Phase II has a rather unusual structure with some molecules in the unit cell ordered and some molecules in other positions in the unit cell dynamically disordered (Figure 9.9).

The structures of the other phases of CH_4 are more complicated still and not yet fully sorted out. A complication here is that the CH_4 molecule is so light that Newtonian mechanics is not appropriate to describe it at such low temperatures, and quantum mechanics must be used to describe the disorder in these solid phases.

The concept of an orientationally disordered solid also can be described from the point of view of thermodynamics. In particular, the idea of an

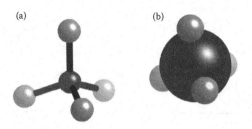

FIGURE 9.8

Two views of the CH_4 molecule. (a) The ball-and-stick model is useful but does not truly represent the space-filling shape of the molecule. (b) A space-filling model of CH_4 more accurately shows methane to be a nearly spherical molecule.

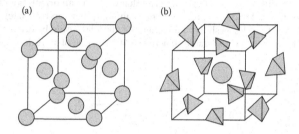

FIGURE 9.9

Structures of two phases of solid CH_4. (a) Phase I, with the CH_4 molecules orientationally disordered (shown as spheres to indicate rotating tetrahedra) on sites of a face-centered cubic lattice. (b) Phase II, with some CH_4 molecules orientationally ordered (orientationally ordered tetrahedra) and some dynamically disordered (shown as a sphere in the center of the unit cell). The latter experience a very low (practically zero) barrier to rotation and a spherically symmetric ground state at low temperature.

orientationally disordered solid leads to the picture of a solid with a molar entropy that is higher than the usual value for a solid. In fact, it has been suggested that one way to characterize an orientationally disordered solid is to describe it as one with an unusually low entropy change on fusion ($= \Delta_{fus}S = S_{liquid} - S_{solid} < 20\,\mathrm{J\,K^{-1}\,mol^{-1}}$ for an orientationally disordered solid), given that S_{liquid} is nearly the same for most liquids and the molar value of S_{solid} is unusually high for an orientationally disordered solid.

An orientationally disordered solid also has been referred to as a *plastic crystal* because of the ease of deformation of this type of material due to the rotating molecules that make it up. The term "orientationally disordered solid" is now preferred over "plastic crystal," as the latter is sometimes too limiting, and the former term accurately describes the physical picture.

Orientational disorder can be understood further in terms of the intermolecular potential: The barrier to rotation of a nearly spherical molecule (see Figure 9.10) can be quite a bit less than the available thermal energy, kT. Therefore, if there is sufficient thermal energy, molecules can rotate on their lattice sites in an orientationally disordered crystal. This situation can be

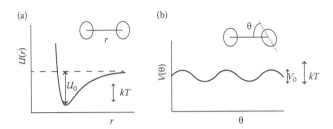

FIGURE 9.10
The energy of interaction of two nearly spherical molecules that are capable of forming an orientationally disordered solid, as a function of (a) intermolecular separation, and (b) orientational angle at fixed separation. If the thermal energy is less than the binding energy ($kT < U_0$ in (a)), the material will be a solid. For a material that can form an orientationally disordered solid, $V_0 \ll U_0$ (i.e., the molecules can rotate more easily than they can translate). If the thermal energy (kT) is greater than V_0, as shown here, then the material would be orientationally disordered. At lower temperatures, the stable structure would be that of an ordered solid.

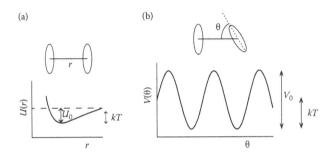

FIGURE 9.11
The energy of interaction of two rod-shaped molecules that are capable of forming a liquid crystalline phase, as a function of (a) intermolecular separation, and (b) orientational angle at fixed separation. For a material that can form a liquid crystal, $U_0 \ll V_0$ (i.e., the molecules can translate more easily than they can rotate). If the thermal energy (kT) is nearly equal to U_0, as shown here, then the material would be liquid crystalline. At lower temperatures, the stable structure would be that of an ordered solid.

contrasted with liquid crystals, where the barrier to reorientation is high due to the molecular shape, but the barrier to translation is relatively low (Figure 9.11). If there is sufficient thermal energy, molecules in a liquid crystal will glide past one another. In both orientationally disordered solids and liquid crystals, when there is less thermal energy (kT decreases), the material forms an ordered solid. As for liquid crystals, orientationally disordered solids are *mesophases*, i.e., intermediate between the liquid and the ordered solid.

Methane is not the only example of a molecular solid exhibiting some aspect of orientational disorder: H_2, N_2, O_2, F_2, CCl_4, neopentane, norbornane, cubane, adamantane, and C_{60} are a few of many other examples. Furthermore, portions of large molecules can be orientationally disordered if there is

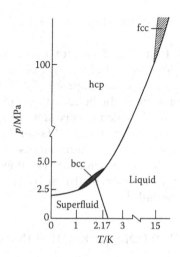

FIGURE 9.12

Phase diagram of the condensed phases of ⁴He. The three solid phases are labeled according to their structure: bcc = body-centered cubic; hcp = hexagonal close-packed; fcc = face-centered cubic. The two liquid phases are labeled as liquid (also called liquid I) and superfluid (also called liquid II).

sufficient energy: typical examples are alkyl chains and methyl groups. The flexibility in a large molecule can be responsible for biological activity, and one reason for the intolerance of living organisms to even relatively small temperature changes is the subtle importance of internal degrees of freedom to their biological functions, and the interplay between thermal energy (kT) and activation barriers to orientational motion.

The phase diagram of helium (Figure 9.12 shows the phase diagram for the condensed phases of the ⁴He isotope) shows other interesting and unique features. For example, there are three solid phases in ⁴He, one with hexagonal close packing, another with body-centered cubic packing, and yet another with face-centered packing. Perhaps more interesting is the fact that helium does not exist as a solid except under applied pressure; it stays as a liquid down to absolute zero. Furthermore, there are two liquid phases! The higher-temperature liquid form, liquid helium I, is the phase that exists at the critical point ($T_c = 5.2$ K), and this is the form of liquid helium that boils at 4.2 K and has wide use as a *cryogen* (a low-temperature fluid).

Liquid helium II has such fascinating properties that it is called a *superfluid*. For example, a suspended beaker filled with liquid helium II will spontaneously lose its contents due to creep up the sides of the beaker, followed by dripping until the beaker is empty. Liquid helium II can flow through the narrowest of openings, without any resistance (i.e., it has zero viscosity). Furthermore, the thermal conductivity of liquid helium II close to the liquid–liquid transition temperature is about 1000 times greater than that of copper (and remember that a liquid has neither phonons nor free electrons to carry heat!). The other He isotope, ^{3}He, has at least three distinct superfluid phases.*

The phase transition from liquid helium I to liquid helium II is often referred to as a λ-transition, owing to the shape of the associated heat capacity anomaly. The source of the unusual properties of liquid helium II is the zero-momentum state of some of the helium atoms in this phase (moving at a speed of about 3×10^{-7} m s^{-1}, which is very slow compared with the speed of the other He atoms which move about 10^9 faster (i.e., 300 m s^{-1} at absolute zero due to zero-point motion). The helium atoms in this phase are said to have undergone a *Bose–Einstein condensation* (i.e., they are "condensed" into their ground state energy), as suggested first by Fritz London in 1938 and now confirmed experimentally.[†]

COMMENT: POLYMORPHISM IN NATURE

$CaCO_3$ exists in two common polymorphs in nature, calcite and aragonite. Calcite is the more thermodynamically stable bulk form at room temperature and pressure. The form produced under biological conditions depends on nucleating macromolecules and on particle size, temperature, and the presence of other ions in solution. For example, the nacre (mother-of-pearl) in shells is typically aragonite "bricks" (~1 μm thick and 10 μm wide and deep) embedded in an organic matrix.

As a final example of an interesting phase diagram of a pure material, we consider polymorphism in carbon. Figure 9.13 shows that the low-pressure stable form is graphite while at higher pressures (such as inside the earth), diamond is the stable form. Since we know that diamonds can exist at room temperature and pressure, we must conclude that they are metastable at ambient conditions and should spontaneously convert to the more thermodynamically stable form, graphite. That diamonds do not convert before our eyes indicates that the activation energy for this conversion is very high. This makes sense in light of the great rearrangement that would be required to go from the three-dimensional diamond lattice (Figure 9.14a) to the two-dimensional layered structure of graphite (Figure 9.14b). The differences between the diamond and the graphite structures are also responsible for the very large differences in their properties: Diamond is hard (due to the three-dimensional structure) whereas graphite is

---

[*] The discovery of superfluidity of ^{3}He at 0.002 K was the topic of the 1996 Nobel Prize in Physics, awarded jointly to David M. Lee (1931–; Cornell University), Douglas D. Osheroff (1945–; Stanford University) and Robert C. Richardson (1937–2013; Cornell University).

[†] The achievement of the Bose-Einstein condensation in dilute gases of alkali atoms was the subject of the 2001 Nobel Prize in Physics jointly to Eric A. Cornell (1961–; University of Colorado), Wolfgang Ketterle (1957–; MIT) and Carl E. Wieman (1951–; University of Colorado).

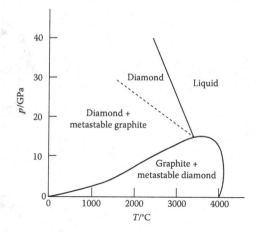

FIGURE 9.13
The pressure–temperature phase diagram of carbon. Solid lines indicate the regions of stable phases; metastability is indicated with a dashed line.

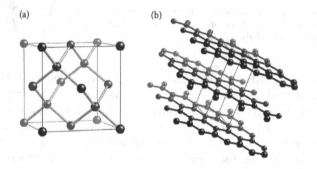

FIGURE 9.14
Structures of two polymorphs of carbon: (a) diamond and (b) graphite. The three-dimensional bonding network of diamond is one of the reasons why diamond is so hard. On the other hand, the layered structure of graphite, with the layers held together with weak van der Waals interactions, makes graphite useful as a lubricant.

so soft that it is the "lead"[*] in a pencil and also a useful lubricant (due to the weak [van der Waals] interaction holding the layers together).

Carbon has other polymorphs, including the family of compounds generally known as *Fullerenes*. The archetypal member of this family is the molecule C_{60}, which was first discovered in 1985.[†] The geometry of the C_{60} molecule is that of a soccer ball (see Figure 9.15), with one carbon atom at each vertex. There can be other molecules with similar structures, and the

---

[*] This is called "lead" because the first deposit of graphite used to make pencils, which was revealed when a tree toppled over, was originally mistaken for the element lead.

[†] This discovery was recognized with the 1996 Nobel Prize in Chemistry, awarded jointly to Robert F. Curl Jr. (1933–; Rice University), Sir Harold W. Kroto (1939–2016; University of Sussex) and Richard E. Smalley (1943–2005; Rice University).

FIGURE 9.15
The molecular structure of C_{60}.

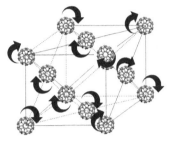

FIGURE 9.16
The structure of C_{60} at room temperature, showing the C_{60} molecules dynamically orientationally disordered on their face-centered cubic lattice sites.

general family is called Fullerenes after R. Buckminster Fuller, the architect who designed geodesic domes as architectural structures. Although first synthesized in the laboratory only a few decades ago, it is now known that Fullerenes exist naturally on earth and even in interstellar space. C_{60} has been found to occur naturally in a 2-billion-year-old impact crater in Sudbury, Canada. The discovery of C_{60} and its family members has led to very exciting new areas of materials research. Some metal atoms can be trapped inside the cage of Fullerenes, designated by the general formula $M@C_n$. For example, $La@C_{82}$ and $Y@C_{82}$ have been prepared. The combination of metals with the insulating Fullerenes can lead to unusual optical and electronic devices: Fullerene derivatives can be made to be *superconducting* (e.g., K_3C_{60}; see Chapter 12 for a discussion of superconductivity). Other molecules with special properties can be added outside or inside the cage, and C_{60} even can be used to produce diamond films.* Pure C_{60} is yellow when in the solid form and magenta when dissolved in benzene, indicating that its electronic energy levels depend on the environment.

The solid form of C_{60} is orientationally disordered at room temperature (see Figure 9.16), with the molecules sitting on the lattice sites of a face-centered cubic (fcc) lattice. Below a solid–solid phase transition at about $T = 250$ K, the C_{60} molecules become more ordered.

---

* For a review, see R. F. Curl and R. E. Smalley, 1991. Fullerenes. *Scientific American*, October 1991, 32.

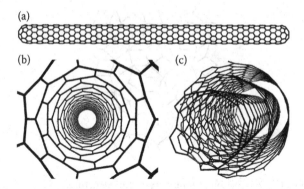

FIGURE 9.17
Some examples of carbon nanotubes. (a) A schematic view of a single-walled carbon nanotube, and (b) a view into its interior. (c) A view of a multiwalled carbon nanotube. (Diagrams courtesy of Chris Kingston.)

COMMENT: CARBON NANOTUBES

Another form of carbon that is attracting considerable attention can be considered as small graphite sheets folded into tubes. (Individual graphite sheets are yet another form of carbon, known as *graphene*.) Since the tubes are made of carbon, and their width is on the nanometer length scale, these are often referred to as carbon nanotubes (or CNTs). CNTs come in many different forms, including single-walled (SWCNT) and multiwalled (MWCNT), as shown in Figure 9.17.

Carbon nanotubes have exceptional properties. For example, they can be as stiff as diamond and more than 10 times stronger than Kevlar (see Chapter 14 for further discussion of mechanical properties). As we shall see in Chapter 12, carbon nanotubes also can have unusual electrical properties, ranging from semiconductors to metals, depending on the exact folding of the tubes.

COMMENT: UNUSUAL OPTICAL PROPERTIES OF C_{60}

C_{60} has an unusual optical property in that, when more intense light shines on it, less light is transmitted. Furthermore, C_{60} becomes a better "light limiter" at longer wavelengths, because the absorption cross sections of the ground state and the excited state both change with wavelength, but in opposite directions.

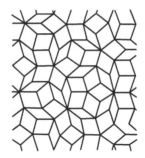

FIGURE 9.18
A two-dimensional representation of a quasicrystalline structure.

COMMENT: QUASICRYSTALS

The 14 crystal types shown in Chapter 1 are the smallest cells that can be regularly and repeatedly translated to produce periodic macroscopic crystalline structures. By contrast, glassy (or amorphous) materials do not have periodic atomic arrangements. For many years it was thought that these were the only possibilities for arrangements of atoms in solids— fully periodic or fully disordered. However, materials such as alloys of aluminum and manganese have been found which have quasiperiodic structures, and they are called *quasicrystals*. These structures are similar to quasiperiodic tiling patterns worked out by the British mathematician Sir Roger Penrose[*]; an example of a Penrose tiling pattern is shown in Figure 9.18. The main feature of the packing arrangement is that all space is filled, yet the arrangement is never exactly repeated.

---

[*] Quasicrystals were discovered by Dan Shechtman (1941–; Technion—Israel Institute of Technology), and for his discovery he was awarded the 2011 Nobel Prize in Chemistry.

9.5 The Phase Rule

There is a very useful rule, called the *phase rule* (also known as Gibbs' phase rule), that can be used to generalize the relationships between the number of free variables specifying the state of a system, the number of chemical components present and the number of phases present. The phase rule can provide very useful information concerning the phases present in a system.

Before deriving the phase rule, it is necessary to introduce a concept that is as important for many-component species as Gibbs energy is for single-component (pure) systems. This concept is *chemical potential*, where the chemical potential for component i, μ_i, is defined as

$$\mu_i = \left(\frac{\partial G}{\partial n_i} \right)_{T, p, n_{j \neq i}}, \qquad (9.13)$$

where the inequality in the constant-variable subscript indicates that the number of moles* of all components except component i are held constant. If the system has only component i (i.e., it is pure species i), then $\mu_i = G_m$ (i.e., the molar Gibbs energy is the same as the chemical potential for a pure material).

The reason for introducing chemical potential is that it is the multicomponent analogue to G. Just as G_m is equal for all phases at equilibrium, the chemical potential of a given species is the same in all equilibrated phases in which it is present. For example, if phase α is in equilibrium with phase β and component i is present in both phases, then

$$\mu_i^\alpha = \mu_i^\beta. \qquad (9.14)$$

We use the symbol F to represent the *number of degrees of freedom*[†] of a system, and it is the aim of the phase rule to determine F, the number of independent variables needed to specify the system. In other words,

$$F = \text{number of variables} - \text{number of relations}. \qquad (9.15)$$

The variables needed to describe the system will involve the composition of each phase and possibly other variables (e.g., temperature and pressure), and we look at these in some detail below.

We use the symbol c to represent the number of *chemical components* in the system. This is the minimum number of chemical species needed to describe the composition of the system. Alternatively, c can be described as the number of independent chemical species. For example, in an aqueous sodium chloride solution, $c = 2$ whether it is considered that H_2O and NaCl are the components or whether the chemical species are considered to be H_2O, NaCl, Na^+, Cl^-, H^+, and OH^-. In the latter case, the number of chemical species is 6 but there are 4 independent relations: $[OH^-] = [H^+]$; $[Na^+] = [Cl^-]$; $H_2O \rightleftarrows H^+ + OH^-$; NaCl $\rightleftarrows Na^+ + Cl^-$. This leaves $6 - 4 = 2$ independent chemical species, that is, $c = 2$.

We let p represent the number of types of *phases* within the system. A phase is considered to be a homogeneous region separated from other homogeneous regions by a phase boundary (or surface of discontinuity). For example, an ice water solution with excess salt has $p = 3$. The discontinuities from one phase to the next can be considered to be the variation in density of the beaker's contents from the top to the bottom of the beaker.

---

* IUPAC recommends the term "amount of substance" in place of "number of moles."
† This parameter is distinct from the "degrees of freedom" used in Chapter 6 to designate the number of spatial and angular coordinates required to describe the motion and position of a molecule, but in both cases the term "degrees of freedom" is used to indicate the variables that are necessary to define the system.

So, using the definition of F (Equation 9.15), we can first consider the number of variables needed to describe the state of the system, and then consider the relations between the variables, to determine a useful relationship for F, that is, the phase rule.

We must know something about the compositions to know the state of the system. Within a given phase, we need to know $(c - 1)$ of the compositions if there are c components present. The value $(c - 1)$, rather than c, arises because the final composition, that is, the composition of the c-th component, can be derived from the other $(c - 1)$ compositions from

$$\sum_i X_i = 1 \tag{9.16}$$

where X_i is the *mole fraction of component i* in the mixture, defined as

$$X_i = \frac{n_i}{n_{tot}} \tag{9.17}$$

where n_i is the number of moles of component i and n_{tot} is the total number of moles in the system. For p phases, this gives a total of $p(c - 1)$ compositional variables that are required to specify the system.

Other variables might also be needed to specify the state of the system. For example, temperature and pressure can vary the state of the system, and, in some circumstances, electric field (e.g., cholesteric liquid crystals) and/or magnetic field can vary the state of the system. We let n represent these other variables; in our examples, often n will be 2 (temperature and pressure) or 1 (temperature or pressure), but we leave the value n unspecified for the general case. Combined with the composition variables, this leads to $n + p(c - 1)$ variables needed to specify the system. However, not all these variables are independent, and we now address this point.

There are chemical potential relations among the phases that serve to reduce the number of variables needed to specify the state of the system. For each component i in the p equilibrated phases, the chemical potentials are all the same, so we can write

$$\mu_i^1 = \mu_i^2 = \ldots = \mu_i^p, \tag{9.18}$$

and for the p phases there are $(p - 1)$ such relations for each component i. Therefore, for c components, there are $c(p - 1)$ such relations, each reducing the number of free variables necessary to specify the state of the system.

Putting this information together, considering the definition of F (Equation 9.15), leads to

$$F = n + p(c - 1) - c(p - 1), \tag{9.19}$$

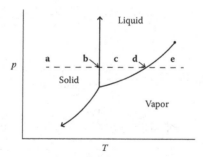

FIGURE 9.19

Phase diagram for a typical pure material, showing the effect of pressure and temperature on phase stability. The solid lines indicate stable phase boundaries. The dashed line indicates heating at constant pressure (see text).

which simplifies to

$$F = c - p + n \qquad (9.20)$$

where Equation 9.20 is the phase rule, first derived in 1875 by J. Willard Gibbs.

The phase rule can be used to describe the number of free variables in a system, and, since we have been discussing single-component phase diagrams, we begin there with the use of the phase rule.

From the typical one-component phase diagram of Figure 9.19, considering that here $c = 1$ and $n = 2$ (temperature and pressure can, in principle, vary), leads to $F = 2$ in the region where only solid exists. This value of F indicates that two variables (temperature and pressure) are needed to specify the state of the system. Another way to express this is that pressure and temperature can vary independently. Similarly, in the liquid region, $F = 2$ and temperature and pressure are independent of each other. In the vapor region, $F = 2$ also.

Along the liquid–solid coexistence line, $c = 1$, $p = 2$, so $F = 1 - 2 + 2 = 1$. This single degree of freedom shows that, although temperature or pressure can be varied along this line, they cannot be varied independently; a new temperature specifies (through the phase diagram or the Clapeyron equation) what the new pressure will be. Similarly, $F = 1$ along the solid–vapor line and along the liquid–vapor line. As long as we are on one of these lines, temperature and pressure are not independent of each other; there is only one degree of freedom.

At the triple point (the coexistence point of solid, liquid, and vapor), $c = 1$, $p = 3$, and $F = 1 - 3 + 2 = 0$. This is said to be an *invariant point,* since neither the temperature nor the pressure can be arbitrarily changed while remaining at the triple point. For this reason, triple points (e.g., $T = 273.16\,K$, $p = 610.6\,Pa$ for H_2O) can be used for calibration of thermometers.

The minimum number of degrees of freedom is 0, as in the invariant triple point, implying that for a situation where $n = 2$, the maximum number of

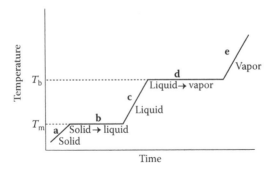

FIGURE 9.20
The warming curve (temperature–time profile) corresponding to isobaric heating of a pure material from solid to liquid to gas (i.e., along the dashed line of Figure 9.19).

phases in equilibrium in a one-component system is 3. (See Figure 9.4 for sulfur, where there are four phases but no more than three phases coexist at any given temperature and pressure.)

If we consider isobarically warming a sample of the material described by the phase diagram of Figure 9.19 along the dashed line shown in the figure, the phase rule can be used to describe the heating curve (temperature as a function of time). As this is a constant pressure situation, $n = 1$. Therefore, $F = 2 - p$ here. In region **a**, where there is one phase (solid), $F = 2 - 1 = 1$, and the temperature can increase. At point **b**, $p = 2$ (solid and liquid), so $F = 0$, and the temperature must be invariant throughout the time it takes to pass through **b**. In **c**, $p = 1$, $F = 1$, so the temperature increases again. At point **d**, $p = 2$ (liquid and vapor), $F = 0$, so the temperature remains constant until the liquid is fully vaporized. In region **e**, $p = 1$ (vapor) and $F = 1$, so the temperature can increase again. The corresponding temperature–time profile is shown in Figure 9.20. The horizontal lines in such a warming curve are called *halts* or *arrest points*, and they are particularly informative, as they indicate regions where $F = 0$. The determination of such warming curves, as we shall see, especially for multicomponent phase diagrams, can be used to construct phase diagrams. We turn next to phase diagrams consisting of two components.

9.6 Liquid–Liquid Binary Phase Diagrams

In the world of materials, mixtures of materials can be somewhat more important than pure materials. The main reason is that mixing materials allows one to tailor-make properties, intermediate between those of the pure materials, or sometimes giving new properties altogether different from those of the pure materials. Although there are a lot of different types of pure materials (over 130,000,000 compounds catalogued in Chemical Abstracts, growing at more than 30,000 per week!), mixtures can be very important.

When a system consists of more than one chemical component, it is referred to as a *multicomponent system*. In the case of two chemical components, it is called a *binary system*. In our discussion of binary systems, we begin by considering a system prepared by mixing two liquids.

Generally speaking, liquids are considered to be *miscible* (fully soluble, e.g., two liquids with similar polarities, such as water and acetone), or *immiscible* (e.g., two liquids with very different polarities, such as CCl_4 and H_2O). In strict terms, at the lowest concentrations, every liquid will mix with every other liquid to some extent, but at higher concentrations, there may be a *miscibility gap* (i.e., a range of concentrations for which two liquid phases coexist).

The miscibility of two liquids, like the solubility of a solid in a liquid, can change as the temperature is changed. Liquids with a miscibility gap can either increase or decrease their miscibility as the temperature is increased.

If the miscibility increases with increasing temperature, this can lead to a temperature above which the two liquids are completely miscible in all proportions. Such as point is referred to as the *upper consolute temperature*, T_{uc}. An example is shown in Figure 9.21.

In contrast, some liquids decrease their miscibility as the temperature is increased, so there is a temperature beneath which the two liquids are miscible in all proportions. An example of such a *lower consolute temperature*, T_{lc}, is shown in Figure 9.22.

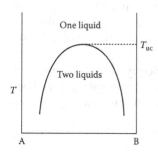

FIGURE 9.21
An example of a liquid–liquid system with increased solubility at increased temperature, showing an upper consolute temperature, T_{uc}, at a composition intermediate between pure A and pure B. An example of such a phase diagram is water–aniline, with $T_{uc} = 441$ K.

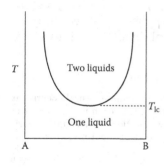

FIGURE 9.22
An example of a liquid–liquid system with increased solubility at decreased temperature, showing a lower consolute temperature, T_{lc}, at a composition intermediate between pure A and pure B. An example of such a phase diagram is water–triethylamine, with $T_{lc} = 291.6$ K.

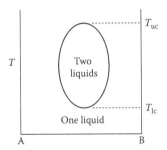

FIGURE 9.23

A liquid–liquid system showing both an upper consolute temperature and a lower consolute temperature, both at compositions intermediate between pure A and pure B. An example of such a phase diagram is water–nicotine, with $T_{uc} = 480$ K and $T_{lc} = 335$ K.

There are a few liquid–liquid systems that show both an upper consolute temperature and a lower consolute temperature. One is shown in Figure 9.23.

9.7 Liquid–Vapor Binary Phase Diagrams

If a liquid is heated sufficiently, it will eventually transform to the vapor phase; this is true for pure liquids and it is also true for binary liquid systems. However, the variation of boiling point with composition can reveal interesting features in a binary liquid system. A few of those features are summarized here because similar features can be seen in some solid–liquid phase diagrams, which are relevant for materials.

If the components of a binary system are sufficiently similar chemically, they will interact much the same with each other as they do with themselves. In this case, the system is said to be an *ideal solution* and p_A, the vapor pressure of component A, is given by *Raoult's law**:

$$p_A = X_A p_A^o, \tag{9.21}$$

where X_A is the mole fraction of component A and p_A^o is the vapor pressure of component A when pure. A similar equation applies to the second component (B), such that the total pressure over an ideal solution, p_{tot}, is given by

$$p_{tot} = X_A p_A^o + X_B p_B^o \tag{9.22}$$

---

* François Marie Raoult (1830–1901) was a leading French experimental physical chemist in the 19th century. He made major experimental contributions to electrochemistry and understanding of solutions, including provision of experimental verification of Arrhenius' theory that salts ionize in aqueous solution. His accurate measurements of vapor pressures led to what we now call Raoult's law.

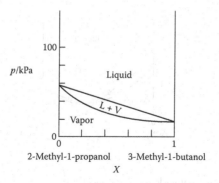

FIGURE 9.24
The pressure-composition diagram for 2-methyl-1-propanol with 3-methyl-1-butanol, $T = 323.1\,K$, which forms a nearly ideal solution. The liquid curve is nearly a straight line, as given by Equation 9.22.

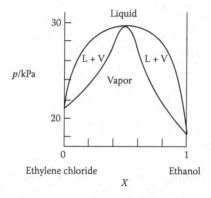

FIGURE 9.25
The liquid–vapor phase diagram for the system ethylene chloride–ethanol at $T = 313.1\,K$, showing pressures greater than ideal pressures and an azeotrope.

where X_B and p_B^0 are the mole fraction of component B and vapor pressure of pure component B, respectively. The resulting vapor pressure as a function of composition for a nearly ideal system is shown in Figure 9.24.

A nonideal binary solution results from the interactions between the component molecules (i.e., A–B interactions) being different from those between like molecules (i.e., A–A and B–B interactions).

If the A–B type interactions are more repulsive than the A–A and B–B interactions, then the vapor pressure of the solution will be higher than the ideal vapor pressure. An example is shown in Figure 9.25, where the increase in vapor pressure has given rise to a maximum in the vapor pressure plot. At the composition of this maximum, the liquid boils at constant temperature, just like a pure component. This composition is called an *azeotrope*, which means "constant boiling mixture."

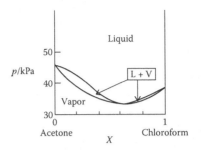

FIGURE 9.26
A liquid–vapor diagram showing a low vapor pressure azeotrope. The components are acetone and chloroform at $T = 308$ K.

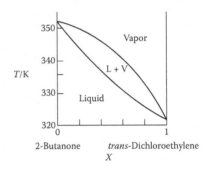

FIGURE 9.27
Temperature-composition boiling point diagram for 2-butanone with *trans*-dichloroethylene at $p = 101$ kPa. This system shows nearly ideal behavior.

On the other hand, if the A–B interactions in a binary solution are more attractive than the A–A or B–B interactions, then the vapor pressure will be lower than in the ideal system. An example is shown in Figure 9.26, where another azeotrope (now at low vapor pressure) occurs.

An ideal binary solution, by virtue of its vapor pressure changing monotonically with composition (Equation 9.22), will have a boiling point that is intermediate between the boiling points of the two components. An example is shown in Figure 9.27. Note that at any composition intermediate between the two pure components, there is a temperature range of coexistence between liquid and vapor.

Deviations from ideality that lead to a vapor pressure that is higher than ideal corresponds to boiling points that are lower than ideal. (A solution boils when the vapor pressure that it exerts equals the external pressure; if the vapor pressure is high, then a lower temperature will achieve boiling.) An example of a boiling point diagram for such a system is shown in Figure 9.28.

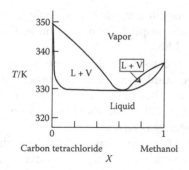

FIGURE 9.28
The boiling point diagram for carbon tetrachloride and methanol at $p = 101\,kPa$. This system exhibits deviations from ideality that lead to a low-boiling azeotrope. In distillation of a mixture, the *distillate* (i.e., the portions that distils) would have the azeotropic composition, making separation of the components by distillation impossible.

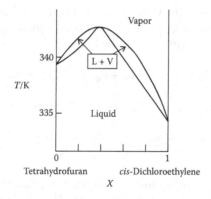

FIGURE 9.29
Boiling point diagram for the system tetrahydrofuran with *cis*-dichloroethylene at $p = 101\,kPa$. The high-boiling azeotrope means that in distillation of any composition other than the azeotropic composition, the azeotrope will remain behind in the still pot. The azeotropic composition would distil at a fixed temperature, much like a pure component.

If the deviations from ideality make the vapor pressure lower than the ideal vapor pressure, then there will be a maximum in the boiling point diagram. An example is shown in Figure 9.29.

9.8 Relative Proportions of Phases: The Lever Principle

Once we enter the realm of multicomponent systems, beginning with binary systems, the phase rule allows coexistence of two or more phases in fixed proportions at fixed temperature and pressure. In some applications, it may be important to know the proportions of the phases that coexist, and it is this matter that is addressed here.

For a binary system at constant temperature and pressure, the number of components is two, the number of important other variables is zero (i.e., $n = 0$), and the minimum number of degrees of freedom (F) is zero. The phase rule (Equation 9.20) shows that the maximum number of phases that can coexist in this circumstance is two.

To consider the relative proportions of these two phases, we can use any phase diagram. Here we make use of the phase diagram for the boiling point of an ideal solution (as in Figure 9.27, redrawn for this purpose as Figure 9.30). At an overall composition X_A' and temperature T^*, as shown in Figure 9.30, the composition of the liquid is given by X_A'' and that of the vapor is given by Y_A. (Note that X is the mole fraction in the liquid, and Y is the mole fraction in the vapor.)

With n_A representing the overall number of moles of component A in the system, n_{liq} representing the total number of moles of liquid (including both A and B), n_{vap} representing the total number of moles of vapor (including both A and B), and n_{tot} representing the total number of moles in the system,

$$n_A = X_A' n_{tot} = X_A' \left(n_{liq} + n_{vap} \right) = X_A'' n_{liq} + Y_A n_{vap}. \tag{9.23}$$

To determine the ratio n_{liq}/n_{vap}, Equation 9.23 can be rearranged

$$\frac{n_{liq}}{n_{vap}} = \frac{Y_A - X_A'}{X_A' - X_A''} = \frac{ab}{bc}, \tag{9.24}$$

where ab and bc refer to the lines in Figure 9.30. Equation 9.24 shows that the mole ratio of the phases present is proportional to the "arms" of a lever (called *tie lines*) with the lever fulcrum at the overall composition. This makes sense; the greater the proportion of a given phase, the shorter the

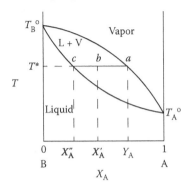

FIGURE 9.30
A boiling point diagram for an ideal solution, for purposes of illustrating the lever principle.

"arm" is to it. The proportionality of Equation 9.24 is called the *lever principle* (or lever rule).

Since no particular suppositions were made concerning the types of phases present, the lever principle applies to any two coexisting phases, regardless of the types of phases, and we will make use of it in the problems at the end of this chapter. An additional useful feature is that the same general form (ratio of tie line lengths) applies to a mass% diagram to give appropriate mass ratios.

9.9 Liquid–Solid Binary Phase Diagrams

By far the most important phases in materials science are condensed phases, including solids (crystalline or amorphous), liquids, and liquid crystals. We concentrate in this section on the phase diagrams of condensed phases of matter that arise when two components are mixed together.

In the case of a single-component (unary) phase diagram, the pressure–temperature profile could be represented in a two-dimensional diagram. When a second component is added, showing the pressure–temperature–composition profile would require a three-dimensional diagram. Although this can be done, it is more common to show a temperature–composition plot, at constant pressure, as laboratory conditions are most commonly isobaric.

The simplest case of mixing two solids would arise if the solids were so much alike that they dissolved in each other completely in the solid state, giving a *solid solution*, the solid analogue of a liquid solution. Examined microscopically, a solid solution is homogeneous. An example of the formation of a solid solution is given in the copper–nickel phase diagram (Figure 9.31). Although

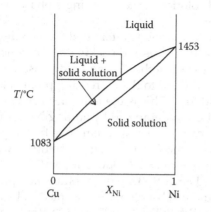

FIGURE 9.31
Melting point diagram for copper with nickel. Note that only at the pure metal compositions is the melting point at one fixed temperature; at all other compositions, there is a temperature range over which liquid and solid coexist.

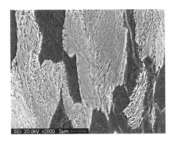

FIGURE 9.32
An image of a solid dispersion, called pearlite, formed by Fe with 0.8 mass% carbon. This structure is formed from bands of ferrite (<0.02 mass% C) and cementite (Fe_3C, with 6.7 mass% C). (See Comment: The Iron Phase Diagram.) (By Michelshock—McGill University, Public Domain, https://commons.wikimedia.org/w/index.php?curid=6464429). Bar is 200 μm.

copper melts at a lower temperature than nickel ($T_m(Cu) = 1083\,°C$; $T_m(Ni) = 1453\,°C$), both Cu and Ni form fcc solids. Since Cu and Ni are similar in size (the atomic radius is 0.128 nm for Cu and 0.125 nm for Ni), they mix completely in the solid state in all proportions. Note that the melting of a mixture of Cu and Ni begins at a temperature intermediate between the melting point of Cu and that of Ni, but an important difference from either pure Cu or pure Ni is that for the mixture, solid and liquid co-exist over a temperature range at all compositions intermediate between pure Cu and pure Ni.

Most pairs of solids (like some pairs of liquids) cannot dissolve in one another; in solids, this often is due to incompatibility of crystal structures. When two solids are immiscible, this leads to a *suspension* or *dispersion* of one solid in the other. On a microscopic scale, a dispersion is *inhomogeneous* (also called *heterogeneous*). A photograph of a dispersion (Figure 9.32) shows that particles of one solid exist in contact with particles of the other solid. In contrast with a solid solution, which is a single phase, a dispersion contains more than one phase.

A schematic example of a solid system that forms a dispersion is shown in Figure 9.33. The molecular sizes and packing of components A and B make it impossible for them to fit together in one solid phase, so they form a dispersion. Note that as A is added to B, the freezing point is lowered; the same is true when B is added to A. This is the familiar *freezing point depression*. (The freezing point of Cu is not depressed by adding Ni in the Cu–Ni phase diagram [Figure 9.31] because one basic premise of freezing point depression is that the pure solid is formed on freezing the solution, whereas in the Cu-Ni case a solid solution is formed.) At low temperatures, the solids formed in the phase diagram of Figure 9.33 are pure solid A and pure B; a dispersion of one in the other would be observed microscopically. Examples of real phase diagrams that are similar to Figure 9.33 include ethylene glycol/water (in its metastable phase diagram; ethylene glycol/water also forms a solid hydrate in its stable phase diagram), and several alloys (As/Au, As/Pb, Au/Si, and Bi/Cd all show almost completely immiscible solids).

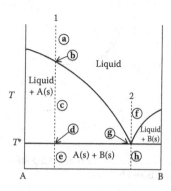

FIGURE 9.33

A freezing point diagram showing completely immiscible solid phases. The constant-composition slices shown correspond to the cooling curves shown in Figures 9.34 and 9.35. The implied label on the x-axis is X_B, ranging from 0 to 1.

Consideration of the numbers of degrees of freedom in a few constant-composition slices of the phase diagram in Figure 9.33 allows plotting temperature–time profiles, to show how cooling (or heating) curves can be used to deduce phase diagrams. From the general form of the phase rule (Equation 9.20) and the fact that $c = 2$ here and $n = 1$ (temperature can vary but pressure is fixed), we find $F = 3 - p$, where p is the number of phases. Slice 1 (Figure 9.33) on cooling gives the phases and degrees of freedom summarized in Table 9.1. The phase(s) in each region are determined by considering what regions an imaginary horizontal line (the tie line) touches. The change from two degrees of freedom to one at point **b** leads to a change in slope in the cooling curve; the cooling is not so fast because the precipitation of solid A is exothermic. This change in slope in a cooling curve is called a *break point*, and it is characteristic of a change in the number of degrees of freedom, with $F \neq 0$. At point **d**, $F = 0$ so temperature is invariant, and there is a *halt point* (also called *arrest*) in the cooling curve corresponding to this temperature. The corresponding cooling curve is shown in Figure 9.34.

TABLE 9.1

The Phases and Degrees of Freedom in the Various Regions of the Phase Diagram along Slice 1, Shown in Figure 9.33[a]

Region	Phase(s)	p (number of phases)	F
a	Liquid	1	2
b	Liquid + A(s)	2	1 (Break)
c	Liquid + A(s)	2	1
d	Liquid + A(s) + B(s)	3	0 (Halt)
e	A(s) + B(s)	2	1

[a] Pressure is 101 kPa; $F = 3 - p$ here, where p is the number of phases.

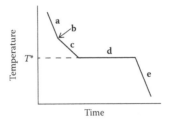

FIGURE 9.34

The cooling curve (temperature as a function of time) along Slice 1 of the phase diagram shown in Figure 9.33. Note the break at **b** and halt at **d**.

TABLE 9.2

The Phases and Degrees of Freedom in the Various Regions of the Phase Diagram along Slice 2, Shown in Figure 9.33[a]

Region	Phase(s)	p (number of phases)	F
f	Liquid	1	2
g	Liquid + A(s) + B(s)	3	0 (Halt)
h	A(s) + B(s)	2	1

[a] Pressure is 101 kPa; $F = 3 - p$ here, where p is the number of phases.

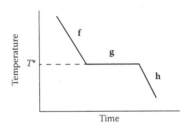

FIGURE 9.35

The cooling curve (temperature as a function of time) along Slice 2 of the phase diagram in Figure 9.33, at a constant pressure of 1 atm. Note the halt at **g**.

Similar consideration of Slice 2 of Figure 9.33 leads to the phases and degrees of freedom summarized in Table 9.2. The corresponding cooling curve is shown in Figure 9.35.

Although cooling curves might provide more reliable experimental information than warming curves (slow cooling from the melt will generally ensure equilibrium as long as there are sufficient nucleation sites, whereas the solid might contain metastable states that could interfere with the equilibrium warming curve), we can consider the effects of warming an equilibrated solid. On warming along the line of Slice 1, the solid (which is a dispersion of solid A in solid B) first begins to melt at temperature T^*. The composition of

TABLE 9.3

Compositions and Melting Points of Eutectic Solders

Melting Point/°C	Composition (mass%)				
	Bi	Cd	Pb	Sn	Hg
60	53.5	—	17	19	10.5
70	45.3	12.3	17.9	24.5	—
80	35.3	9.5	35.1	20.1	—
91.5	51.6	8.1	40.2	—	—
96	52.5	—	32	15.5	—
100	50	—	32.2	17.8	—
124	55.5	—	44.5	—	—
138	57	—	—	43	—
154	14	—	43	43	—
177	—	32	—	68	—
188	—	—	38	62	

the first liquid that appears, considering the tie line, will be the composition of point **g**. Note that this is not the same as the overall composition of the system. As the sample is heated further, the composition of the liquid will change, becoming richer in A as the temperature is increased. The liquid and solid continue to co-exist throughout the range from point **d** to point **b**, and above **b** only liquid exists.

Note that the point **g** on Slice 2 is a special point. At any temperature higher than this point, the system is completely liquid. At any temperature lower than this point, the system is completely solid. Therefore, on heating at this composition, the system melts to give a liquid of the same composition as the solid. This special composition is called the *eutectic composition*, which means that it melts at a single temperature without change in composition. The corresponding temperature ($T*$ in Figure 9.33) is called the *eutectic temperature*.

The existence of a eutectic can be useful in choosing materials with a particular melting point. For example, although both tin and lead melt too high to be useful as solders, a 26 atomic % Pb mixture of tin and lead (see Figure 9.36) can be useful as a solder, in that it melts at a fixed temperature (188 °C) without change in composition. Some useful low-melting eutectic alloys are listed in Table 9.3.

Just as liquids can show partial miscibility, so can some solids. The tin–lead phase diagram (Figure 9.36) shows *partial miscibility*. The two solids formed, S_1 and S_2, are *solid solutions* of variable composition, rich in tin and lead, respectively. The composition of S_1 or S_2 can be read from the phase diagram (e.g., S_2 is 93 atomic % lead at $T = 100$ °C).

Cooling of Slice A or Slice B (Figure 9.36) of the Sn–Pb phase diagram will lead to cooling curves as in the phase diagram of Figure 9.33, but Slice C will

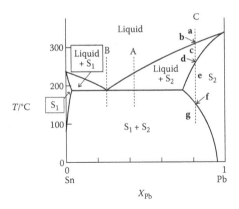

FIGURE 9.36

Melting point diagram of tin with lead. Note the eutectic at 26 atomic % Pb, which corresponds to tin-lead solder, melting at 188 °C. S_1, sometimes expressed as (Sn), is solid rich in tin; S_2, also written as (Pb), is solid rich in lead.

TABLE 9.4

Degrees of Freedom in Slice C of the Tin–Lead Phase Diagram (Figure 9.36)[a]

Region	Phase(s)	p (number of phases)	F
a	Liquid	1	2
b	S_2 + liquid	2	1 (Break)
c	S_2 + liquid	2	1
d	S_2 + liquid	2	1 (Break)
e	S_2	1	2
f	$S_1 + S_2$	2	1 (Break)
g	$S_1 + S_2$	2	1

[a] $F = 3 - p$ here, where p is the number of phases.

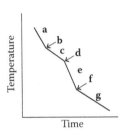

FIGURE 9.37

Cooling curve (temperature as a function of time) for Slice C of the tin-lead phase diagram (Figure 9.36). Note the breaks at **b**, **d**, and **f**. At **b** it begins to cool more slowly because of the exothermicity of precipitation of solid S_2.

be slightly different. Again, $F = 3 - p$ here, which gives the degrees of freedom listed in Table 9.4. The corresponding cooling curve is shown in Figure 9.37. Note that in this slice there is no invariant point ($F = 0$ point) and there is no halt in the cooling curve.

COMMENT: THE IRON PHASE DIAGRAM

The stable form of pure iron at room temperature has a body-centered cubic (bcc) structure and it is called α-*iron*. Below 760 °C, it is *ferromagnetic* (i.e., permanently magnetized by an applied magnetic field; see Chapter 13); above 760 °C, there is too much thermal energy for ferromagnetism to be maintained, but the structure remains bcc until 906 °C when it transforms to a fcc structure (γ-*iron*). At 1401 °C, pure iron transforms to another bcc phase, δ-*iron*, which is stable up to the melting point of 1530 °C.

Additions of small amounts (<2 mass%) of carbon to iron leads to the formation of *steel*. Higher concentrations of carbon give *cast iron*. For this reason, the Fe–C phase diagram, a portion of which is shown in Figure 9.38, is among the most important phase diagrams known. As can be seen from the diagram, below 2 mass% C, heating can lead to a solid solution (γ-iron, which can be hot-rolled, pressed, drawn, or otherwise shaped), but above 2 mass% C (i.e., cast iron) heating does not achieve a malleable solid state and therefore shaping is done from the melt by casting.

The Fe–C phase diagram shows a number of features, including the eutectic at about 1150 °C and 4.2 mass% C (liquid stable at higher temperatures) and the *eutectoid* (this is like a eutectic but with a solid solution stable at higher temperatures) at 723 °C and about 1 mass% C. The formation of the compound Fe_3C plays an important role in the mechanical properties of steel and cast iron, and its formation can be controlled by heat treatment. The presence of Fe_3C leads to brittle iron that can be cast into its final form. However, Fe_3C is metastable with respect to graphite, and in some treatments, the presence of graphite allows the production of malleable irons that can be worked into shape.

Figure 9.38 shows that α-iron dissolves only a little carbon (0.05 mass% carbon), whereas γ-iron dissolves up to 2 mass% carbon. This difference is because there is more room for carbon at *interstitial octahedral sites* (i.e., spaces within the lattice that have octahedrally placed neighbors) in the fcc structure of γ-iron than in the *tetrahedral interstitial sites* (spaces in the lattice with tetrahedral neighbors) of bcc α-iron.

The solid solution of carbon in γ-iron is called *austentite*; that for α-iron is called α-ferrite. If austentite is cooled slowly below the eutectoid at 723 °C, it phase separates to α-ferrite and Fe_3C (*cementite*); this phase separation gives a heterogeneous material called *pearlite* (see Figure 9.32).

If austentite is cooled rapidly, it forms a metastable structure called *martensite* in which carbon atoms are at the interstitial sites of a bcc lattice, but the crystal structure is tetragonal (expanded along one axis but with 90° angles retained). Martensite is the hardest component in quenched (rapidly cooled) steels because the tetragonal structure of the crystallites does not pack well in the cubic arrangements of their neighboring materials. This is one way in which steel is hardened by heat treatment.

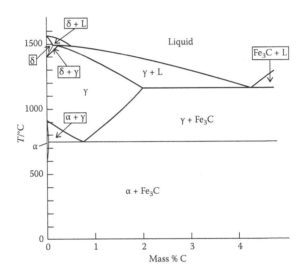

FIGURE 9.38
A portion of the Fe–C phase diagram at atmospheric pressure, showing the Fe-rich region.

9.10 Compound Formation

So far, in considering binary systems, we have only taken into account physical interactions, not stronger interactions. A strong interaction between the two components can lead to compound formation, and, in terms of their phase diagrams, compounds can be distinguished as congruently melting or incongruently melting.

A *congruently melting compound* is one in which the liquid formed on melting has the same composition as the solid compound. A generic example is shown in Figure 9.39, where AB is the 1:1 compound of components A and B. This diagram can be considered to be the sum of two other simple phase diagrams—the AB/B diagram (right side of Figure 9.39) and the A/AB

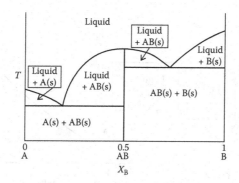

FIGURE 9.39
The melting point diagram in the binary A-B system, at constant pressure. The compound AB melts congruently.

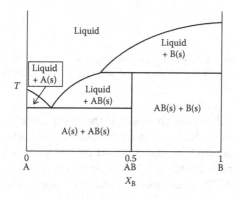

FIGURE 9.40
Temperature-composition phase diagram for the system A-B at constant pressure, showing the incongruently melting compound AB. When AB(s) is heated, it melts first to give liquid and B(s), and then, at a higher temperature, gives liquid only.

diagram (left side of Figure 9.39).* When looked at this way, the features are not new. The compound AB melts to give a liquid of composition 50 mol% A, 50 mol% B, the same as the composition of the solid.

An *incongruently melting compound* gives, on melting, a liquid and a solid both with different compositions from the compound. This situation is shown schematically for compound AB in Figure 9.40. An example of an incongruently melting compound is KNa_2, which decomposes on melting to give a solid rich in Na and a liquid that is depleted in Na relative to KNa_2. The Na–K phase diagram is shown in Figure 9.41.

---

* The only difference from two side-by-side phase diagrams is that the slope of the melting point around the composition of AB is zero, whereas it is nonzero for the pure components, A and B. The slope is nonzero only when the liquid is the same chemical species as the solid, and A melts to give liquid A, and B melts to give liquid B, whereas in the binary diagram shown above, AB melts to give A(liq) miscible with B(liq), not AB(liq).

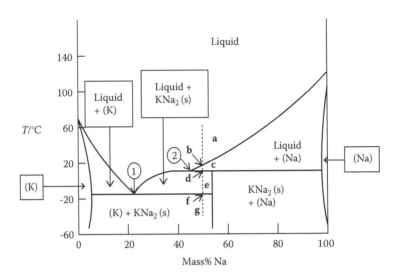

FIGURE 9.41
The K–Na temperature–composition phase diagram at constant pressure. The compound KNa$_2$ melts incongruently at 7°C. (K) is a solid solution rich in potassium and (Na) is a solid solution rich in sodium.

The K–Na phase diagram shows that the composition at 54.0 mass% Na (which corresponds to KNa$_2$) begins to melt at 7°C, giving solid solution rich in Na, designated (Na), and a liquid depleted in Na relative to KNa$_2$. As the temperature is increased, the proportion of liquid (which contains K and Na) and solid (Na-rich solid solution) changes, yielding more liquid as the temperature goes up. This liquid also becomes richer in Na as the temperature is increased, because it is forming from (Na). Eventually all the solid will become liquid. A major distinguishing feature between congruent and incongruent melting is that for congruent melting the vertical line at the compound composition directly reaches the liquid region; for incongruent melting, the vertical line at the compound composition stops short of the liquid region, usually at a horizontal line in the phase diagram.

The K–Na phase diagram shows a eutectic (point ① in Figure 9.41) and also another special point (point ② in Figure 9.41), called a *peritectic point*. The peritectic point corresponds to the liquid composition that is formed at the temperature of melting of an incongruently melting compound. This temperature is referred to as the *peritectic temperature*.

Cooling at constant composition along the dashed line in Figure 9.41 leads to the phases and degrees of freedom (from $F = 3 - p$), as noted in Table 9.5. The corresponding cooling curve is shown in Figure 9.42. The lower-temperature halt, from the phase diagram, is a *eutectic halt*, while the higher temperature halt is a *peritectic halt*. It is not possible to distinguish whether a halt is eutectic or peritectic (or something else that gives $F = 0$) from one cooling diagram alone; several cooling curves are necessary to

TABLE 9.5

Phases and Degrees of Freedom in the K–Na System[a]

Region	Phase(s)	p (Number of phases)	F
a	Liquid	1	2
b	Liquid + (Na)[b]	2	1 (Break)
c	Liquid + (Na)[b]	2	1
d	Liquid + (Na)[b] + KNa_2(s)	3	0 (Peritectic halt)
e	Liquid + KNa_2(s)	2	1
f	Liquid + KNa_2(s) + (K)[b]	3	0 (Eutectic halt)
g	(K)[b] + KNa_2(s)	2	1

[a] Regions correspond to those labeled in Figure 9.41; $F = 3 - p$ here.
[b] (Na) indicates a solid solution rich in Na but containing some K, and (K) indicates a solid solution rich in K but containing some Na.

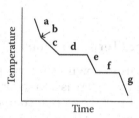

FIGURE 9.42
The cooling curve (temperature as a function of time) for cooling at constant composition in the K–Na phase diagram shown in Figure 9.41. Note the break at **b** going to a slower cooling rate due to precipitation of (Na), the peritectic halt at **d** and the eutectic halt at **f**.

determine the phase diagram. Indeed, this is often how phase diagrams are determined experimentally.

It can be important to know whether a compound melts congruently or incongruently, especially if there is some probability that the compound will melt in an application. Incongruent melting can be particularly worrisome as the original material might not be recoverable from solidification of the corresponding liquid, since cooling this liquid will first form another solid ((Na) in the K/Na case), which might not be able to rearrange itself on further cooling to provide the desired equilibrium compound (KNa_2, in the K/Na case).

The main factor determining whether a compound melts congruently or incongruently is the stability of the liquid phase relative to the pure solid components at the temperature at which there is too much thermal energy for the lattice to hold together (i.e., at the melting point). If it would lower the molar Gibbs energy to produce one of the solid components rather than just liquid, melting will be incongruent (to give the liquid and the more stable solid). Of course, the phase stability could change with a change in another parameter such as pressure. This situation is shown schematically in Figure 9.43.

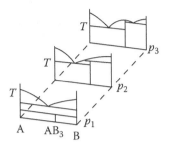

FIGURE 9.43
Various isobaric temperature-composition diagrams at different pressures. The effect of increasing pressure is to increase the stability of compound A_3B, such that it transforms to A(s) and B(s) at low pressure (p_1); it melts incongruently at intermediate pressure (p_2); it melts congruently at high pressure (p_3).

9.11 Three-Component (Ternary) Phase Diagrams

In the single component (also known as unary) phase diagrams, we were able to plot the phase stability region as a function of temperature and pressure, in a two-dimensional plot. For a two-component (binary) phase diagram, in order to add the additional dimension of composition, the phase diagram was either three-dimensional (p, T, composition; see Figure 9.43) or two-dimensional (p and composition at constant temperature as in Figure 9.25, or T and composition at constant pressure as in Figure 9.28).

For three components, there are two independent composition variables (X_1 and X_2 where the third, X_3, the mole fraction of component 3, is given by the relation $X_1 + X_2 + X_3 = 1$) as well as pressure and temperature to consider. Although it is possible to include a diagram that shows composition and temperature or composition and pressure, it is most common in three-component systems to show composition diagrams at constant temperature and constant pressure. The usual way to do this is graphing using triangular coordinates, as shown in Figure 9.44.

In a triangular coordinate system, each vertex in the graph represents a pure component, A, B, or C, as shown in Figure 9.44. The advantage of triangular coordinates is that the three components are considered in a symmetric way. The side AB is as far as possible from C, so it contains no C, and has varying A–B compositions from 100% A at point A, to 100% B at point B. Similarly, the side AC represents the A–C binary phase diagram (no B), and the side B–C represents the binary B–C phase diagram (no A).

The interior of the triangle presents intermediate compositions containing A, B, and C. For example, the composition 10% A is found along the "10" line parallel to side BC. (These lines start at 0% A on BC and work up to 100% A at point A.) If the B composition is 20%, this is on the 20% line parallel to

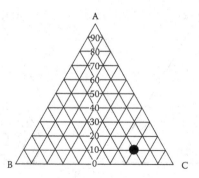

FIGURE 9.44

Triangular coordinates, to represent the constant-temperature, constant-pressure phase diagram of the generic A-B-C ternary system. Each vertex represents the pure component (A, B, or C, as labeled), and each side presents a binary phase diagram (A-B, opposite vertex C; A-C, opposite vertex B; B-C, opposite vertex A). The interior of the triangle represents nonzero compositions of all three components. The numbers shown correspond to concentration of A in percent (mass or mole). The point marked ● represents 10% A, 20% B, 70% C. Note that a ternary diagram can represent mass% or mole%.

FIGURE 9.45

Ethanol–methanol–water phase diagram at $T = 25\,°C$, $p = 101\,kPa$. All three liquid components are miscible in all proportions.

side AC. The composition 10% A, 20% B is at the point where the 10% A line intersects the 20% B line, as highlighted in Figure 9.44. This meeting point also crosses the 70% C line, as it must since %C = 100% – %A – %B = 70% in this case. The additivity of the overall composition to 100% serves as a useful check in plotting points on ternary graphs.

A simple example of a ternary phase diagram at constant pressure and temperature is the ethanol–methanol–water phase diagram. At $T = 25\,°C$ and $p = 101\,kPa$, ethanol, methanol, and water are soluble in one another in all proportions, so the entire interior of the triangle represents same phase, one miscible liquid. This phase diagram is shown in Figure 9.45.

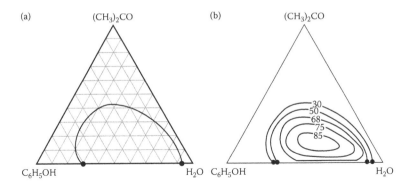

FIGURE 9.46
The ternary phase diagram for phenol (C_6H_5OH)—water (H_2O)—acetone ((CH_3)$_2CO$): (a) at room temperature and (b) as a function of temperature. The heavy points on the $C_6H_5OH-H_2O$ axis indicate the limits of solubility, and at room temperature, there are two liquid phases present at compositions poor in (CH_3)$_2CO$ and intermediate in C_6H_5OH and H_2O. From (b) we can see that by 68 °C, C_6H_5OH and H_2O are soluble in all proportions, but the addition of (CH_3)$_2CO$ still leads to two liquid phases. Above about 90 °C, the upper consolute temperature, there is only one liquid phase, independent of the proportions.

When water, acetone, and phenol are mixed at room temperature and pressure, the situation is not quite so simple. Water and acetone are miscible in all proportions, so there are no phase boundaries on the water–acetone axis. Similarly, acetone and phenol are miscible in all proportions, so there are no phase boundaries on the acetone–phenol axis. However, water and phenol exhibit only limited miscibility. When only a little phenol is added to water, the two mix, but beyond a certain amount the solution separates into two phases, one rich in water and the other rich in phenol. Similarly, if a little water is added to phenol, it will dissolve, but if more water is added so that the miscibility is exceeded, the solution will separate into two layers, one rich in phenol and the other rich in water. (Note that neither layer is pure as the other component will still dissolve to some extent.) The corresponding water–phenol–acetone phase diagram is shown in Figure 9.46, where the temperature–dependence (showing an upper consolute temperature) is included.

The 25 °C water–K_2CO_3–methanol phase diagram is shown in Figure 9.47. If sufficient K_2CO_3 is added to water, past the solubility limit, the result would be a saturated solution with excess K_2CO_3 as the hydrate $K_2CO_3 \cdot 3/2H_2O$ (solid). Similarly, K_2CO_3 has only limited solubility in methanol. These solubility limits are represented by the positions at which the phase boundary lines cross the $H_2O–K_2CO_3$ and $CH_3OH–K_2CO_3$ axes. At regions close to the $H_2O–CH_3OH$ axis, there will be one solution, since these liquids are soluble in each other in all proportions when there is nothing else present. However, starting with a 1:1 (by mass) mixture of H_2O and CH_3OH and adding a salt like K_2CO_3 can have a rather startling effect: Addition of sufficient salt causes the formation of two separable

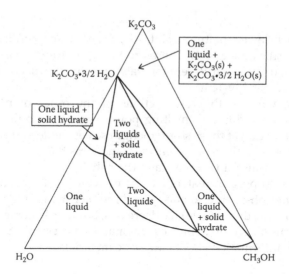

FIGURE 9.47
The H_2O–K_2CO_3–CH_3OH ternary phase diagram at $T = 25\,°C$ and $p = 101$ kPa.

liquids. One of these liquids is rich in water, and the other is rich in methanol; they separate because the salt (K_2CO_3) ties up water in ion hydration spheres, so there is less water available to dissolve the methanol. Furthermore, because the

See the videos "Ternary Phase Diagram: Simple Case" and "Ternary Phase Diagram: Partial Miscibility" under Student Resources at PhysicalPropertiesOfMaterials.com

ionic strength of the aqueous solution is now higher, methanol is no longer miscible. This process is used to cause separation of otherwise miscible solvents by the addition of a salt, and it is termed *salting out*. In the phase diagram it is represented by a region in which two immiscible liquid phases coexist. At higher K_2CO_3 concentrations (i.e., closer to the K_2CO_3 vertex), there are two liquids and excess K_2CO_3 is in the hydrate form, $K_2CO_3{\cdot}3/2H_2O$ (solid).

COMMENT: SIZE EFFECTS AND THERMAL STABILITY

The discussions of phase stability in this chapter have focused on bulk materials (i.e., materials with [at least] millions of atoms). However, thermal stability can be quite different when the number of atoms becomes small or the system is confined in certain ways.

As an example of the latter, consider the melting of water. Pure, bulk water melts at 273 K. However, water in confined environments can have a much lower melting point, e.g., 214 K, when confined in the channels of silicate MCM-41. Therefore, the equilibrium phase diagram of water (Figure 9.5) is different for water in confined spaces, including water in biological systems.

Similarly, wires with cross sections of the order of nanometers (nanowires) have different melting points from their bulk materials. The ease of motion at these shorter length scales can lead to significant diffusion between the components in nanowires that are composed of different materials for their core and their shell.

TiO_2 is an important material with many applications and at least three polymorphs: rutile, anatase, and brookite. While rutile is the stable bulk form, as the particle size decreases, brookite becomes most stable, and then, at very small particles, anatase is most stable. See Chapter 10 for more information about nanoscale materials.

9.12 A Tongue-in-Cheek Phase Diagram: A Tutorial

The phase diagram of Figure 9.48 has been proposed as an imaginative answer to the age-old problem "How do they get the caramel in the Caramilk™ bar?"

The idea is that the appropriate composition of chocolate and caramel (about 50–50) is heated to a processing temperature, as noted on the diagram, and then cooled to room temperature (see Figure 9.48). This would cause phase separation, into solid chocolate and a liquid phase rich in caramel.

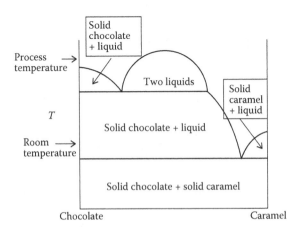

FIGURE 9.48
Proposed chocolate–caramel binary phase diagram.

a. Does this proposed phase diagram seem to answer the question? Why or why not?

b. Speculate on how one could successfully create a liquid caramel phase inside a solid chocolate structure.

9.13 Applications of Supercritical Fluids: A Tutorial

Supercritical fluids are fluids at temperatures and pressures beyond the critical conditions. In this region, liquid and gas are no longer distinguishable.

a. Describe some expected physical properties of a supercritical fluid, such as density (compared with ordinary gas) and viscosity (compared with the liquid).

b. How could the critical point (temperature, pressure, molar volume) for a given compound be determined experimentally?

c. Would you expect the solubility of a solute in a supercritical fluid to be the same as in the same solvent in the liquid state? How could this be used to advantage? As an example, consider the extraction of caffeine from green coffee beans using supercritical CO_2 ($p_c = 7353\,kPa$, $T_c = 304.2\,K$), reducing the caffeine in the beans from its initial range of 1%–3% to as low as 0.02% without removing other flavor and aroma components. Following caffeine extraction, the CO_2 solvent is removed to be used again and the caffeine can be used for addition to other products. This process is now used commercially to decaffeinate millions of kilograms of coffee beans annually.

d. Some mixtures of high boiling materials cannot easily be separated by fractional distillation. Can supercritical fluids be used for a suitable extraction procedure? Consider that supercritical fluids can be used as the carrier in chromatographic separations.

e. In some cases, such as SiO_2 dissolved in water, the solubility is very low, only a few % even when the water is in its supercritical phase ($p_c = 22.14\,MPa$, $T_c = 647.3\,K$). However, the viscosity of the solution changes considerably at the critical temperature: for SiO_2 in H_2O, it is about $10^{-2}\,P$ (poise) at room temperature and near $10^{-4}\,P$ in the supercritical region. This change in viscosity is very important in the growth of SiO_2 crystals from supercritical water.

   i. Why can SiO_2 not be grown as single crystals from the melt? (See problem 9.9.)

   ii. Why is the viscosity an important consideration in growth of rather insoluble materials from solution? Consider that large

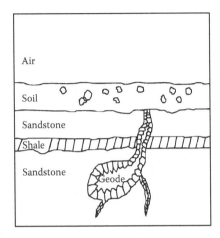

FIGURE 9.49
In certain geological conditions, pockets of supercritical aqueous solution allow the growth of quartz crystals in geodes.

single crystals of SiO_2 are required (e.g., for the electronics industry), and nonequilibrium growth of crystals can lead to dendrites rather than large single crystals.

iii. What laboratory conditions would be necessary to grow SiO_2 crystals by *hydrothermal crystal growth* methods (i.e., from supercritical water)?

iv. Can these conditions be achieved in nature? Consider the formation of a geode when a hydrothermal solution saturated with silicon dioxide fills a cavity in a rock matrix, as shown schematically in Figure 9.49.

9.14 Learning Goals

- Origins of condensation of gases
- Phase equilibria in pure materials
- Clapeyron equation
- Polymorphism
- Allotropes
- Orientationally disordered phases
- Fullerenes
- Carbon nanotubes
- Quasicrystals

- Phase rule; cooling and warming curves
- Liquid–liquid binary phase diagrams, including miscibility
- Liquid–vapor binary phase diagrams, including ideality and nonideality
- The lever principle
- Liquid–solid binary phase diagrams, including solid solutions, immiscibility, eutectic, partial miscibility, compound formation (congruent and incongruent melting), peritectic
- Fe–C phase diagram (steel and cast iron)
- Ternary phase diagrams
- Supercritical fluids
- How to draw phase diagrams from cooling (or warming curves) and vice versa

9.15 Problems

9.1 At $20\,°C$, the vapor pressure of benzene is $13.3\,kPa$ and that of octane is $2.7\,kPa$. If $1.00\,mol$ of octane is dissolved in $4.00\,mol$ of benzene and the resulting solution is ideal, calculate:

   a. the total vapor pressure;

   b. the composition of the vapor;

   c. the new vapor composition if the vapor is collected, re-condensed and then allowed to re-equilibrate.

9.2 A sample of methane is known to be contaminated with some $N_2(g)$. To determine the extent of the contamination, the sample is cooled in liquid nitrogen to the temperature of the normal boiling point of N_2 (i.e., its boiling point at $p = 101\,kPa$). At this temperature, the methane is fully condensed and can be considered to have negligible vapor pressure. The observed pressure is found to be $1.5\,kPa$. What is the purity of the original sample (in mol% methane)?

9.3 CBr_4 forms four solid phases. Phases II and III coexist at $T = 350\,K$, $p = 0.1\,GPa$, and also at $T = 375\,K$, $p = 0.2\,GPa$, with phase II stable at lower temperatures. Which phase is denser: II or III? Explain, using the Clapeyron equation.

9.4 Glycerol and m-toluidine form a binary phase diagram that has an upper and lower consolute temperature. Mixtures of various mass% m-toluidine (W) were found on warming to separate into two phases at T_1 (i.e., become turbid) and lose their turbidity at T_2 as follows:

W	$T_1/°C$	$T_2/°C$
18	48	53
20	18	90
40	8	120
60	10	118
80	19	83
85	25	53

a. Plot the phase diagram (T versus mass% m-toluidine) and find the upper and lower consolute temperatures.

b. Use the phase diagram to explain what happens when m-toluidine is added dropwise to glycerol at 60 °C.

c. At 40 mass% m-toluidine at $T = 70\,°C$, what phases exist and in what proportion (by mass)?

9.5 From the Clapeyron equation, derive the *Clausius*–Clapeyron equation*:

$$\ln\left(\frac{p_2}{p_1}\right) = \left(\frac{\Delta_{vap}H}{R}\right)\left(\frac{1}{T_1} - \frac{1}{T_2}\right). \tag{9.25}$$

Clearly state and justify all assumptions made in the derivation. The Clausius–Clapeyron equation is useful in determining the variation of vapor pressure with temperature.

9.6 The following data were obtained by cooling solutions of Mg and Ni.

X_{Ni}	0.00	0.044	0.139	0.202	0.383	0.669	0.752	1.00
$T_{break}/°C$	—	608	—	—	1050	—	—	—
$T_{break}/°C$	651	510	510	510&770	770	1180	1080	1450

Two compounds form in this system; they are Mg_2Ni and $MgNi_2$. Plot and label the phase diagram for the Mg–Ni system.

9.7 In the thermal decomposition of $BaCO_3(s)$, which results in the formation of $BaO(s)$ and $CO_2(g)$, it is found that the equilibrium pressure is a function of temperature only. (This applies whether one starts with $BaCO_3$ alone, or two of $BaCO_3$, BaO and/or CO_2, or all three, i.e., $BaCO_3$, BaO and CO_2.) Does the equilibrium system consist of pure crystals of $BaCO_3$ and pure crystals of BaO or does it contain a solid solution of these compounds? Only one of these alternatives is allowed thermodynamically. Explain. Hint: Consider the phase rule.

---

* Rudolph Julius Emmanuel Clausius (1822–1888) was born in Prussia and was a professor of mathematical physics in Switzerland and Germany. Clausius made many contributions to thermodynamics, kinetic theory (he introducèd the concept of mean free path), and electricity and magnetism.

9.8 The following information was obtained from a set of cooling curves for Sb–Cd:

X_{Cd}	Break/°C	Halt/°C
0	—	630
0.188	550	410
0.356	461	410
0.412	—	410
0.480	419	410
0.600	—	439
0.683	400	295
0.925	—	295
1	—	321

Sketch the T versus X_{Cd} phase diagram. Label all areas.

9.9 The binary phase diagram for SiO_2 (quartz) and Al_2O_3 shows several interesting features. Pure SiO_2 crystallizes from the melt at 1735 °C as *cristobalite*. At 1475 °C, cristobalite undergoes a first-order phase transition to another quartz polymorph characterized by its small (often triplet) crystals, *tridymite*. Aluminum oxide crystallizes at 2020 °C to produce *corundum*, a very hard mineral often used as an abrasive. In the SiO_2/Al_2O_3 phase diagram, at 1540 °C there is a eutectic. There is one compound in the binary phase diagram, and this is *mullite*, a rare clay mineral of disputed empirical formula, most likely Al_6SiO_{11}. Mullite melts incongruently at 1805 °C. Sketch and label the SiO_2/Al_2O_3 binary phase diagram (with the abscissa as mass% Al_2O_3). Al_6SiO_{11} is 83.5 mass% in Al_2O_3.

9.10 Cooling curves for liquid mixtures of magnesium and zinc give the following results:

Mass% Zn	Freezing Start/°C	Freezing End/°C
0	651	651
10	623	346
20	580	346
30	520	346
40	443	346
50	360	346
60	437	346
70	520	346
80	577	346
84.3	595	595
90	557	368
95	456	368
97.5	379	368
100	419	419

   a. Draw and label the phase diagram, temperature versus composition (mass% Zn) as accurately as possible.
   b. If 100 g of a liquid mixture containing 75.0 mass% Zn is equilibrated at 460 °C, what phase(s) will be formed and in what amounts?

9.11 The following data for the ternary system $Li_2SO_4/(NH_4)_2SO_4/H_2O$ show the mass% composition of the liquid phase in equilibrium with solids as indicated, at 30 °C.

Liquid Phase		Solid Phase(s)
% Li_2SO_4	% $(NH_4)_2SO_4$	
25.1	0.0	$Li_2SO_4 \cdot H_2O$
23.7	6.7	$Li_2SO_4 \cdot H_2O$
23.0	9.4	$Li_2SO_4 \cdot H_2O$
21.9	12.4	$Li_2SO_4 \cdot H_2O + NH_4LiSO_4$
16.6	19.3	NH_4LiSO_4
10.8	28.7	NH_4LiSO_4
7.8	35.6	NH_4LiSO_4
6.5	39.5	$NH_4LiSO_4 + (NH_4)_2SO_4$
2.9	41.4	$(NH_4)_2SO_4$
0.0	44.1	$(NH_4)_2SO_4$

   a. Draw the phase diagram using triangle coordinates. Label each area.
   b. Consider a binary mixture of Li_2SO_4 and $(NH_4)_2SO_4$ containing 80.0 mass% Li_2SO_4. If this mixture is dissolved in water and then the resulting solution is evaporated to yield a ternary mixture containing 25% water by mass, what phase(s) will be present in the remaining mixture? Indicate this mixture on your phase diagram with ⋆. (Hint: As long as the $Li_2SO_4/(NH_4)_2SO_4$ ratio is the same when the water is added, at all proportions of water, this solution will fall on a straight line going from the original mixture to the water vertex.)

9.12 In a publication[*] concerning work on the high-temperature superconductor $YBa_2Cu_3O_{6.5}$ (which is made from Y_2O_3, BaO, and CuO), the authors state "(a)t most three phases can coexist in this system when the temperature and partial pressure of O_2 (0.21 atm [21 kPa]) are fixed." Justify this statement. Keep in mind that this is a ternary system.

[*] H. Steinfink, J. S. Swinnea, Z. T. Suí, H. M. Su, and J. B. Goodenough, 1987. *Journal of the American Chemical Society*, 109, 3348.

9.13 Sketch the Ag–Sn phase diagram (T versus X_{Sn}) based on the following information. Pure Sn melts at 350°C; pure Ag melts at 980°C; Ag$_3$Sn is the only intermetallic compound formed and it melts incongruently at 440°C to give liquid and S$_1$, where S$_1$ is a solid solution rich in Ag; there is one eutectic ($X_{Sn} = 0.91$) and it melts at 210°C; Sn and Ag$_3$Sn are insoluble in the solid state; Ag$_3$Sn and S$_1$ are insoluble in the solid state. Label the phase diagram.

9.14 The pressure–temperature coexistence line for solid–vapor equilibrium (sublimation) is always less steep in a p versus T plot for a pure material than is the melting line. Explain why. A diagram could be helpful.

9.15 The quotation at the start of Part III of this book reads, "*Passed through the fiery furnace, As gold without alloy...*" The implication is that "gold without alloy" (i.e., pure gold) can pass through a fiery furnace without damage, whereas gold "with alloy" (i.e., impure gold) would be damaged in a furnace. Provide an explanation. Include a phase diagram.

9.16 The system Na$_2$SO$_4$/Na$_2$MoO$_4$/H$_2$O was studied at 25°C by W.F. Linke and J.A. Cooper (1956, *Journal of Physical Chemistry*, 60, 1662). They determined that the following solid phases just precipitate out of a saturated liquid at these compositions (compositions given are mass% in the liquid):

Liquid Phase		
Mass% Na$_2$SO$_4$	Mass% Na$_2$MoO$_4$	Solid Phase(s)
32.90	0.0	Na$_2$SO$_4$
28.14	5.89	Na$_2$SO$_4$
21.66	14.15	Na$_2$SO$_4$
15.40	22.73	Na$_2$SO$_4$
10.50	30.00	Na$_2$SO$_4$
9.47	31.89	Na$_2$SO$_4$ + NaMoO$_4$·2H$_2$O
9.42	31.92	Na$_2$SO$_4$ + NaMoO$_4$·2H$_2$O
9.46	31.90	Na$_2$SO$_4$ + NaMoO$_4$·2H$_2$O
7.28	33.73	Na$_2$MoO$_4$·2H$_2$O
3.88	36.65	Na$_2$MoO$_4$·2H$_2$O
0.0	39.90	Na$_2$MoO$_4$·2H$_2$O

The hydrate Na$_2$MoO$_4$·2H$_2$O is 85.11 mass% Na$_2$MoO$_4$.

a. Plot the points for a ternary graph and then complete the ternary phase diagram using triangular coordinates.

b. Consider a dry sample (no water) that is 40% by mass Na$_2$SO$_4$ (i.e., 60 mass% Na$_2$MoO$_4$). Imagine this sample to be completely dissolved in water and describe what happens as the water present is allowed to evaporate.

9.17 Polymorphism can be a considerable problem when dealing with chromophores. For example, a compound can have different optical properties in different polymorphs. Explain how temperature and pressure can influence the stability of a phase and the resulting optical properties of a compound.

9.18 The silver–lanthanum isobaric binary phase diagram has three compounds. LaAg melts congruently at 886 °C; $LaAg_2$ melts incongruently at 864 °C; $LaAg_3$ melts congruently at 955 °C. In the La–Ag phase diagram, there are also three eutectics: 11 at% La and 778 °C; 43.5 at% La and 741 °C; 71 at% La and 518 °C. Pure silver melts at 960.5 °C and pure lanthanum melts at 812 °C. Sketch and label the La–Ag binary phase diagram.

9.19 In the gold–lead phase diagram, there are two compounds, Au_2Pb, which melts incongruently at 418 °C, and $AuPb_2$, which melts incongruently at 254 °C. The peritectic compositions are 44 atomic% Pb and 72 atomic% Pb, respectively. In addition, there is a eutectic at 215 °C, 84.4 atomic% Pb. Pure gold melts at 1063 °C and pure lead melts at 327 °C. Sketch and label the Au–Pb binary phase diagram.

9.20 A bar of a material has a small concentration of impurity (initial value C_0), shown as a function of distance along the bar, x, in Figure 9.50. The portion of the melting point diagram near the pure material, showing the influence on the melting point of the addition of a small amount of this impurity, is also shown in Figure 9.50. The bar will be heated with a small heater, such that successive regions of the bar will be melted, as shown schematically in Figure 9.51.

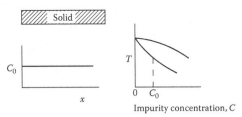

FIGURE 9.50
A bar of metal has an initial impurity concentration constant at C_0 along its length. The melting diagram of the material at low impurity concentration, C, also is shown.

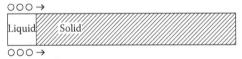

FIGURE 9.51
A heater passes along the bar, melting material in the heater region.

a. At the region of the molten material, is the solid purer or less pure than the liquid? Use the phase diagram in Figure 9.50 to answer this question.

b. As the heater moves along the sample from left to right, the purity profile (C versus x) changes. Sketch the final C versus x relationship, when the heater has travelled the full length of the bar and then the bar has fully solidified.

c. This procedure is called *zone refinement* and it is used to purify materials such as silicon. How can the purity of the final bar (or regions of it) be improved further?

9.21 Compound A is *optically active*; that is, it comes in two molecular forms that are chemically identical except that they are nonsuperimposable mirror images of each other. One form (*d-A*) rotates polarized light in one direction and the other form (*l-A*) rotates polarized light in the other direction. In the solid state, there are (at least) two possible arrangements for the *racemic mixture* (equimolar *d-A* and *l-A*): The two isomers are totally insoluble in the solid state, or the two isomers form a racemic compound (*d-A + l-A* in a 1:1 mole ratio). Sketch binary phase diagrams (temperature versus mole fraction of *d-A*) for each of these possibilities.

9.22 In the phase diagram for H_2O/Fe_2Cl_6, six chemically distinct species are possible: Fe_2Cl_6, $Fe_2Cl_6 \cdot 4H_2O$, $Fe_2Cl_6 \cdot 5H_2O$, $Fe_2Cl_6 \cdot 7H_2O$, $Fe_2Cl_6 \cdot 12H_2O$, and H_2O. Use the definition of "component" to show why the $Fe_2Cl_6 - H_2O$ phase diagram is a binary (two-component) phase diagram.

9.23 The Gibbs energy change for the mixing of two liquids is given by:

$$\Delta_{mix}G = \Delta_{mix}H - T\Delta_{mix}S \qquad (9.26)$$

and if $\Delta_{mix}H$ is independent of temperature, from Equation 6.54, it follows that

$$\frac{d\Delta_{mix}G}{dT} = -\Delta_{mix}S. \qquad (9.27)$$

a. For a binary system with an upper consolute temperature, what is the sign of $(d\Delta_{mix}G/dT)$? What does this imply about $\Delta_{mix}S$? Explain.

b. For a binary system with a lower consolute temperature, what is the sign of $(d\Delta_{mix}G/dT)$? What does this imply about $\Delta_{mix}S$? Explain.

9.24 The slope of the solid–liquid line in the phase diagram of H_2O (Figure 9.5a) at atmospheric pressure is $-13.7\,MPa\,K^{-1}$. Consider an ice skater whose mass is 80 kg with a contact surface between the skate blades and the ice of 20 cm^2.

   a. Calculate the pressure exerted on the ice by the skater.

   b. Determine how much this pressure lowers the melting point of ice.

   c. From (b), do you think that the pressure-induced formation of a liquid layer under the skate blades is a plausible explanation for the ice skating phenomenon?

9.25 The compounds 4,4′-dimethoxyazoxybenzene (DMAB) and 4,4-diethoxyazoxy-benzene (DEAB) both form liquid crystalline phases. Above 135 °C, DMAB is an isotropic liquid, as is DEAB above 160 °C. In the isotropic liquid phase, DEAB and DMAB mix with each other in all proportions. At 135 °C, pure DMAB transforms to a nematic liquid crystalline phase, as does pure DEAB at 160 °C. The DMAB and DEAB nematic phases are miscible in each other in all proportions, and the isotropic liquid to nematic liquid crystal transition temperature changes smoothly as DEAB is added to DMAB. At 118 °C, pure DMAB freezes to a crystalline solid. Similarly, pure DEAB freezes at 132 °C. DMAB and DEAB are not soluble in the solid state, and they do not form compounds. However, they do form a eutectic at 97 °C and $X_{DEAB} = 0.45$. (All data are for $p = 101\,kPa$.)

   a. Sketch and label the isobaric T versus X_{DEAB} phase diagram.

   b. Deduce and draw the cooling curve (temperature versus time) from 160 °C to 90 °C for $X_{DEAB} = 0.45$.

9.26 The refractive index of a typical liquid is quite different from that of its vapor, mostly due to differences in density. Therefore, there is a discontinuity in refractive index on passing from liquid to vapor. At and near the critical point, there are large fluctuations in refractive index, leading to a phenomenon known as *critical opalescence*: The material appears milky and opaque. Use a diagram showing the light path to explain the origin of critical opalescence of a material in which both the liquid and vapor are transparent and colorless.

9.27 The following data are from Lightfoot and Prutton's 1947 study of the $CaCl_2$-KCl-H_2O system at 75 °C.

Saturated Solution (Mass%)		
CaCl$_2$	KCl	Solid Phase(s)
0	33.16	KCl
11.73	21.62	KCl
18.27	16.00	KCl

(*Continued*)

Saturated Solution (Mass%)		
CaCl$_2$	KCl	Solid Phase(s)
28.47	9.62	KCl
37.65	6.77	KCl
47.65	8.43	KCl
50.19	10.32	KCl,2KCl·CaCl$_2$·2H$_2$O
50.92	9.36	2KCl·CaCl$_2$·2H$_2$O
53.85	6.21	2KCl·CaCl$_2$·2H$_2$O
56.33	4.51	2KCl·CaCl$_2$·2H$_2$O
57.62	3.60	2KCl·CaCl$_2$·2H$_2$O,CaCl$_2$·2H$_2$O
57.77	2.56	CaCl$_2$·2H$_2$O
58.58	0	CaCl$_2$·2H$_2$O

a. Sketch and label the isothermal isobaric phase diagram.

b. State what phases exist on evaporation of a solution that is initially 80 mass% H$_2$O, 10 mass% CaCl$_2$ and 10 mass% KCl, until no water remains.

9.28 Label the phases in the following phase diagrams:

a. Mg–Pb (Figure 9.52);

b. Al–Zn (Figure 9.53).

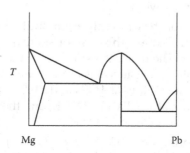

FIGURE 9.52
The Mg–Pb phase diagram.

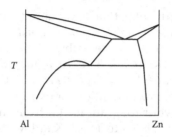

FIGURE 9.53
The Al–Zn phase diagram.

9.29 Consider the binding forces within a carbon nanotube and also the data for thermal conductivity of graphite (Table 8.1). On this basis, would you expect the thermal conductivity of carbon nanotubes to be exceptionally high or low? Explain.

9.30 Consider the lack of periodic structure in quasicrystals (Figure 9.18) and comment on how this could influence their thermal conductivity relative to crystalline materials with similar compositions. Consider all the factors in the Debye equation.

9.31 Graphite, diamond, and graphene (a single-atom thick layer of graphite) are all allotropes of carbon. At ambient conditions, graphite is the most stable form (see Figure 9.13). Graphene is very high in energy due to its high surface area (see Chapter 10). Based on this information, sketch a Gibbs energy diagram that includes these three allotropes of carbon, analogous to that shown in Figure 6.9, for ambient conditions, and label each allotrope.

9.32 HfB_2 has been proposed by NASA scientists as a useful ultra-high-temperature ceramic for aerospace applications due to its high melting point (melts congruently at 3380 °C) and high thermal conductivity. Pure hafnium melts at 2230 °C and B melts at 2076 °C. In the binary Hf–B phase diagram, there are two compounds: HfB_2 and also HfB, which melts incongruently at 2100 °C. Name two advantages of HfB_2 over HfB.

9.33 The element boron has recently been studied experimentally at high pressure. There are three main solid phases (allotropes) of boron: α, β, and γ. The phase densities are in the order $\gamma > \alpha > \beta$. The triple point (α–β–γ coexistence) is at 8 GPa and 1850 K. The phases α and β are in equilibrium at 4 GPa and 1400 K; the phases α and γ are in equilibrium at 11 GPa and 1420 K; and the phases β and γ are in equilibrium at 10 GPa and 2300 K.

   a. Use this information to sketch the p (y-axis) $vs.$ T (x-axis) phase diagram of boron. Include the pressure range from 0 to 15 GPa and the temperature range from 0 to 2500 K. Label the phase(s) in each region. No explanation is required.

   b. Recent experimental thermodynamic studies (M. A. White et al., 2015. *Angewandte Chemie International Edition*, 54, 3626) show that β-boron is the stable phase at ambient pressure from $T \sim 2000$ K to $T < 3$ K. At ambient pressure and very low temperature, $\Delta_{trs}H(\alpha \rightarrow \beta) \sim 0$ kJ mol^{-1}. Use this information and the Clapeyron equation to extrapolate the α-β equilibrium line in your diagram for (a) from the lowest measured temperature (1400 K) to $T \rightarrow 0$ K at low pressure. Show your reasoning.

9.34 The mineral leucite, $KAlSi_2O_6$, forms an incongruently melting compound with SiO_2, called *potash feldspar*, of formula $KAlSi_3O_8$.

a. What is the mass% of SiO_2 in $KAlSi_3O_8$? Show your calculation.

b. $KAlSi_3O_8$ melts incongruently at 1150 °C to give a liquid that is 40 mass% SiO_2. This is the only compound formed in the $KAlSi_2O_6/SiO_2$ binary phase diagram. There is a eutectic at 990 °C and 55 mass% SiO_2. Pure leucite melts at 1685 °C and pure SiO_2 melts at 1713 °C. There is one more complication in this phase diagram; SiO_2 has two polymorphs, cristobalite (stable above 1475 °C) and tridymite (stable below 1475 °C). Draw the $KAlSi_2O_6/SiO_2$ binary phase diagram, expressed as a function of mass% SiO_2. (Note that not all details of the phase diagram are known from the above information, but the main features should be consistent with the above.) Label all the areas in the phase diagram.

9.35 The ternary phase diagram for a hypothetical system A/B/C, given in mole %, is shown by the heavy lines in Figure 9.54.

a. State which binary mixtures (A/B, A/C, and/or B/C) show binary compounds and give the approximate composition.

b. As for binary diagrams (see Figure 9.30), a lever rule can be used to give the proportions of various components in ternary diagrams. Consider the point labeled ★ in Figure 9.54. At this composition, there are three phases: B, b, and a. The relative molar proportions are given as follows, where the ℓ values are as shown on Figure 9.54: $n_b/(n_B + n_a) = \ell_2/\ell_1$ and $n_B/n_a = \ell_4/\ell_3$. Use this information to estimate the numbers of moles of B, a, and b if the total number of moles is 1.

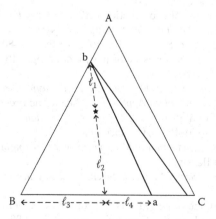

FIGURE 9.54
A generic ternary phase diagram, expressed as mole%.

Further Reading

General References

M. F. Ashby and D. R. H. Jones, 2005. *Engineering Materials 2: An Introduction to Microstructures, Processing and Design*, 3rd ed. Butterworth-Heinemann, Oxford.

D. R. Askeland, P. P. Fulay, and W. J. Wright, 2010. *The Science and Engineering of Materials*, 6th ed. Cengage-Engineering, Stamford, CT.

H. Baker, Ed., 1990. *ASM Handbook. Vol. 3: Alloy Phase Diagrams*. ASM International, Materials Park, OH.

A. Barton, 1997. *States of Matter: States of Mind*. Taylor & Francis, Washington, DC.

R. J. Borg and G. J. Dienes, 1992. *The Physical Chemistry of Solids*. Academic Press, Oxford.

F. Case, 2013. Alloys: Magical mixtures of metals. *Chemistry World*, November, 54.

W. D. Callister, Jr. and D. G. Rethwisch, 2013. *Materials Science and Engineering: An Introduction*, 9th ed. John Wiley & Sons, Hoboken, NJ.

D. G. Clerc and D. A. Cleary, 1992. Spinodal composition and an interesting example of the application of several thermodynamic principles. *Journal of Chemical Education*, 72, 112.

M. de Podesta, 2002. *Understanding the Properties of Matter*, 2nd ed. CRC Press, Boca Raton, FL.

E. G. Ehlers, 1972. *The Interpretation of Geological Phase Diagrams*. Dover, New York.

A. Findlay, revised by A. N. Campbell, 1938. *The Phase Rule and Its Applications*. Longmans, Green and Co, London.

B. Fultz, 2014. *Phase Transitions*. Cambridge University Press, Cambridge, UK.

G. G. Hall, 1991. *Molecular Solid State Physics*. Springer-Verlag, New York.

R. A. Higgins, 1994. *Properties of Engineering Materials*, 2nd ed. Industrial Press, New York.

J. M. Honig, 2007. *Thermodynamics: Principles Characterizing Physical and Chemical Processes*. Academic Press, Oxford.

K. Knapman, 2000. Polymorphic predictions. *Modern Drug Discovery*, March, 53.

W. J. Moore, 1967. *Seven Solid States*. W.A. Benjamin, New York.

N. Mott, 1967. The solid state. *Scientific American*, September, 80.

A. Navrotsky, 1997. Thermochemistry of new, technologically important inorganic materials. *MRS Bulletin*, May, 35.

J. B. Ott and J. Boerio-Goates, 2000. *Chemical Thermodynamics. Vol. I: Principles and Applications; Vol. II: Advanced Applications*. Academic Press, Oxford.

C. N. R. Rao and J. Gopalakrishnan, 1997. *New Directions in Solid State Chemistry*. Cambridge University Press, Cambridge.

J. F. Shackelford, 2008. *Introduction to Materials Science for Engineers*, 7th ed. Prentice-Hall, Upper Saddle River, NJ.

W. F. Smith, 1990. *Principles of Materials Science and Engineering*. McGraw-Hill, New York.

N. O. Smith, 1997. The Gibbs energy basis and construction of melting point diagrams in binary systems. *Journal of Chemical Education*, 74, 1080.

P. C. Taylor, 2016. Exotic forms of silicon. *Physics Today*, December, 34.

R. S. Treptow, 1993. Phase diagrams for aqueous systems. *Journal of Chemical Education*, 70, 616.

L. H. Van Vlack, 1989. *Elements of Materials Science and Engineering*, 6th ed. Prentice-Hall, Upper Saddle River, NJ.

Fullerenes, Carbon Nanotubes, Graphene and Related Materials

Graphene fundamentals and functionalities. Special issue. W. Lu, P. Soukiassian, and J. Boeckl, Eds. *MRS Bulletin*, December 2012, 1119–1321.

P. Ajayan, P. Kim, and K. Banerjee, 2016. Two-dimensional van der Waals materials. *Physics Today*, September, 38.

R. Baum, 1994. Endohedral fullerenes purified by porphyrin-silica. *Chemical and Engineering News*, October 10, 37.

J. Cartwright, 2015. Graphene band gap heralds new electronics. *Chemistry World*, September 29.

M. W. Cole, V. H. Crespi, M. S. Dresselhaus, G. Dresselhaus, J. E. Fischer, H. R. Gutierrez, K. Kojima, G. D. Mahan, A. M. Rao, J. O. Sofo, M. Tachibana, K. Wako, and Q. H. Xiong, 2010. Structural, electronic, optical and vibrational properties of nanoscale carbons and nanowires: A colloquial review. *Journal of Physics—Condensed Matter*, 22, 334201.

R. Dagani, 1999. Building with buckyballs. *Chemical and Engineering News*, October 18, 54.

R. Dagani, 2001. Molecular surgery on C_{60}. *Chemical and Engineering News*, April 23.

M. Davenport, 2015. Much ado about small things. *Chemical and Engineering News*, June 8, 10.

C. Day, 2015. Predicting pentagonal graphene. *Physics Today*, April, 17.

M. S. Dresselhaus, G. Dresselhaus, and P. C. Eklund, 1996. *Science of Fullerenes and Carbon Nanotubes*. Elsevier, Amsterdam.

T. W. Ebbesen, 1996. Carbon nanotubes. *Physics Today*, June, 26.

R. J. Fitzgerald, 2014. Graphene's newest cousin, germanene. *Physics Today*, November, 20.

A. Hildebrand, U. Hilgers, R. Blume, and D. Wiechoczek, 1996. Playing with the soccer ball—an experimental introduction to fullerene chemistry. *Journal of Chemical Education*, 73, 1066.

D. R. Huffman, 1991. Solid C_{60}. *Physics Today*, November, 22.

M. Imboden and D. Bishop, 2014. Top-down nanomanufacturing. *Physics Today*, December, 45.

M. Jacoby, 2017. Manganese thiophosphate joins the 2-D materials club. *Chemical and Engineering News*, October 30, 7.

M. Jacoby, 2017. 2-D materials stack up. *Chemical and Engineering News*, May 29, 36.

J. L. Miller, 2014. Tailor-made molecules grow into identical carbon nanotubes. *Physics Today*, October, 14.

J. L. Miller, 2015. Two-dimensional materials let protons pass. *Physics Today*, February, 11.

J. F. Niven, M. B. Johnson, S. M. Juckes, M. A. White, N. T. Alvarez, and V. Shanov, 2016. Influence of annealing on thermal and electrical properties of carbon nanotube yarns. *Carbon*, 99, 485.

P. Patel, 2017. Strong graphene aerogels bounce back. *Chemical and Engineering News*, July 17, 7.

R. Rawls, 2000. Fullerenes are out of this world. *Chemical and Engineering News*, March 27, 8.

A. Scott, 2016. Graphene's global race to market. *Chemical and Engineering News*, April 11, 28.

E. Stoye, 2015. Graphene beyond the hype. *Chemistry World*, June 19.

R. Van Noorden, 2011. The trials of the new carbon. *Nature*, 469, 14.

A. P. E. York, 2004. Inorganic fullerenes, onions and tubes. *Journal of Chemical Education*, 81, 673.

Helium

B. Bertman and R. A. Grujer, 1967. Solid helium. *Scientific American*, August, 65.

R. J. Donnelly, 1988. Superfluid turbulence. *Scientific American*, November, 100.

R. Hallock, 2015. Is solid helium a supersolid? *Physics Today*, May, 30.

Ice and Water

L. Glasser, 2004. Water, water, everywhere: Phase diagrams of ordinary water substance. *Journal of Chemical Education*, 81, 414; erratum 81, 645.

J. Kemsley, 2017. In clouds, cubic-structure ice. *Chemical and Engineering News*, July 24, 9.

M. J. Shultz, 2018. Crystal growth in ice and snow. *Physics Today*, February, 34.

H. E. Stanley, 1999. Unsolved mysteries of water in its liquid and glass states. *MRS Bulletin*, May, 22.

Liquids

R. E. Hummel, 2004. *Understanding Materials Science*, 2nd ed. Springer, New York.

S. R. Logan, 1998. The behavior of a pair of partially miscible liquids. *Journal of Chemical Education*, 75, 339.

Orientationally Disordered Solids

N. G. Parsonage and L. A. K. Staveley, 1978. *Disorder in Solids*. Oxford University Press, Oxford.

Phase Diagrams and Phase Transitions

F. S. Bates and G. H. Fredrickson, 1999. Block copolymers—designer soft materials. *Physics Today*, February, 32.

P. W. R. Bessonette and M. A. White, 1999. Thermodynamic curves describing a second order phase transition. *Journal of Chemical Education*, 76, 220.

N. Blagden and R. Davey, 1999. Polymorphs take shape. *Chemistry in Britain*, March, 44.

D. Calvert, M. J. Smith, and E. Falcão, 1999. Equipment for a low-cost study of the naphthalene-biphenyl phase diagram. *Journal of Chemical Education*, 76, 668.

M. C. Flemings, 1998. Solidification science and engineering practice. *MRS Bulletin*, May, 30.

S. J. Hawkes, 1999. There is no perceptible inflection at the triple point. *Journal of Chemical Education*, 76, 226.

J. M. Honig, 1999. Mean field theory of phase transitions. *Journal of Chemical Education*, 76, 848.

M. Jacoby, 2017. Why it's so hard to make metallic hydrogen. *Chemical and Engineering News*, June 12, 16.

W. B. Jensen, 2006. The origin of the term allotrope. *Journal of Chemical Education*, 83, 838.

W. J. Nellis, 2000. Making metallic hydrogen. *Scientific American*, May, 84.

M. R. Ranade, A. Navrotsky, H. Z. Zhang, F. F. Banfield, S H. Elder, A. Zaban, P. H. Borse, S. K. Kulkarni, G. S. Doran, and H. J. Whitfield, 2002. Energetics of nanocrystalline TiO_2. *Proceedings of the National Academy of Science*, 22, 6476.

R. Trautman. B. J. Griffin, and D. Scharf, 1998. Microdiamonds. *Scientific American*, August, 82.

M. A. White, A. Cerqueira, C. A. Whitman, M. B. Johnson, and T. Ogitsu, 2015. Determination of Phase Stability of Elemental Boron. *Angewandte Chemie International Edition*, 54, 3626.

E. Wilson, 2001. Tin's last waltz. *Chemical and Engineering News*, November 27, 4.

J. K. Woodworth, J. C. Terrance, and M. M. Hoffmann, 2006. Using nuclear magnetic resonance spectroscopy for measuring ternary phase diagrams. *Journal of Chemical Education*, 83, 1064.

Quasicrystals

Special issue on quasicrystals. D. J. Sordelet and J. M. Dubois, Eds. *MRS Bulletin*, November 1997, 34–74.

C. Day, 2001. Binary quasicrystals discovered that are stable and icosahedral. *Physics Today*, February, 17.

J.-M. Dubois, 2005. *Useful Quasicrystals*. World Scientific.

R. A. Dunlap, 1990. Periodicity and aperiodicity in mathematics and crystallography. *Science Progress* (Oxford), 74, 311.

M. Jacoby, 1999. Quasicrystals: A new kind of order. *Chemical and Engineering News*, March 15, 44.

D. R. Nelson, 1986. Quasicrystals. *Scientific American*, August, 42.

A. L. Pope and T. M. Tritt, 2004. Thermal conductivity of quasicrystals. In *Thermal Conductivity*, T. M. Tritt, Ed. Springer, Boston.

P. W. Stephens and A. I. Goldman, 1991. The structure of quasicrystals. *Scientific American*, April, 44.

H. C. von Baeyer, 1990. Impossible crystals. *Discover*, February, 69.

Supercritical Fluids

T. Clifford and K. Bartle, 1993. Chemistry goes supercritical. *Chemistry in Britain*, June, 499.

F. Hensel, 1988. Critical point phenomena of fluid metals. *Chemistry in Britain*, May, 457.

R. A. Laudise, 1987. Hydrothermal synthesis of crystals. *Chemical and Engineering News*, September 26, 30.

S. G. Mayer, J. M. Gach, E. R. Forbes, and P. J. Reid, 2001. Exploring phase diagrams using supercritical fluids. *Journal of Chemical Education*, 78, 241.

C. L. Phelps, N. G. Smart, and C. M. Wai, 1996. Past, present, and possible future applications of supercritical fluid extraction technology. *Journal of Chemical Education*, 73, 1163.

M. Poliakoff and S. Howdie, 1995. Supercritical chemistry: Synthesis with a spanner. *Chemistry in Britain*, February, 118.

K. Shakesheff and S. Howdle, 2003. Gently does it. *Chemistry in Britain*, December, 30.

R. W. Shaw, T. B. Brill, A. A. Clifford, C. A. Eckert, and E. U. Franck, 1991. Supercritical water. *Chemical and Engineering News*, December 23, 26.

E. K. Wilson, 2000. Materials made under pressure. *Chemical and Engineering News*, December 18, 34.

P. York, 2005. Supercritical fluids: realizing potential. *Chemistry World*, February, 50.

K. Zosel, 1978. Separation with supercritical gases: Practical applications. *Angewandte Chemie International Edition England*, 17, 702.

Websites

For links to relevant websites, see PhysicalPropertiesOfMaterials.com

10

Surface and Interfacial Phenomena

10.1 Introduction

Although we have concentrated up until now on bulk properties of materials, the surfaces of materials can have special and important properties that are distinct from those inside the bulk of the solid or liquid. For a simple example, consider that in the interior of an ionic crystal the charges balance out when all directions are averaged, whereas at the surface there can be a layer of ions all of the same charge, and this situation will greatly influence the response of other ions to the surface.

One important property of a surface is its ability to interact with other atoms and molecules. Molecules may be *absorbed* into the interior of the material, as in the case of *graphite intercalates* (also called graphite inclusion compounds, i.e., graphite with other types of atoms between the graphitic layers), an example of which is shown in Figure 10.1a.

Molecules could also be *adsorbed* onto the surface, as shown schematically in Figure 10.1b. The strength of the interaction between the *adsorbate* (adsorbed species) and the surface determines the type of adsorption. Weak interaction ($\Delta H \approx -20 \, kJ \, mol^{-1}$, due to weak van der Waals interactions) is termed *physisorption* (or physical adsorption). Stronger interaction ($\Delta H \approx -200 \, kJ \, mol^{-1}$, due to covalent bonding) is termed *chemisorption* (or chemical adsorption).

Understanding the properties of surfaces can be very important in chemistry. For example, many reactions are catalyzed on surfaces that allow energetically favorable reaction pathways, primarily through chemisorption of reactants.

In the study of surfaces, it might be important to have a large surface area and/or to have very well defined surfaces. The large surface area can be particularly important in the case of a property that scales with the amount of surface area, such as the enthalpy change associated with adsorption. However, the surface must be regular for measurements to be meaningful.

A very important surface, both for its homogeneity and its rather large surface area, is that of *exfoliated graphite*. This material, which is composed of graphitic particles, is produced from intercalated graphite (e.g., with bromine between the layers) that has been reacted to remove the intercalating

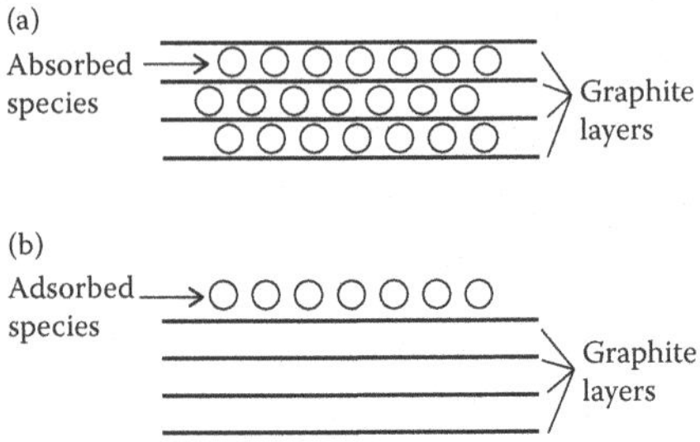

(a)

Absorbed $\longrightarrow$ 　　　　　　　　　　　　　　　　　　　　　　Graphite
species　　　　　　　　　　　　　　　　　　　　　　　　　　layers

FIGURE 10.1
Cross-sectional views of absorption into and
adsorption onto graphite. (a) Atoms or molecules absorbed into a graphite structure to
form a graphite intercalate. (b) Atoms or molecules also could be adsorbed onto a graphite
structure.

(b)

Adsorbed $\longrightarrow$ 　　　　　　　　　　　　　　　　　　Graphite
species　　　　　　　　　　　　　　　　　　　　　　layers

species, blowing apart the structure and leaving behind islands of graphitic
structure with surface areas of several thousand $m^2\,g^{-1}$. Exfoliated graphite
is used in products as diverse as automotive gaskets, electrodes, fire retar-
dants, and materials to absorb oil spills. Note that exfoliated graphite still
has several layers of graphitic carbon. When the carbon sheets are only one-
atom thick, this material is called *graphene*.

COMMENT: GRAPHENE

Graphene is an allotrope of carbon, consisting of a single layer of
carbon atoms (Figure 10.2). Rather famously, Geim* and co-workers
extracted graphene from graphite using tape to peel away the single-
carbon-thick layer. Graphene has a high surface area, $\sim\!3000\,m^2\,g^{-1}$,
and unusual optical, thermal, electronic, and mechanical properties.
For example, it shows nonlinear optical properties, exceptionally high
thermal conductivity and melting point, and high strength. Graphene
shows promise for applications in solar cells, light-emitting diodes, and
touch panels.

---

* Andre Geim (1958–) and Konstantin Novoselov (1974–) both of the University of
Manchester were awarded the 2010 Nobel Prize in Physics for groundbreaking experiments regarding the two-dimensional material graphene.

FIGURE 10.2
View of the 2D structure of graphene. Note that graphene is not infinite in all directions, and
edges and ripples can impart different properties. By Alexander AIUS - Own work, CC BY-SA
3.0, https://commons.wikimedia.org/w/index.php?curid=11294534.

10.2 Surface Energetics

The Gibbs energy of a surface is always higher than that of the bulk material because the surface atoms are not able to achieve bonds in all directions as the interior atoms can. In this sense surfaces are unstable relative to the bulk material.

One consequence is that if there is flexibility within the surface, it will choose to have the minimum surface area. For example, free-floating bubbles are spherically shaped; any other shape would require more surface area.

Another consequence is that it may be more favorable energetically for a material to be covered with another material. For a given mass, a group of small crystals is of higher energy (i.e., less favorable) than one large crystal due to the decreased surface area of the large crystal. This fact can be used to grow large crystals from small ones, as follows. If the small crystals are placed in a saturated solution and the temperature is increased, more material will dissolve (provided the temperature coefficient of solubility is positive, as is the usual situation). If the temperature is lowered, material will precipitate out due to the lowered solubility, and the most energetically favored way for the precipitation to occur is onto the largest crystals present. Subsequent heating–cooling cycles can eventually transform many small crystals into a few larger crystals. Such a process is shown schematically in Figure 10.3.

The energy at a surface is always higher than the bulk but the *surface energy*, G_s, depends on the material in question. The data in Table 10.1 show one aspect of the driving force of the rusting process: Metal oxide has a lower surface energy than pure metal. Coating the metal surface with an organic material lowers the surface energy below that of the metal oxide, and this is one reason why undercoating a car prevents rust formation although the dominant effect is keeping oxygen away from the metal.

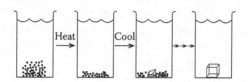

FIGURE 10.3
This diagram shows schematically how a group of small crystals (e.g., a powder) can be transformed to one large crystal, if (as is usual) the material is more soluble when the solution is heated. Heat leads to greater solubility, and cooling causes preferential precipitation on a seed crystal. After a number of cycles, powder can be transformed into a single crystal. This process is thermodynamically favored on the basis of reduced Gibbs energy for the sample with reduced surface area for a given mass.

TABLE 10.1

Surface Energies, G_s, of Some Materials

Material	$G_s/(J\ m^{-2})$
Metal	~1
Metal oxide	~0.3
Ice	~0.1
Wax	~0.07

Source: R. G. Linford, 1972. *Chemical Society Reviews*, 1, 445.

10.3 Surface Investigations

Surfaces can be investigated by many means. For example, the enthalpies of adsorption of species can be important in gaining understanding of the surface–adsorbate interactions. Many microscopic techniques can be used to investigate surfaces, and one family is highlighted here: the *scanning tunneling microscope* (STM) and the related *atomic force microscope* (AFM), both of which are scanning probe microscopies.

The first STM was described in 1986.* Its principle involves the overlap of the wavefunctions of a very sharp (atomic resolution) "tip" that scans the surface, with the wavefunctions of the surface atoms. When the overlap is sufficient, electrons can tunnel between the surface and the tip, completing an electrical circuit. The tunneling current is very sensitive to the tip–substrate separation distance and, with fine control of the tip (0.01 nm), atomic resolution can be achieved. The STM is shown schematically in Figure 10.4a. An STM image of Si is shown in Figure 10.4b.

In STM, the imaging relies on measurement of the tunneling current between the tip and a conducting substrate. For insulating materials, it is more useful to determine the topography by measuring the attractive (van der Waals) force between the tip and the surface. The force will be repulsive if the distance between the tip and surface is very close (the so-called contact mode), or attractive if the distance is longer (noncontact mode). The force can be determined very accurately by the deflection of laser light hitting a cantilever that has the tip at one end. It also is possible to use an AFM tip to move individual atoms and image the result.

---

* Ch. Gerber et al., 1986. *Review of Scientific Instruments*, 5 7, 221. Gerd Binnig (1947–) and Heinrich Rohrer (1933–2013) were researchers at IBM Zurich, Switzerland, and co-winners of the 1986 Nobel Prize in Physics, for their design of the scanning tunneling microscope; their Nobel Prize was shared with Ernst Ruska (1906–1988), a researcher from the Fritz–Haber Max Planck Institute in Berlin, who was cited for his design of the first electron microscope.

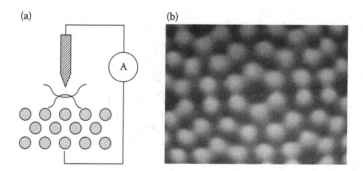

FIGURE 10.4

(a) A schematic view of the scanning tunneling microscope. The overlap of a wave function of a surface atoms with the scanning tip completes the circuit and allows a measurable tunneling current. The fine control of the tip position is accomplished with a piezoelectric material. (b) An STM image of Si, courtesy of M. Jericho.

10.4 Surface Tension and Capillarity

As in the solid state, the molecules at the surface of a liquid are at higher energies than those in the bulk of the liquid. This situation is shown schematically in Figure 10.5, where the molecules in the bulk are shown to be attracted equally in all directions, whereas the molecules at the surface have a net inward attraction to their neighbors. This situation again shows that the surface is high in energy with respect to the bulk, and the most favorable situation would be to have as little surface as possible. Therefore, very small drops of liquid are very nearly spherical (although larger ones may be flattened by their weight): The spherical shape minimizes the surface area.

The energy required to increase the surface area leads to the definition of surface tension, γ, as*

$$\gamma = \frac{dG}{dA} \tag{10.1}$$

where G is the Gibbs energy and A is the surface area, so SI units of γ are J m^{-2} (for a liquid, surface tension is the same as surface energy).

The presence of surface tension leads to some interesting facts. For example, surface tension is a major factor in allowing insects to exist on the surface of a pond; although gravitational forces would dictate that the insect (which is denser than water) should be at the bottom of the pond, the cost in increasing the surface area, as the "skin" of the pond is ripped and the

---

* The most important equations in this chapter are designated with ▌ to the left of the equation.

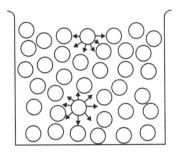

FIGURE 10.5
The attractive interaction of molecules with each other in liquids leads to no net force on an interior molecule but a net inward force on surface molecules.

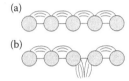

FIGURE 10.6
(a) Surface tension normally keeps water from flowing through cotton or canvas. This is the basis of its water repellency. (b) Once a pathway has been established for a leak in the fabric by lowering the surface tension, the water will continue to pour through.

insect sinks, is too high. The energetic cost of increasing the surface area of the water in the pond is related to the contact angle of the insect with the water and the degree of wetting of the solid phase (insect) by the liquid. The former can be seen by example: Careful placement of a needle horizontally on a water surface will allow it to float, but more vertical placement allows the needle to fall to the bottom while only making a tiny (low-energy cost) slit. The latter effect can be responsible for the initial water repellency of cotton or canvas. If touched, your finger (or, more accurately, the oils on your finger) will lower the surface tension, and the water will then leak continuously where you touched. This situation is shown schematically in Figure 10.6.

For a spherical drop of radius R, the total surface Gibbs energy is $4\pi R^2 \gamma$; a decrease in radius by dR would lower the surface Gibbs energy by $8\pi R \gamma dR$. At equilibrium, the tendency to shrink is balanced by a pressure difference, Δp, between the interior of the drop and its exterior such that the work against this pressure difference, $\Delta p 4\pi R^2 dR$, equals the difference in surface Gibbs energy, $8\pi R \gamma dR$, which simplifies to

$$\Delta p = \frac{2\gamma}{R}. \tag{10.2}$$

TABLE 10.2

Surface Tension, γ, for Some Liquids at $T = 25\,°C$

Liquid	$\gamma/(10^{-3}\,J\,m^{-2})$
Br_2	40.95
H_2O	71.99
Hg	485.48
CS_2	31.58
CH_3OH	22.07
CH_3NH_2	19.15
CH_3CH_2OH	21.97

Source: J. J. Jasper, 1972. Journal of Physical and Chemical Reference Data, 1, 841.

Equation 10.2 shows the relationship between the radius of curvature of a surface and surface tension; this equation can be used to determine surface tension experimentally.

Some typical values of surface tensions are given in Table 10.2. They can be seen to vary rather widely, and account for the large variation in sizes of free-falling drops for different liquids. The higher the surface tension, the larger the drops, as a large drop will have less surface area than many smaller drops with the same total volume. The size of the drop can be used to quantify surface tension, γ, using Tate's law:

See the videos "Surface Tension", "Cube Bubble" and "Spiral Bubble" under Student Resources at PhysicalPropertiesOfMaterials.com

---

$$m_{ideal} = 2\pi r \gamma / g \qquad (10.3)$$

where m_{ideal} is the ideal mass of a drop falling from an opening of a tube of radius r and g is the gravitational acceleration. In this context, m_{ideal} is related to the observed mass, m_{obs}, by

$$m_{obs} = f\, m_{ideal} \qquad (10.4)$$

where f is a correction factor, given in Table 10.3, that accounts for residual liquid left behind when the drop detaches, as shown in Figure 10.7.

Another way to determine surface tension is the *capillary effect*. Water rises in small openings (capillaries) because the energy reduction on wetting the surface (a favorable effect) is balanced by the gravitational energy cost. The smaller the capillary diameter, the higher the water rises (less gravitational cost for a given increase in height due to the decreased diameter). If a liquid

TABLE 10.3

Correction Factors for the Drop Weight Method to
Determine Surface Tension[a]

$r/V^{1/3}$	f
0.30	0.7256
0.40	0.6828
0.50	0.6515
0.60	0.6250
0.70	0.6093
0.80	0.6000
0.90	0.5998
1.00	0.6098
1.10	0.6280
1.20	0.6535

Source: A. W. Adamson, 1967. *Physical Chemistry of
        Surfaces,* 2nd ed. John Wiley & Sons, Hoboken, NJ.
[a] r is the tube radius and V is the volume of the drop.

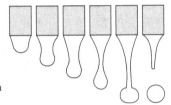

FIGURE 10.7
The sequence of shapes for a drop falling from an open
tube. Note the residual liquid after the drop is detached.

wets the walls of a capillary (e.g., water on glass), as shown in Figure 10.8, the
radius of curvature, R, is related to the capillary radius, r:

$$\Delta p = \frac{2\gamma}{R} = \frac{2\gamma\cos\theta}{r} \tag{10.5}$$

where now $\Delta p = p_1 - p_2$ and p_1 and p_2 are pressures outside and inside the liquid, respectively, and θ quantifies the curvature of the meniscus (see Figure
10.8). The value $(p_1 - p_2)$ corresponds to the change in the hydrostatic pressure
of the liquid, ρgh, where ρ is the density, g is the gravitational acceleration,
and h is the capillary rise. These forces balance at equilibrium and therefore

$$\rho gh = \frac{2\gamma\cos\theta}{r} \tag{10.6}$$

and surface tension can be determined by direct measurement of capillarity.
     Water wets a glass surface, making $h > 0$, whereas for mercury, $\theta \approx 140°$ (see
Figure 10.9), which leads to $h < 0$, that is, capillary depression results.
     The *capillary rise* for water at room temperature would be 2.8 cm for a
1-mm-diameter tube and 28 cm for a 0.1-mm-diameter tube. The rise of water

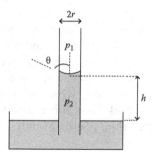

FIGURE 10.8
Rise of a fluid such as water in a glass capillary.

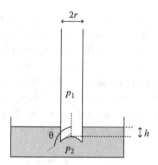

FIGURE 10.9
Mercury is depressed in a glass capillary.

in a capillary accounts for the transport properties of fluids in plants and animals. It also partially accounts for the maximum possible height of a tree: It is limited by the minimum possible diameter of the capillaries involved, based on the molecules from which they are structured.

Typical *capillary depression* for mercury is −0.79 cm for a 2-mm-diameter tube and −15.7 cm for a 0.1-mm-diameter tube. Although today pressures in laboratories are seldom measured using mercury manometers, when such manometers are used, the effects of capillarity on the "true" height of the mercury need to be taken into account for accurate pressure determination, especially in small capillaries.

COMMENT: NONSTICK COATING

A coating made by copolymerization of esters of fluoroalkyl alcohols with functionalized vinyl monomers gives a nonstick material that cannot be wetted by solvents. Its cross-linked structure makes it very durable, and its nonstick properties arise from the orientation of the perfluoro groups on the surface. Although the nonstick surface has some similarities to Teflon (also a fluorinated polymer), there is a major difference, in that Teflon must be heated to get a good coating, and this material is heat sensitive. However, the surface of this material is so nonstick that magic markers, which will wet Teflon, cannot wet this surface. (S. Borman, 1994. Uses for nonwettable, nonstick coating probed. *Chemical and Engineering News*, March 7, 1994, 6.)

COMMENT: SURFACE WETTING

When a small drop of liquid is placed on a flat surface, it can either wet the surface (contact angle, θ_c, is zero) or partially wet the surface ($\theta_c > 0$). For partial wetting, at the contact line of the droplet, which is typically a circle on the surface, there are three phases in equilibrium: the solid, S; the liquid, L; and the vapor, V. Each interface has a certain Gibbs energy per unit area, given by γ_{SL}, γ_{SV}, and γ_{LV}, where the last is the familiar surface tension, usually represented by γ. The British scientist, Thomas Young, showed that, at equilibrium, the following relationship holds:

$$\gamma_{SV} - \gamma_{SL} - \gamma \cos\theta_c = 0. \tag{10.7}$$

Therefore, the measurement of the contact angle of a liquid droplet of known surface tension on a solid surface gives the difference in surface energetics of the S – V and S – L interfaces. Alternatively, we can predict the contact angle of a liquid droplet on a solid surface from knowledge of the three surface energies involved.

An example of the determination of the contact angle of water on a thin film of carbon nanotubes is shown in shown in Figure 10.10.

COMMENT: ANTIFOGGING

Fogging of windows and glasses occurs when very small water droplets condense on the surface. The water droplets scatter light and make the surface appear translucent. They form when the humidity is high (e.g., the mirror in a bathroom when the shower is running), or when warm, high-humidity air hits a cold surface (e.g., eyeglasses when you come in from the cold on a winter's day). Coatings can prevent fogging. An example is coatings composed of alternating layers of silica nanoparticles and a polymer. The nanoparticles in the coating are super-hydrophilic, leading to much smaller contact angles than on the uncoated surface. Instead of light-scattering droplets, the antifog coating leads to a uniform, transparent sheet of water on the surface.

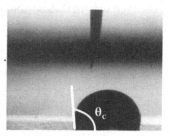

FIGURE 10.10
The contact angle, θ_c, of a 3-mm-diameter water droplet on a carbon nanotube thin film is about 100°. The water was dropped from a syringe that can be seen at the top of the image. (Photo courtesy of M. Jakubinek.)

10.5 Liquid Films on Surfaces

When *amphiphilic* molecules* (molecules that are both *hydrophobic* [water-hating] and *hydrophilic* [water-loving]) are dispersed on a water surface, they tend to orient themselves with their hydrophobic ends (e.g., alkyl tails) away from the water and their hydrophilic ends (e.g., polar head groups) pointing into the water. This situation is shown schematically at low surface coverage in Figure 10.11a.

If the water surface is compressed, the surface layer becomes more ordered until eventually the surface coverage is exactly one molecule thick, with each molecule occupying the smallest surface area that it can (Figure 10.11b); this is called a *monolayer*. As the water surface is compressed further, its structure will collapse (Figure 10.11c).

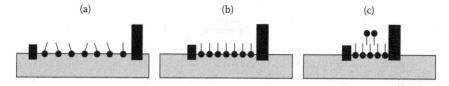

(a) (b) (c)

FIGURE 10.11
(a) A schematic view of surfactant molecules at low surface coverage on water. The polar head group points into the water and the nonpolar tail points out of the water. (b) As the pressure is increased (i.e., the surface area is decreased), a one-molecule-thick film (monolayer) is formed. (c) At higher pressures, the film collapses. Here the pressure is defined as the force per unit length of the edge of the monolayer. The nonpolar tails of the surfactant molecules would be dynamically disordered.

---

* As we shall see in Chapter 11, amphiphilic molecules influence surface tension and therefore are usually called *surface active agents* often abbreviated as *surfactants*.

The surface pressure (which has units of N m^{-1}, compared with N m^{-2} for bulk pressure) can be measured by the force on a floating barrier, as shown in the trough in Figure 10.12. As the surface area is decreased, the pressure increases, until the monolayer is formed; with the collapse of the monolayer, the pressure falls, as shown in the phase diagram in Figure 10.13. The difference in surface tension above and below what we now know to be monolayer coverage was first described by Pockels.*

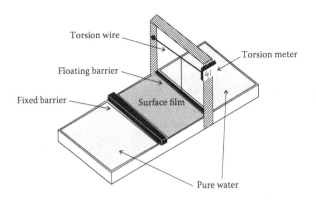

FIGURE 10.12
A schematic diagram of a floating-barrier trough for measuring the pressure of surface films. This is commonly referred to as a Langmuir–Blodgett trough.

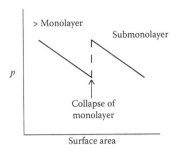

FIGURE 10.13
Two-dimensional pressure versus area relationship for a Langmuir–Blodgett film showing the collapse of the film as the surface area is decreased.

---

* Agnes Pockels (1862–1935) was born in Vienna and raised in Germany. Despite her lack of formal training (she had only high school education), she carried out seminal work related to surface films, starting with her first investigations at age 18. She carried out her work in her home, examining the influence of films on water. Ignored by most German scientists, she was encouraged by her brother, a professor of physics, to write to Lord Rayleigh about her results, and Lord Rayleigh aided her in publishing her work in *Nature* in 1891.

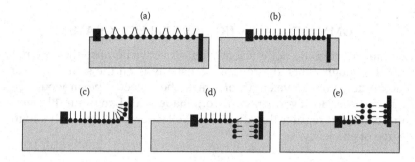

FIGURE 10.14
Deposition of a Langmuir–Blodgett film on a solid substrate by successive passes, (a) to (e), of a moveable barrier. In this case, the layers are all head-to-head and tail-to-tail.

Coherent (i.e., highly organized) films of this type are usually referred to as *Langmuir–Blodgett films.*[*] They have many special properties due to the highly regular arrangement of molecules one molecular layer thick. Although such films themselves are rather fragile, they can be transferred to solid substrates, giving rise to many applications. The process of the transfer of one or more layers is shown schematically in Figure 10.14. This transfer allows stacking of up to thousands of monolayers on a substrate in either pure or doped forms.

Langmuir–Blodgett films can lead to interesting electrical and optical properties, which can be tailor-made by changing the composition of the film. For example, insulating films can be placed on semiconducting substrates, enhancing their electrical properties for use as photovoltaic devices. The *microlithographic* (literally "small printing") possibilities of these films can lead to "wires" of 2–10 nm width. Special optical devices with second- or third-harmonic generation (recall that these take light of a certain frequency ν and produce light of frequency 2ν or 3ν, respectively; see Chapter 5) are possible by coating fibers with coherent films.

The highly organized Langmuir–Blodgett films also can be of use in adhesion, catalysis, and in the field of biosensors. Polymerization within layers after deposition can lead to increased mechanical stability.

The important factors that lead to the utility of Langmuir–Blodgett films are nanometer control over the thickness of the film, the uniformity of the film, and the orientation and architecture of the molecules involved.

---

[*] Irving Langmuir (1881–1957) was an American chemical physicist who spent most of his career as a research scientist at the General Electric Company in Schenectady, New York where he worked on gases, films, and electrical discharges. He was awarded the 1932 Nobel Prize in Chemistry for contributions to surface science. Katherine Burr Blodgett (1898–1979), also was a distinguished scientist working at General Electric. Blodgett was the inventor of antireflective coating for glass.

COMMENT: CAN MIRRORS REFLECT X-RAYS?

Ordinary mirrors do not efficiently reflect x-rays because they are not smooth enough. Since the roughness usually is on the same order of magnitude as the wavelength of x-rays, the reflected beam would be quite diffuse. However, x-ray mirrors made with Langmuir–Blodgett films can be so regular that they allow reflection of x-rays with a high degree of efficiency.

COMMENT: LOW-EMISSIVITY GLASS

Low-emissivity coatings reduce the transparency of glass to ultraviolet and infrared light, without substantially reducing the amount of visible light that passes through. When heat or light is absorbed by glass, it can be re-radiated by the glass surface, in a process known as emissivity. Emitted energy increases unwanted heat transfer through a window by heat loss from the building in winter and heat gain from the outside in the summer. Low-emissivity (low-E) coatings reduce that unfavorable energy transfer: About 85% of long-wavelength light hitting an uncoated window can be transmitted, but transmission can be reduced to 2% with a low-E coating. Typical low-E coatings are either layers of metals, metal oxides, and metal nitrides, deposited by a technique known as magnetron sputter vacuum deposition or metal oxides that are applied to the surface by chemical vapor deposition (CVD). Low-E coatings are very thin, of the order of 10^{-8} m, and often made of many layers. These low-E coatings are similar in many ways to Langmuir–Blodgett films; the main distinction is that Langmuir–Blodgett films are composed of molecular layers and generally more ordered than sputtered or CVD-deposited metals and oxides.

COMMENT: COMMAND SURFACES

When a light-sensitive material is coated on a surface, this layer can be used to modify the properties of the assembly. The azobenzene layer shown in Figure 10.15 changes orientation in the presence of light, which in turn changes the alignment of the outer liquid–crystalline layer, as shown. The resultant change in optical and/or electrical properties can be controlled "on command." (K. Ichimura et al., 1993. *Langmuir*, 9, 3298.)

> ## COMMENT: SELF-ASSEMBLED MONOLAYERS
>
> Self-assembled monolayers (SAMs) are formed from bi-functional organic molecules that spontaneously attach themselves to surfaces by chemisorption. The SAM is generally highly organized, as shown schematically in Figure 10.16. The SAM structure is similar to a Langmuir–Blodgett film, the main distinction being that it self-assembles from solution. The driving force is the strong interaction of the head group (e.g. containing sulfur) with the substrate surface (e.g. gold). SAMs can be inexpensive to produce, leading to versatile coatings with controlled wetting and adhesion and other chemical and physical properties. SAMs also can be used to functionalize nanostructured materials.

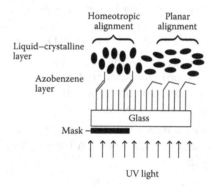

FIGURE 10.15
Light induces a change in the orientation of the azobenzene layer in this coating, which in turn changes the orientation of the liquid–crystalline layer and the optical and electrical properties of this assembly in the area in which the light is not masked.

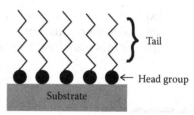

FIGURE 10.16
A schematic structure of a self-assembled monolayer (SAM), showing the bifunctional adsorbate molecule, with a head group that adheres to the substrate surface, and a tail that does not.

10.6 Nanomaterials: A Tutorial

Many materials and devices have their properties associated with the extremely small dimensions of the repeat unit; as discussed already, these are often referred to as *nanomaterials* if at least one of their dimensions is less than 100 nm.

a. The band structure of a metal relies on the electrons being delocalized over ~10^{23} atoms. If a "nanostructure" is composed of atoms that in their bulk phase are metallic, do you expect the nanomaterial to be metallic? Explain.

b. Do you expect the band structure of a semiconductor material in a nanostructure to be the same as that for the bulk material? Explain. Give an indication of the utility of a semiconductor nanostructure.

c. There are several methods that can be used to produce nanostructures. Langmuir–Blodgett deposition is one; others include chemical vapor deposition (CVD), molecular beam epitaxy (MBE; growth of a material from a molecular beam onto a substrate), and lithographic techniques. Explain how scanning tunneling microscopy (STM) and atomic force microscopy (AFM) could be used to characterize these structures.

d. On the basis of the principles by which they operate, consider the required atmosphere for STM compared with AFM. Can either be used in open air, or with liquids present? Discuss.

e. What is the advantage of having such small-scale materials for the storage of information? Consider that, for example, in digital magnetic tape storage devices, the magnetic field orientation at each domain is oriented either "ON" or "OFF."

f. Propose uses for nanostructured materials in optical devices. Consider interference effects and nonlinear optical effects and any other pertinent effects.

g. How would you expect the heat capacity of a nanostructured material to compare with its bulk counterpart? Explain.

h. How would you expect the intrinsic thermal conductivity of a nanostructured material to compare with the same material as a single crystal? Explain.

i. Explain whether a nanomaterial would be expected to be more or less stable thermodynamically than its bulk counterpart. Could your answer influence long-term stability of nanomaterials?

10.7 Learning Goals

- Absorption vs. adsorption
- Chemisorption vs. physisorption
- Graphene

- Surface energy
- STM and AFM
- Surface tension and capillarity
- Films on surfaces: monolayer, Langmuir–Blodgett films
- Nanomaterials
- Self-assembled monolayers

10.8 Problems

10.1 Calculate (a) the ideal mass and (b) the observed mass of a drop of water falling from an opening of radius:

   i. 1 mm

   ii. 3 mm

10.2 If a given surfactant molecule has a cross-sectional area of 0.06 nm², calculate:

   a. The number of such molecules that will be dispersed in a monolayer film of dimensions 15 cm × 30 cm.

   b. The volume of 0.1 M solution (where 1 M ≡ 1 mole of solute per liter of solution) of this surfactant in a nonpolar volatile solvent to produce the film in (a). (This can be dispersed dropwise on the surface and the solvent will evaporate, leaving behind the surfactants on the surface of the water.)

10.3 STM and AFM are both very useful techniques for studying surfaces. Most researchers in this field agree that "the tip is everything." Explain this comment.

10.4 Often, after a shower, there is still some water in the shower head. This water may stay for some time, but if there is some force applied to the shower head (which can be as little as the air pressure changing on closing the door to the bathroom), the water will drain away. Explain this phenomenon. Include energetics in your explanation.

10.5 Nippon Telephone & Telegraph (NTT) experienced trouble with snow sticking to wires and antennas in the Hida Mountains of Japan. As a result, NTT developed a coating to which snow would not stick. This coating, which can be painted on, is a fluorine-containing compound with a low surface Gibbs energy, which only weakly interacts with water. As a result, when applied to a roof with a slope of at least 45°, snow does not accumulate.

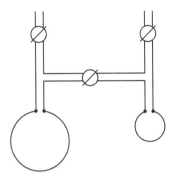

FIGURE 10.17
Two soap bubbles, one larger than the other, are attached to glass tubes as shown here.

     a. Can you suggest other uses for a low water–surface interaction coating? Consider applications where materials must move through water.

     b. The NTT material does not absorb light in the visible range and its fine particles on the surface disperse white light. What color is the coating? Does this lead to any practical applications that would be ruled out if it were another color?

10.6 Two soap bubbles, one smaller than the other, are each attached to ends of glass tubes, and the tubes are connected *via* a stopcock, as shown in Figure 10.17. When that stopcock is opened, will the bubbles change in size, and, if so, which will grow? Explain your answer.

10.7 Given the surface tension of water (7.2×10^{-2} J m^{-1}) at room temperature, calculate the minimum energy required to break a 1.0-cm-diameter sphere of water into 10^{-6} m diameter droplets. (This is the minimum energy because breaking a sphere will produce droplets of the required size and also smaller droplets.)

10.8 A clever way to determine surface energies of solids has been devised (F. H. Buttner et al., 1952. *Journal of Physical Chemistry*, 56, 657). A number of wires of the material, all of the same initial diameter and length, have weights of various masses attached to them. The wires are heated to about 90% of the absolute temperature of their melting point to increase atomic mobility. The gravitational pull on the wire will tend to elongate the wires, but with increased atomic mobility, the surface tension will tend to decrease the length of the wire. The first factor will dominate with large weights and the second will dominate with small weights. At some intermediate weight, the two forces balance such that no elongation is observed. Derive a relationship between the surface energy (G_s), the wire radius (r), and the mass of the weight that causes no elongation (m).

10.9 a. A bubble has a volume of 10 cm³. Calculate its surface area if its shape is a

   i. Cube;

   ii. Tetrahedron; and

   iii. Sphere.

   b. Which shape gives the smallest surface area?

10.10 AgNO₃ and NaCl are both quite soluble in water, whereas AgCl is virtually insoluble. When aqueous solutions of AgNO₃ and NaCl are mixed, a fine precipitate of AgCl forms. This precipitate has a much higher surface area than a large crystal of AgCl with an equivalent mass and the large crystal would have a lower energy. Why is a large crystal of AgCl not formed?

10.11 a. A spherical bubble of volume V breaks into N smaller spherical bubbles, each of the same size. If A is the surface area of the original bubble and A_N is the total surface area of the smaller bubbles, derive a relationship for A/A_N.

   b. What is A/A_N for:

   i. $N = 2$?

   ii. $N = 10$?

   iii. $N = 1000$?

   iv. $N = 10^6$?

10.12 Magnesium oxide (MgO) easily forms a nanomaterial, with cubes of the order of nanometers formed by the combustion of magnesium metal. Calculate:

   a. The surface area of:

   i. One MgO cube with 1 cm sides;

   ii. The same total volume of MgO divided into cubes each with 40 nm sides.

   b. The proportion of surface atoms on the cubes from each situation in (a), expressed in percent of the total number of atoms. The density of MgO is 3.6 g cm⁻³.

10.13 a. Take a clean beaker or glass and fill it with water. Sprinkle pepper on the surface of the water. Now put a tiny bit of soap on the end of your finger and dip it into the water. What happens? Why?

   b. If you repeat the same experiment using oil in place of water, are your results the same? Explain.

10.14 The surface tension of water is more than 3 times that of methanol (see Table 10.2). Why?

10.15 After a snowfall, under certain weather conditions, the flakes of snow consolidate into larger ice crystals. What is the driving force for this transformation?

10.16 A low-emissivity fibrous material has been prepared (Y. Cui et al., *Advanced Materials* 2018, 30, 1705374) and, when woven, it reflects the heat of a body back to the body. The process is similar to low-E glass used for insulation and the fibers are modelled on polar bear hair. How is this structure of advantage to a polar bear?

10.17 Ruby glass is a glass that derives its color from the presence of about 0.01% gold in ordinary glass (see Chapter 3). If the glass is cooled relatively quickly, the gold is fairly uniformly distributed in the glass, and it is colorless. If the glass is heated to about 920 K for several hours and then cooled, the gold aggregates to form nanoparticles (40–140 nm). What is the driving force for gold aggregation?

10.18 TiO_2 polymorph stability depends on both the structure and on how finely divided the material is. Relative to bulk rutile, bulk brookite and bulk anatase are 0.71 and 2.61 kJ mol^{-1} higher in energy, respectively. The surface energies of rutile, brookite, and anatase are 2.2, 1.0, and 0.4 J m^{-2}, respectively.

a. Use this information to draw a diagram that shows the energy of each of these three TiO_2 polymorphs as a function of surface area from zero surface area (i.e., bulk) to 15,000 m^2 mol^{-1} and indicate by a heavy line which polymorph is most stable as a function of surface area.

b. The energetics described above are enthalpies. What additional information would be required to expect that the phase stability you showed in (b) as Gibbs energies? Would the result be expected to be much different? Explain.

Further Reading

General References

Many physical chemistry textbooks include information on surface science. In addition, the following address many of the general points raised in this chapter.

Diamond films: Recent developments. Special issue. D. M. Gruen and I. Buckley-Golder, Eds. *MRS Bulletin*, September 1998, 16–64.

Polymer surfaces and interfaces. Special issue. J. T. Koberstein, Ed. *MRS Bulletin*, January 1996, 16–53.

Solid-liquid interfaces: Molecular structure, thermodynamics, and crystallization. Special issue. M. Asta, F. Spaepen, and J. F. van der Veen, Eds. *MRS Bulletin*, December 2004, 920–966.

V. K. Agarawal, 1988. Langmuir-Blodgett films. *Physics Today*, June, 40.

F. J. Almgren Jr. and J. E. Taylor, 1976. The geometry of soap films and soap bubbles. *Scientific American*, July, 82.

A. E. Anwander, R. P. J. S. Grant, and T. M. Letcher, 1988. Interfacial phenomena. *Journal of Chemical Education*, 65, 608.

D. J. Barber and R. Loudon, 1989. *An Introduction to the Properties of Condensed Matter.* Cambridge University Press, Cambridge.

A. Barton, 1997. *States of Matter: States of Mind.* Taylor & Francis, Washington, DC.

F. E. Condon and F. E. Condon Jr., 2001. The floating needle. *Journal of Chemical Education,* 78, 334.

G. Cooke, 1997. Laying it on thin. *Chemistry in Britain,* April, 54.

V. S. J. Craig, A. C. Jones, and T. J. Senden, 2001. Contact angles of aqueous solutions on copper surfaces bearing self-assembled monolayers. *Journal of Chemical Education,* 78, 345.

P. De Gennes, 1985. Wetting: Statics and dynamics. *Reviews of Modern Physics,* 57, 827.

M. Dionisio and J. Sotomayor, 2000. A surface chemistry experiment using an inexpensive contact angle goniometer. *Journal of Chemical Education,* 77, 59.

G. Ertl and H.-J. Freund, 1999. Catalysis and surface science. *Physics Today,* January, 32.

F. Frankel and G. M. Whitesides, 1997. *On the Surface of Things: Images of the Extraordinary in Science.* Chronicle Books, San Francisco, CA.

H. D. Gesser, 2000. A Demonstration of surface tension and contact angle. *Journal of Chemical Education,* 77, 58.

L. C. Giancarlo, H. Fang, L. Avila, L. W. Fine, and G. W. Flynn, 2000. Molecular photography in the undergraduate laboratory: Identification of functional groups using scanning tunneling microscopy. *Journal of Chemical Education,* 77, 66.

M. Gross, 2000. The incredible nanoplotter. *Chemistry in Britain,* October, 25.

B. Halford, 2016. Natural trifecta inspires water-harvesting surface. *Chemical and Engineering News,* February 29, 10.

G. G. Hall, 1991. *Molecular Solid State Physics.* Springer-Verlag, New York.

U. Henricksson and J. C. Eriksson, 2004. Thermodynamics of capillary rise: Why is the meniscus curved? *Journal of Chemical Education,* 81, 150.

L. V. Interrante, L. A. Casper, and A. B. Ellis, Eds., 1995. Materials chemistry: An emerging discipline. In *Advances in Chemistry,* Series 245. American Chemical Society, Washington, DC.

M. Jacoby, 2001. Custom-made biomaterials. *Chemical and Engineering News,* February 5, 30.

M. Jacoby, 2012. Anti-icing surfaces. *Chemical and Engineering News,* July 23, 31.

K. Kabza, J. E. Gestwicki, and J. L. McGrath, 2000. Contact angle goniometry as a tool for surface tension measurements of solids, using Zisman plot method. *Journal of Chemical Education,* 77, 63.

M. Lagally, 1993. Atom motion on surfaces. *Physics Today,* November, 24.

J. Lahann and R. Langler, 2005. Smart materials with dynamically controllable surfaces. *MRS Bulletin,* March, 185.

L. C. McKenzie, L. M. Huffman, K. E. Parent, and J. E. Hutchison, 2004. Patterning self-assembled monolayers on gold. *Journal of Chemical Education,* 81, 545.

M. Meyyappan and M. K. Sunkara, 2010. *Inorganic Nanowires.* Taylor & Francis, New York.

T. Munguia and C. A. Smith, 2001. Surface tension determination through capillary rise and laser diffraction patterns. *Journal of Chemical Education,* 78, 343.

Y. Pomeau and E. Villermaux, 2006. Two hundred years of capillarity research. *Physics Today,* March, 39.

T. Richardson, 1989. Langmuir-Blodgett films. *Chemistry in Britain,* December, 1218.

A. J. Rosenthal, 2001. Demonstration of surface tension. *Journal of Chemical Education,* 78, 332.

A. M. Rouhi, 1999. Contemporary biomaterials. *Chemical and Engineering News*, January 18, 51.

G. A. Somorjai and G. Rupprechter, 1998. The flexible surface: Molecular studies explain the extraordinary diversity of surface chemical properties. *Journal of Chemical Education*, 75, 162.

P. Spanoghe, J. Cocquyt, and P. Van de Meeren, 2001. A low-cost dynamic surface tension meter with a LabVIEW interface and its usefulness in understanding foam formation. *Journal of Chemical Education*, 78, 338.

A. Swift, 1995. Skimming the surface. *Chemistry in Britain*, November, 887.

Atomic Microscopy

Atomic probe tomography. Special issue. Y. Amouyal and G. Schmitz, Eds. *MRS Bulletin*, January 2016, 13–52.

Scanning probe microscopy in materials science. Special issue. E. Meyer, S. P. Jarvis, and N. D. Spencer, Eds. *MRS Bulletin*, July 2004, 443–487.

R. Baum, 1994. Chemical force microscopy. *Chemical and Engineering News*, October 3, 6.

C. Bustamante and D. Keller, 1995. Scanning force microscopy in biology. *Physics Today*, December, 32.

H. Carmichael, 1997. In search of atomic resolution. *Chemistry in Britain*, November, 28.

R. Dagani, 1996. Putting the nano finger on atoms. *Chemical and Engineering News*, December 2, 20.

D. M. Eiger and E. K. Schweizer, 1990. Positioning single atoms with a scanning tunneling microscope. *Nature*, 344, 524.

F. J. Giessibl and C. F. Quate, 2006. Exploring the nanoworld with atomic force microscopy. *Physics Today*, December, 44.

W. F. Heinz and J. H. Hoh, 2005. Getting physical with your chemistry: Mechanically investigating local structure and properties of surfaces with the atomic force microscope. *Journal of Chemical Education*, 82, 695.

C. M. Henry, 2001. Spectroscopy gets up close and personal. *Chemical and Engineering News*, April 30.

M. Jacoby, 2017. Scanning probes investigate nanoscale phenomena. *Chemical and Engineering News*, February 27, 25.

J. Leckenby, 1995. Probing surfaces. *Chemistry in Britain*, March, 212.

C. Lieber, 1994. Scanning tunneling microscopy. *Chemical and Engineering News*, April 18, 28.

S. Magonov and Y. Godovsky, 1998. Atomic force microscopy, Part 7: Studies of thermal phase transitions in polymers. *American Laboratory*, November, 15.

M. M. Maye, J. Luo, L. Han, and C.-J. Zhong, 2002. Chemical analysis using scanning force microscopy. An undergraduate laboratory experiment. *Journal of Chemical Education*, 79, 207.

G. Meyer and K. H. Rieder, 1998. Lateral manipulation of single absorbates and substrate atoms with the scanning tunneling microscope. *MRS Bulletin*, January, 28.

C. O'Driscoll, 2001. Atoms in the spotlight. *Chemistry in Britain*, December, 20.

D. Rugar and P. Hansma, 1990. Atomic force microscopy. *Physics Today*, October, 23.

M. Shusteff, T. P. Burg, and S. R. Manalis, 2006. Measuring Boltzmann's constant with a low-cost atomic force microsope: An undergraduate experiment. *American Journal of Physics*, 74, 863.

Nanomaterials

Advances in carbon nanotubes. Special issue. M. S. Dresselhaus and H. Dai, Eds. *MRS Bulletin*, April 2004, 237–285.

Nanomaterial combines low density, stiffness, elasticity. *Chemical and Engineering News*, December 11, 2000, 43.

Nanoscale characterization of materials. Special issue. E. T. Yu and S. J. Pennycock, Eds. *MRS Bulletin*, August 1997, 17–61.

Optics of nanostructures. Special issue. D. S. Chemia, Ed. *Physics Today*, June 1993, 22–73.

Silicon nanocrystals amplify light. *Chemical and Engineering News*, November 27, 2000, 21.

Toward applications of ceramic nanostructures. Special issue. S. Seal and M.-I. Baraton, Eds. *MRS Bulletin*, January 2004, 9–47.

S. Ashley, 2001. Nanobot construction crews. *Scientific American*, September 2001, 84.

D. Bradley, 2018. Thousands of 2D materials are just waiting to be discovered. *Chemistry World*, February 15.

S. Casalani, C. A. Bortolotti, F. Loenardi and F. Biscarini, 2017. Self-assembled mono-layers in organic electronics. *Chemical Society Reviews*, 46, 40.

C. Q. Choi, 2006. Crystal steer. *Scientific American*, April, 26.

M. W. Cole, V. H. Crespi, M. S. Dresselhaus, G. Dresselhaus, J. E. Fischer, H. R. Gutierrez, K. Kojima, G. D. Mahan, A. M. Rao, J. O. Sofo, M. Tachibana, K. Wako, and Q. H. Xiong, 2010. Structural electronic, optical and vibrational properties of nanoscale carbons and nanowires: A colloquial review. *Journal of Physics—Condensed Matter*, 22, 334201.

E. Corcoran, 1990. Diminishing dimensions. *Scientific American*, November, 122.

R. Dagani, 1998. Architectural tour of the nano world. *Chemical and Engineering News*, September 21, 70.

R. Dagani, 2000. Nanotubes spun into long fibres. *Chemical and Engineering News*, November 20, 9.

R. Dagani, 2001. Nanotube magic. *Chemical and Engineering News*, April 16, 6.

R. Dagani, 2001. Polymers "worms" line up on silicon. *Chemical and Engineering News*, April 2.

R. Dagani, 2001. Slimming down inorganic tubes. *Chemical and Engineering News*, April 23.

P. Day, 1996. Room at the bottom. *Chemistry in Britain*, July, 29.

C. Day, 2016. Boron nitride nanotubes reinforce polymer materials. *Physics Today*, March, 20.

C. Dekker, 1999. Carbon nanotubes as molecular quantum wires. *Physics Today*, May, 22.

J. T. Dickinson, 2005. Nanotribology: Rubbing on a small scale. *Journal of Chemical Education*, 82, 734.

K. E. Drexler, 2001. Machine-phase nanotechnology. *Scientific American*, September, 74.

B. D. Fahlman, 2002. Chemical vapor deposition of carbon nanotubes. An experi-ment in materials chemistry. *Journal of Chemical Education*, 79, 203.

M. Freemantle, 2001. 3-D Images of nanoparticles. *Chemical and Engineering News*, May 28, 9.

M. Freemantle, 2006. Exploiting polymer nanostructures. *Chemical and Engineering News*, April 24, 39.

K. J. Gabriel, 1995. Engineering microscopic machines. *Scientific American*, September, 150.

D. Gammon and D. G. Steel, 2002. Optical studies of single quantum dots. *Physics Today*, October, 36.

J. E. Greene, 2014. Organic thin films: From monolayers on liquids to multilayers on solids. *Physics Today*, June, 43.

B. Halford, 2005. Inorganic menagerie. *Chemical and Engineering News*, August 29, 30.

J. R. Heath, 1995. The chemistry of size and order on the nanometer scale. *Science*, 270, 1315.

M. Jacoby, 1999. Nanotubes coil, story unfolds. *Chemical and Engineering News*, October 4, 31.

M. Jacoby, 2001. Boron nitride nanotubes formed. *Chemical and Engineering News*, March 19, 9.

M. Jacoby, 2001. Wires for a nanoworld. *Chemical and Engineering News*, January 1, 28.

M. Jacoby, 2001. Writing between the lines. *Chemical and Engineering News*, February 12, 10.

M. Lagally, 1998. Self-organized quantum dots. *Journal of Chemical Education*, 75, 277.

J. Leckenby and R. Eby, 2004. Toward industrial nanotechnology using a nanoscale patterning technique. *American Laboratory*, January, 11.

S. Lindsay, 2005. Single-molecule electronic measurements with metal electrodes. *Journal of Chemical Education*, 82, 727.

J. Newton, 2017. Quantum dots and a bright future. *Chemistry World*, November 19.

G. Ozin and A. Arsenault, 2005. *Nanochemistry*. Royal Society of Chemistry, Cambridge.

P. M. Petroff, A. Lorke, and A. Imamoglu, 2001. Epitaxially self-assembled quantum dots. *Physics Today*, May 2001, 46.

M. J. Pitkethly, 2004. Nanomaterials—the driving force. *Nanotoday*, December, 20.

M. C. Roco, 2006. Nanotechnology's future. *Scientific American*, August, 39.

A. M. Rouhi, 2001. From membranes to nanotubules. *Chemical and Engineering News*, June 11.

R. W. Siegel, 1993. Exploring mesoscopia: The bold new world of nanostructures. *Physics Today*, October, 64.

N. Tao, 2005. Electrochemical fabrication of metallic quantum wires. *Journal of Chemical Education*, 82, 720.

M. Thayer, 2003. Nanomaterials. *Chemical and Engineering News*, September 1, 15.

R. Trager, 2016. Boron nitride nanotube composites outperform carbon cousins. *Chemistry World*, January 14.

G. M. Whitesides, 2001. The once and future nanomachine. *Scientific American*, September, 78.

E. Wilson, 1995. Microlithography uses metastable argon atoms. *Chemical and Engineering News*, September 4, 8.

R. M. Wilson, 2018. Silicon-based quantum dots have a path to scalable quantum computing. *Physics Today*, April, 17.

X. Ye and C. M. Wai, 2003. Making nanomaterials in supercritical fluids: A review. *Journal of Chemical Education*, 80, 198.

P. Zurer, 2001. Nanostructures: Shells and tubes. *Chemical and Engineering News*, March 12, 12.

Websites

For links to relevant websites, see PhysicalPropertiesOfMaterials.com

11

Other Phases of Matter

11.1 Introduction

Many of the phases of matter discussed so far, such as gases, liquids, pure solids, solid solutions, liquid crystals, and orientationally disordered crystals, are *homogeneous* (i.e., uniform throughout). Both in nature and in the laboratory, materials can be combined (by alloying, coating, bonding, dispersion, lamination, etc.) to give new materials with different properties. There are important states of matter that are *heterogeneous* (not uniform throughout); a Langmuir–Blodgett film is but one example. The film is different from the substrate, making it heterogeneous. Other examples of heterogeneous materials are described in this chapter.

11.2 Colloids

A *colloid* is a system of semi-organized particles too small to be visible in an optical microscope, typically 10–1000 nm in size. A colloid can be stable (e.g., a macro-molecular solution or a micellar system; see Section 11.3) or metastable (e.g., a colloidal dispersion). The word "colloid" is derived from the Greek for "gluelike," as glue is a colloid. An important characteristic of colloids is that they do not pass through membranes that are permeable to (smaller) dissolved species.

A colloidal dispersion has its particles (*dispersed phase*) held in a *dispersing medium*. Gas, liquid, or solid can be either the dispersed phase or the dispersing medium, and this distinction leads to eight types of colloidal dispersion, listed with examples in Table 11.1.

In principle, colloidal dispersions are metastable with respect to the separated forms. This lack of stability is apparent to anyone who has let whipped cream sit too long before eating it! However, colloidal dispersions can be kinetically stabilized, and can last for a long time. Gold sols prepared by

TABLE 11.1

Colloidal Dispersion Compositions and Examples

Dispersed Phase	Dispersing Medium	System Name	Examples
Solid	Solid	Solid dispersion	Colored glass
Solid	Liquid	Sol or suspension	Gold in water, ink
Solid	Gas	Aerosol of solid particles	Smoke
Liquid	Solid	Solid emulsion	Jelly
Liquid	Liquid	Emulsion	Milk, mayonnaise
Liquid	Gas	Aerosol of liquid droplets	Fog
Gas	Solid	Solid foam	Pumice, Styrofoam
Gas	Liquid	Foam	Whipped cream

Michael Faraday* more than 150 years ago are still in their dispersed form today; they are on display in the British Museum in London, England. Some colloidal dispersions can be kinetically stabilized by the presence of surface charges on the dispersed particles; they do not *flocculate* (coalesce) due to the mutual repulsion between particles.

Suspended charged particles can be made to move in electrical fields, and this is the basis for the separation process called *electrophoresis*. This process can also be used preparatively, as an electrical field applied across a colloidal suspension of rubber latex can cause flocculation at one electrode. The flocculated latex particles can take on the shape of the electrode, and this is how rubber gloves, for example, are produced.

COMMENT: COLLOIDAL SPONGES

Colloidal microgels are polymeric materials with cells of dimensions in the range 10^{-9} to 10^{-6} m, swollen with solvent molecules. One example is poly(N-isopropylacrylamide), which, in aqueous solution, shrinks on heating due to conformational changes in the polymer, making the polymer–solvent interaction less favorable at high temperature. This reversible transformation reduces the volume by a factor of 60 on heating from 25 °C to 50 °C. The gels also are sensitive to pH and the presence of other ions and molecules. Colloidal microgels have applications in drug delivery, oil recovery, and as viscosity modifiers.

---

* Michael Faraday (1791–1867) was an English scientist who gained his scientific training when Sir Humphrey Davy hired him as an assistant during his temporary loss of sight due to a laboratory accident. Faraday made many major contributions to the theories of electricity and magnetism.

COMMENT: PROCESSING OF CERAMICS

Sol-gel processing can be used to prepare ceramics with considerable control, as schematically illustrated in Figure 11.1. The interactions between ceramic particles can be controlled by coating particles with surfactant (reducing the attraction) or with polymeric surface modifiers (reducing the repulsion). Sol-gel processing also is used to deposit thin films for chemical sensors and other sensors such as pH electrodes, and also to prepare nanoscale ceramic materials.

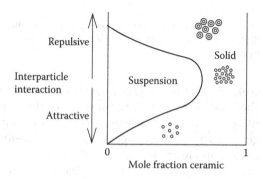

FIGURE 11.1
The balance between attractive and repulsive forces can determine the packing of ceramic particles in a gel phase, with packing fractions ranging from very low values to about 70%, depending on the interparticle interactions.

11.3 Micelles

Micelles are another heterogeneous phase of matter involving aggregates. At low concentrations of amphiphilic molecules (e.g., soap), individual ions are dispersed in aqueous solution. Above a certain temperature (the Krafft temperature, T_K) and above a certain concentration (the *critical micellar concentration*, abbreviated *cmc*) the solution will consist of ions and aggregates called *micelles*. A micelle is a three-dimensional structure which, in a polar solvent, usually has the hydrophilic ends of the molecules pointing out toward the solvent, and the hydrophobic ends pointing inward. Examples of micellar structures are shown schematically in Figure 11.2. For $T > T_K$, the hydrophilic part (hydrocarbon tail) of the surfactant molecule is mobile and dynamically disordered in *trans* and *gauche* configurations, entropically stabilizing the micellar phase. For $T < T_K$, the stabilizing entropy effect is diminished, and the molecules are more rigid, making the crystalline form more stable than the micelles.

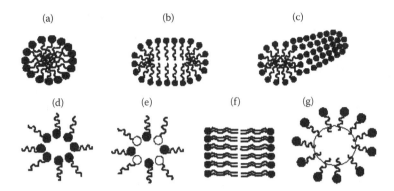

FIGURE 11.2
Schematic view of organized surfactant structures: (a) spherical micelle, (b) oblong micelle, (c) tubular micelle, (d) reverse micelle, (e) mixed reverse micelle, (f) lamellar micelle, and (g) oil droplet solubilized by micelle.

A micelle, typically consisting of 50 to a few hundred monomer ions,* is a dynamic entity, both with respect to the motion of the hydrophobic ends of the constituent species and with respect to exchange with the free ions in solution. A generalized phase diagram of a micelle-forming solution is shown schematically in Figure 11.3.

The physical proof of the existence of micelles comes from many different types of experiments. Because the particle dimensions are of the same order of magnitude as the wavelength of visible light, light scattering can be used to determine the average particle size. A typical light-scattering set-up is shown in Figure 11.4.

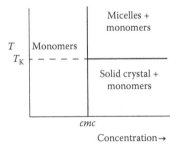

Concentration→

FIGURE 11.3   A schematic temperature–composition phase diagram for an aqueous solution of surfactant molecules indicating the region of micelle formation. T_K is the Krafft temperature, and for $T > T_K$ and concentrations higher than the critical micellar concentration, *cmc*, micelles form.

---

* Typical *cmc* values for aqueous solutions at 20 °C are: $CH_3(CH_2)_{11}N(CH_3)_3Br$, 14.4 mol·L^{-1}, ≈ 50 monomers per micelle; $CH_3(CH_2)_{11}SO_4Na$, 8.1 mol·L^{-1}, ≈ 62 monomers per micelle; $CH_3(CH_2)_{11}(OCH_2)_6OH$, 0.1 mol·L^{-1}, ≈ 400 monomers per micelle.

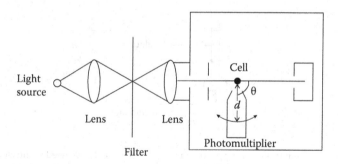

A schematic diagram of a light-scattering apparatus for determination of particle size.

From Chapter 4, we know that the intensity of scattered light of wavelength λ at angle θ, I_θ, and distance d from the sample is related to the incident light intensity, I_i, by:

$$\frac{I_\theta}{I_i} = \frac{8\pi^4\alpha^2\left(1+\cos^2\theta\right)}{d^2\lambda^4} \qquad (11.1)$$

where α is the polarizability. The scattered light depends on the interaction of light with the sample, which depends on the refractive index of the pure solvent (n_0) and the refractive index of the solution (n). Equation 11.1 can be rewritten* in terms of these quantities and the concentration (c, expressed in mass per unit volume) and the *mass-averaged molecular mass*[†] of the solute, $\bar{M}_m$:

$$\frac{I_\theta}{I_i} = \frac{2\pi^2 n_0^2\left(1+\cos^2\theta\right)(n-n_0)^2\,\bar{M}_m}{cN_A d^2\lambda^4} \qquad (11.2)$$

where N_A is Avogadro's number. From Equation 11.2, the mass-averaged molar mass of the solute (micelle) can be determined.

The presence of the micelle can be determined by comparison with the calculated properties of a solution with the same concentration of surfactant molecules that are not aggregated: An ionic micellar solution will have lower

---

* See R. A. Alberty, 1987. *Physical Chemistry*, 7th ed. John Wiley & Sons, Hoboken, NJ, p. 475, for a derivation.
† It is important to note that this is an *average* molecular mass as not all micelles in a solution will have exactly the same composition. There will be a distribution of molecular masses, and this particular average is weighted by the mass of the solute particles.

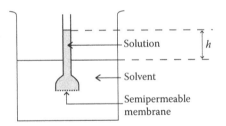

FIGURE 11.5
A schematic representation of a simple osmometer, which is a device used to measure osmotic pressure. The solvent flows through the semipermeable membrane into the solution to increase the pressure of the solution by Π, the osmotic pressure. This flow is reflected in h, the difference between the levels of the solution and the surrounding solvent at equilibrium.

electrical conductivity (it drops dramatically at the *cmc* due to aggregation of ions and some counter ions) and less effective colligative properties such as freezing point depression, boiling point elevation, vapor pressure lowering, and osmotic pressure, due to the aggregation.

These properties can be used to determine the *average molecular mass*[*] of the aggregate in the micelle. For example, the *osmotic pressure*, Π, is the pressure that is developed inside a solution to bring it into equilibrium with pure solvent that is separated from it by a membrane (see Figure 11.5). The membrane is semipermeable; it allows passage of solvent but not solute. As the solvent passes into the solution, the pressure in the solution increases by Π, until equilibrium is attained. The increase in pressure can be reflected in the solution rising up a tube as shown in Figure 11.5, and the rise, h, will be proportional to Π:

$$\Pi = \rho g h \tag{11.3}$$

where g is the gravitational acceleration and ρ is the density of the solution. The osmotic pressure, Π, is directly related to the *number-averaged molecular mass*,[†] $\bar{M}_n$:

$$\Pi = \frac{\bar{M}_n RT}{V_m} \tag{11.4}$$

where V_m is the molar volume of the solvent.

---

[*] This average can be by mass, that is, $\bar{M}_m$, or by number, that is, $\bar{M}_n$, where the former is determined by light scattering and the latter is determined by osmotic pressure.
[†] That is, the average weighted in proportion to the probability of having a certain molecular mass.

11.4 Surfactants

Knowledge of the properties of surfactant molecules can lead to the stabilization of emulsions. As can be seen from Table 11.1, an *emulsion* results from a colloidal suspension of one liquid in another. Oil and water will form two layers if poured gently together, but if shaken vigorously (with the mechanical work going to increase the surface area of the droplets), an emulsion can be formed.

Although an emulsion will tend to separate out again over time, an amphiphilic material added to the emulsion will act as a *surfactant* (which is an abbreviation for *surface active agent*) by lowering the interfacial tension between the oil and water. The surfactant does this by solubilizing one end in the oil and the other in the water, thus stabilizing the emulsion. Because surfactants lower the surface tension, they can lead to spreading, wetting, solubilization, emulsion and dispersion formation, and frothing. For example, foams are metastable but can be made to last longer by the addition of a surfactant. For these reasons, surfactants are common additions to household products such as foods (e.g., as emulsifiers in salad dressing) and cleansing agents.

COMMENT: BIOMIMETIC GELS

Polymer gels that mimic some biological activities have been developed. For example, a moving gel loop can be made to move by simple manipulations: In an electric field, surfactant molecules collect on the strand's top surface and electrostatic forces cause the gel to shrink; with a change in field polarity, the surfactant enters the aqueous phase again and the strand extends to a new position. In this way, the gels can act as artificial muscles. (Y. Osada and S. B. Ross-Murphy, 1993. Intelligent gels. *Scientific American*, May 1993, 82.)

COMMENT: WAITING FOR THE PAINT TO DRY

Latex paints contain a mixture of water, small polymer particles, and a surfactant. The surfactant stabilizes the polymer particles and prevents them from aggregating in the aqueous solution; surfactant molecules surround the polymer particles, in a micellar form, with their nonpolar parts facing inward to solubilize the polymers, and their polar headgroups facing outward into the aqueous solution. After application of the paint, over time the water evaporates, leaving behind the film of latex particles.

COMMENT: SURFACE TENSION IN LUNGS

It has been shown that low surface tension is important in our lungs, to keep them from collapsing. In premature babies, the absence of the surfactant that reduces the surface tension in their lungs has been related to their respiratory difficulties.

11.5 Inclusion Compounds

Inclusion compounds are multicomponent materials in which one type of species forms a host in which other species reside. They can exist in solution or in the solid state. In the latter, the range of topologies of the host lattice allows several different types of structures, as shown in Figure 11.6. Inclusion compounds belong to the larger family of materials known as *supramolecular materials*; these are defined by properties that stem from *assemblies* of molecules. In the case of inclusion compounds, many physical and chemical properties are different in the supramolecular assembly than for the pure host or guest species. Supramolecular assemblies are generally held together by noncovalent forces such as hydrogen bonding, van der Waals interactions, and Coulombic forces. Industrial applications of inclusion compounds include fixation of volatile fragrances and drugs and inclusion of pesticides to make them safer to handle.

A *clathrate* is a material in which the host lattice totally traps the guest species. The term clathrate was coined in 1948 by H. M. Powell,* and a

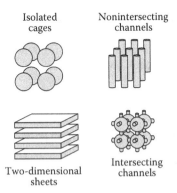

Isolated cages
Nonintersecting channels
Two-dimensional sheets
Intersecting channels

FIGURE 11.6
Schematic representations of possible inclusion compound topologies in the solid state. Isolated cages (e.g., clathrates), nonintersecting channels (e.g., urea compounds), sheets (e.g., graphite intercalates), and intersecting channels (e.g., zeolites) give extended structure in zero, one, two, and three dimensions, respectively.

---

* H. M. Powell, 1948. *Journal of the Chemical Society*, 61.

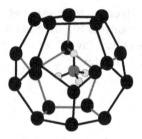

FIGURE 11.7
A portion of the structure of methane clathrate hydrate. Only the oxygen atoms of the H_2O molecules that form the cage are shown; the hydrogen atoms of the water molecules are dynamically disordered. The methane molecule (CH_4) resides as a guest in the cage.

number of materials are now known to form clathrates. Examples are *clathrate hydrates* (H_2O molecules form the host lattice with a number of possible small molecules as guests; see Figure 11.7 and also the tutorial "Applications of Inclusion Compounds" at the end of this chapter).

The case of intersecting channels can be illustrated with the structures of zeolites. *Zeolites* are aluminosilicates that can both occur naturally (and are then usually given mineral names after the location of their discovery) and can be synthesized in the laboratory. They are important for their ability to selectively take up small molecules (e.g., molecular sieves take up water to dry solvents) and also can be used as catalysts to crack petroleum. Some zeolite structures are illustrated in Figure 11.8.

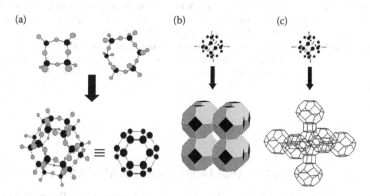

FIGURE 11.8
(a) A schematic representation of the building blocks of a zeolite. Four- and six-membered rings composed of tetrahedral TO_4 units, with T = Si or Al, come together to form cages. For simplicity, only the T atoms are shown at the vertices in the lower right diagram and the O atoms are omitted. The polyhedron is known as a sodalite cage. (b) Additions of other sodalite cages onto the four-membered rings of the central sodalite cage provide a three-dimensional structure with interconnected cages, known as the sodalite structure. (c) Additions of a short channel then other sodalite cages onto the four-membered rings of the sodalite cage provide a three-dimensional structure with cages and channels, known as zeolite A. The lowest structures in (b) and (c) are common forms of representation of zeolite structures.

If the host lattice forms layers with guests between the layers, this is an *intercalate* structure. For example, alkali metals can be intercalated into graphite. Another example is the clay minerals. The structure of one type of clay mineral is shown in Figure 11.9.

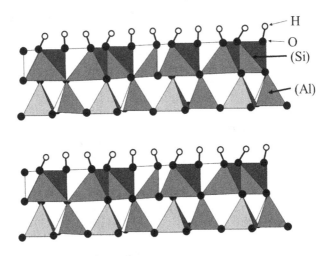

FIGURE 11.9
The structure of kaolinite, a clay mineral of idealized chemical composition $Al_2Si_2O_5(OH)_4$, also sometimes written as $Al_2O_3 \cdot 2(SiO_2) \cdot 2(H_2O)$ to indicate its propensity to lose water at high temperatures. Its name comes from the location Kao-Ling, Jianxi, China. Kaolinite has a layered structure formed of tetrahedral silicate sheets linked through oxygen atoms to sheets of alumina octahedra called *gibbsite layers*. The layered structure of kaolin and the weak silicate-gibbsite interaction allows it to include other species between the layers. Rocks that are rich in kaolinite are known as china clay or kaolin, and they are mined for use in fine china and glossy paper.

COMMENT: MOLECULAR SELF-ASSEMBLY

In the self-assembly shown schematically in Figure 11.10, the π-electron-rich 1,5-dioxynaphthalene unit combines with the π-electron deficient tetracationic cyclophane; this structure can be likened to threading a bead on a wire. The assembly can be monitored optically, as the resulting material has a charge-transfer band in the visible region. (R. Ballardini et al., 1993. *Angewandte Chemie International Edition in English*, 32, 1301.) Molecular level manipulation of structures is important in nanoscale devices.

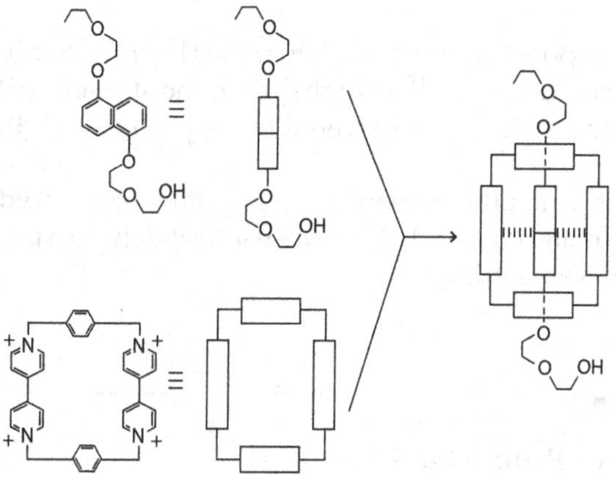

FIGURE 11.10
This molecular self-assembly can be likened to threading a bead on a wire. For the design and synthesis of molecular machines using such techniques, the 2016 Nobel Prize in Chemistry was awarded jointly to Jean-Pierre Sauvage (1944–) of the University of Strasbourg, Sir J. Fraser Stoddart (1942–) of Northwestern University, and Bernard L. Feringa (1951–) of the University of Groningen.

COMMENT: COMPOSITES

Composites are a very important category of heterogeneous materials. Composite materials are composed of one component in a matrix of the other, but the compositions and forms can vary widely from fiber-reinforced polymers to concrete. The main features of composites are that the components are readily distinguishable (i.e., the material is heterogeneous), but the properties of the overall composite are distinct from the properties of the constituents. The matrix can range from plastic, to metal, or ceramic. The other constituent, sometimes called a reinforcement, can be in the form of powder, fiber, or even nanowires, and it can be oriented within the matrix or randomly dispersed.

The properties of composites usually do not obey the *rule of mixtures*, in which property P would be given by*:

$$P = \sum_i f_i P_i \tag{11.5}$$

where f_i is the fraction of component i which has property value, P_i. For example, in a two-component composite, it is not unusual for a property such as strength to be very different from the weighted strength

---

* The most important equations in this chapter are designated with ▌ to the left of the equation.

of the two components, given by Equation 11.5. In fact, the strength of a composite could very well be higher than the strength of the stronger component! Examples are considered in the problems at the end of this chapter.

Composites can offer superior performance compared with pure (monolithic) materials, and they are used widely in sporting goods, bridges, and even cutting tools.

11.6 Hair Care Products: A Tutorial

A large portion of the consumer economy is devoted to hair care products. In many ways, these are related to materials science; the material involved is the biomaterial we know as hair.

Several factors are considered when assessing hair quality: thickness of fiber (i.e., diameter), fiber density (number of hairs/cm^2), stiffness, luster, configuration (curly or straight), and static charge. Some of these properties can be controlled, while others (e.g., density) cannot.

Hair is a protein fiber, with an outer scaly cuticle and an inner cortex. Most of the properties of the hair are concerned with the state of the cuticle.

a. Proteins have both acidic and basic groups and so does hair. Because of this, hair is amphoteric. The acidic groups are $-COOH$, and the basic groups are $-NH_2$. The whole fiber appears neutral at pH = 5.5, and the cuticle is neutral at pH = 3.8. Why is the pH of a shampoo important? What happens to the functional groups of hair at low pH? What happens at high pH?

b. Hair conditioners generally are *cationic surfactants* (i.e., the surfactant ions carry a positive charge), usually with long chain fatty (hydrocarbon) portions. Would these be expected to attach to hair better at low pH or at high pH? Explain.

c. The fatty portion of hair conditioner can act as a lubricant. Is this good or bad? How might the molecular mass of the surfactant play a role?

d. Hair conditioners can leave a film on hair. Is this a desirable feature or not? Consider the reflective properties of films and also their surface energies.

e. At high pH, hair will swell and high molecular weight molecules can be absorbed into the cortex. Can you see an application for this?

f. One of the most annoying things in winter, as anyone with long hair can tell you, is the build-up of static electricity in hair. Why is this not usually a problem in summer? How could hair conditioner help prevent static, flyaway hair?

g. If hair is curled when wet and then allowed to dry in a curled configuration, it retains its curl. Why?

h. How can curl in hair made "permanent"?

11.7 Applications of Inclusion Compounds: A Tutorial

Inclusion compounds can be synthesized in the laboratory, although many occur naturally.

a. Water, when mixed with suitable small molecules and cooled to an appropriate temperature and sometimes also requiring applied pressure, can form ice-like structures called *clathrate hydrates*, in which the water molecules form hydrogen-bonded cages in which the "guest" molecules reside. When the free guest is a gas, such as methane, the resulting materials are called *gas hydrates*. They can exist naturally, and have been found at the bottom of the ocean along the margins of the continents. It has been estimated that there is more combustible hydrocarbon mass in the hydrate form than in the terrestrial oil and gas reserves, but the challenge is to recover the hydrates efficiently. Clathrate hydrates can melt congruently or incongruently, depending on the guest species, the composition, and the overpressure of gas. The organic molecule tetrahydrofuran (THF) is a liquid at room temperature and mixes with water in all proportions. At a mole ratio of 1:17 with water, it forms a clathrate hydrate that melts congruently at $5\,°C$. Given that pure ice melts at $0\,°C$ and pure THF melts at $-108\,°C$ and no compounds other than $THF·17H_2O$ form, sketch and label the H_2O–THF binary phase diagram.

b. Decomposition of naturally occurring clathrate hydrates has been proposed as an explanation for the Bermuda Triangle (i.e., the mysterious disappearance of ships and aircraft in that region). Explain.

c. Explain how an inclusion compound could be used to produce a nonlinear optical material. Discuss advantages and disadvantages.

d. Explain how an inclusion compound could be used to produce a material with special electronic properties. Especially consider nanomaterials.

e. The few inclusion compounds that have been investigated to date appear to have thermal conductivities that increase with temperature at all investigated temperatures.[*] This characteristic is usually associated with glassy materials, but the inclusion compounds are crystalline. Explain this apparent contradiction.

---

[*] See, for example, D. Michalski and M. A. White, 1995. *Journal of Physical Chemistry*, 99, 3774.

11.8 Learning Goals

- Colloids and their various types
- Micelles: structures and conditions for formation; determination of sizes by light scattering and osmotic pressure
- Surfactants
- Inclusion compounds (structures and applications)
- Composites, rule of mixtures

11.9 Problems

11.1 Ethanol and water form a water-rich clathrate hydrate that melts incongruently at about –75 °C. Sketch and label the H_2O–CH_3CH_2OH phase diagram.

11.2 Choosing two phases each from a list of three phases would lead to nine combinations, but only eight types of colloids are listed in Table 11.1. Why is there no entry for a gas dispersed in a gas?

11.3 *Aerogels* are an interesting phase of matter. They are a subset of aerosols, composed of a suspension of fine particles in air. They were first formed in the 1930s by the preparation of a gel phase (from sodium silicate and hydrochloric acid), followed by the removal of the solvent under supercritical conditions. (Supercritical conditions are required because if the solvent is removed as a liquid, the gel collapses due to surface tension considerations.) Recent syntheses have produced aerogels with more than 99% air and densities as low as 3×10^{-3} g cm^{-3}. (The density of air is 1.2×10^{-3} g cm^{-3}.)

   a. One of the main uses of aerogels is in thermal insulation. Explain why an aerogel would conduct heat much less efficiently than the corresponding bulk material (e.g., silica aerogel compared with pure silica).

   b. How could one determine the particle size within the aerogel?

   c. Suggest other uses for aerogels.

11.4 When potassium atoms are intercalated between the layers of carbon atoms in graphite, the extent of intercalation can be followed by color changes. For example, stage 2 intercalates are blue, whereas stage 1 intercalates are gold colored. (The designation "stage n" indicates that there are n layers of graphite between each pair of intercalant layers.) On the basis of the colors, do you expect potassium–graphite intercalates to be semiconductors or metals? Explain.

11.5 An advertisement for NonScents claims that it is a "100% natural odor magnet", "a zeolite," which, by virtue of its "negative molecular ion charge," attracts airborne odor molecules that have a "free ride on dust particles, which have a positive molecular ion charge". The advertisement further claims that "millions of tiny micropores... give the material a great adsorbent surface area." Is NonScents more likely to be useful to trap large organic molecules or small organic molecules? Explain.

11.6 An emulsion of two transparent liquids (e.g., oil and water) leads to a milky-white liquid if the refractive indices of the two phases differ. Why does the emulsion not appear transparent?

11.7 Figure 11.8 shows two zeolite structures, sodalite and zeolite A, that can be considered in terms of building on the four-membered rings on the sodalite cage. Another zeolitic form could be built up by adding hexagonal prisms onto the six-membered rings on the sodalite cage and then adding another sodalite cage at the end of the hexagonal prisms. This mechanism forms the zeolite known as faujasite. How would you expect faujasite to differ from sodalite and zeolite A in terms of the molecules that can be included inside the structure? Explain.

11.8 TiO_2 is often added to paints to make them opaque by scattering light. However, when TiO_2 particles are close together, they scatter less light. Furthermore, TiO_2 particles are expensive and have a high embodied energy. New Dow products reduce the amount of TiO_2 required for opaque paints by using a process in which binder polymer molecules self-assemble around the TiO_2 particles. Not only does this process reduce the amount of TiO_2 required, but without this process, TiO_2 can agglomerate, making a less homogeneous paint. Why is a more homogeneous paint preferable?

11.9 If an inclusion compound has chiral channels, suggest how this material could be used to separate two enantiomers that fit in the channels.

11.10 Ivory is a nanostructured composite material, composed mostly of collagen (40%) and the mineral hydroxyapatite (60%).

a. The room-temperature specific heat of dehydrated collagen is 1.3 J K^{-1} g^{-1} and that of pure hydroxyapatite is 0.77 J K^{-1} g^{-1}. Use Equation 11.5 to predict the room-temperature specific heat of ivory based on the rule of mixtures and compare it to the observed value of 1.15 J K^{-1} g^{-1}. Comment on the magnitude of the difference and possible origins.

b. The room-temperature thermal conductivity of pure collagen is 0.56 W m^{-1} K^{-1} and that of pure hydroxyapatite is 2.0 W m^{-1} K^{-1}. Use Equation 11.5 to predict the room-temperature thermal conductivity of ivory based on the rule of mixtures and compare it to the observed value of 0.37 W m^{-1} K^{-1}. Comment on the magnitude of the difference and possible origins.

c. The elastic modulus (see Chapter 14) of ivory is 10.4 GPa, while that of collagen is about 2 GPa and that of hydroxyapatite is about 4 GPa. Again, compare the rule-of-mixtures value, now for elastic modulus, with the observed value, and comment on any differences.

Further Reading

General References

Supramolecular materials. Special issue. J. S. Moore, Ed. *MRS Bulletin*, April 2000, 26–58.

P. Ball, 1994. *Designing the Molecular World.* Princeton University Press, Princeton, NJ.

A. Barton, 1997. *States of Matter: States of Mind.* Institute of Physics Publishing, Bristol.

R. J. Hunter, 2001. *Introduction to Modern Colloid Science*, 2nd ed. Oxford University Press, New York.

L. F. Lindoy and I. M. Atkinson, 2000. *Self-Assembly in Supramolecular Systems.* Cambridge University Press, Cambridge.

T. E. Mallouk and H. Lee, 1990. Designer solids and surfaces. *Journal of Chemical Education*, 67, 829.

J. B. Ott and J. Boerio-Goates, 2000. *Chemical Thermodynamics. Vol. I: Principles and Applications; Vol. II: Advanced Applications.* Academic Press, Oxford.

B. D. Wagner, P. J. MacDonald, and M. Wagner, 2000. A visual demonstration of supramolecular chemistry: Observable fluorescence enhancement upon host-guest inclusion. *Journal of Chemical Education*, 77, 178.

Colloids and Surfactants

New developments in colloid science. Special issue. D. A. Weitz and W. B. Russel, Eds. *MRS Bulletin*, February 2004, 82–106.

From dynamics to devices: Directed self-assembly of colloidal materials. Special issue. D. G. Grier, Ed. *MRS Bulletin*, October 1998, 21–50.

C. P. Ballard and A. J. Fanelli, 1993. Chapter 1: Sol-gel route for materials synthesis. In Chemistry of Advanced Materials, C. N. R. Rao, Ed. Blackwell Scientific Publications, Oxford.

M. J. L. Castro, H. Ritacco, J. Kovensky, and A. Fernández-Cirell, 2001. A simplified method for the determination of critical micelle concentration. *Journal of Chemical Education*, 78, 347.

J. A. Clements, 1962. Surface tension in the lungs. *Scientific American*, December, 121.

R. Dagani, 1997. Intelligent gels. *Chemical and Engineering News*, June 9, 26.

C. Day, 2006. Colloid particles crystallize in an increasingly wide range of structures. *Physics Today*, June, 15.

V. Garbin, 2013. Colloidal particles: Surfactants with a difference. *Physics Today*, October, 68.

A. P. Gast and W. B. Russel, 1998. Simple ordering in complex fluids. *Physics Today*, December 1998, 24.

M. Hair and M. D. Croucher, Eds., 1982. Colloids and Surfaces in Reprographic Technology. In *ACS Symposium Series 200*, Washington.

B. Halford, 2014. Colloids yield full color palette. *Chemical and Engineering News*, April 28, 28.

M. Jacoby, 2002. 3-D structures from stable gels. *Chemical and Engineering News*, July 1, 7.

C. D. Keating, M. D. Musick, M. H. Keefe, and M. J. Natan, 1999. Kinetics and thermodynamics of Au colloid monolayer self-assembly. *Journal of Chemical Education*, 76, 949.

C. A. Murray and D. G. Grier, 1995. Colloidal crystals. *American Scientist*, 83, 23.

Y. Osada and S. B. Ross-Murphy, 1993. Intelligent gels. *Scientific American*, May 1993, 82.

E. Pefferkorn and R. Varoqui, 1989. Dynamics of latex aggregation. Modes of cluster growth. *Journal of Chemical Physics*, 91, 5679.

L. L. Schramm, 2005. *Emulsions, Foams, and Suspensions: Fundamentals and Applications*. Wiley-VCH, Hoboken, NJ.

P. C. Schulz and D. Clausse, 2003. An undergraduate physical chemistry experiment on surfactants: Electrochemical study of commercial soap. *Journal of Chemical Education*, 80, 1053.

D. R. Ulrich, 1990. Chemical processing of ceramics. *Chemical and Engineering News*, January 1, 28.

H. B. Weiser, 1949. *A Textbook of Colloid Chemistry*. John Wiley & Sons, Hoboken, NJ.

K. I. Zamaraev and V. L. Kuznetsov, 1993. Chapter 15: Catalysts and adsorbents. In Chemistry of Advanced Materials, C. N. R. Rao, Ed. Blackwell Scientific Publications, Oxford.

Composites

H. R Clauser, 1973. Advanced composite materials. *Scientific American*, July, 36.

R. F. Gibson, 2010. A review of recent research on mechanics of multifunctional composite materials and structures. *Composite Structures*, 92, 2793.

M. B. Jakubinek, C. Samarasekera, and M. A. White, 2006. Elephant ivory: A low thermal conductivity, high strength nanocomposite. *Journal of Materials Research*, 21, 287.

A. Kelly, 1967. The nature of composite materials. *Scientific American*, September, 160.

Inclusion Compounds

J. L. Atwood et al., Eds., 1996. *Comprehensive Supramolecular Chemistry*, 11 Volume Series. Pergamon Press, Oxford.

J. L. Atwood, J. E. D. Davies, and D. D. MacNicol, Eds., 1984. *Inclusion Compounds, Vol. 1: Structural Aspects of Inclusion Compounds Formed by Inorganic and Organometallic Host Lattices*. Academic Press, Oxford.

J. L. Atwood, J. E. D. Davies and D. D. MacNicol, Eds., 1984. *Inclusion Compounds, Vol. 2: Structural Aspects of Inclusion Compounds Formed by Organic Host Lattices.* Academic Press, Oxford.

J. L. Atwood, J. E. D. Davies, and D. D. MacNicol, Eds., 1984. Inclusion Compounds, Vol. 3: Physical Properties and Applications. Academic Press, Oxford.

J. L. Atwood, J. E. D. Davies, and D. D. MacNicol, Eds., 1991. Inclusion Compounds, Vol. 4: Key Organic Host Systems. Oxford University Press, New York.

J. L. Atwood, J. E. D. Davies, and D. D. MacNicol, Eds., 1991. Inclusion Compounds, Vol. 5: Inorganic and Physical Aspects of Inclusion. Oxford University Press, New York.

T. Bein, 2005. Zeolitic host–guest interactions and building blocks for the self-assembly of complex materials. *MRS Bulletin*, October, 713.

S. Borman, 1995. Organic "Tectons" used to make networks with inorganic properties. *Chemical and Engineering News*, January 2, 21.

S. Borman, 1997. Nanoporous sandwiches served to order. *Chemical and Engineering News*, April 28.

J. F. Brown Jr., 1962. Inclusion compounds. *Scientific American*, July, 82.

S. Carlino, 1997. Chemistry between the sheets. *Chemistry in Britain*, September, 59.

J. E. D. Davies, 1977. Species in layers, cavities and channels (or trapped species). *Journal of Chemical Education*, 54, 536.

J. M. Drake and J. Klafter, 1990. Dynamics of confined molecular systems. *Physics Today*, May, 46.

J. Haggin, 1996. New large-pore silica zeolite prepared. *Chemical and Engineering News*, May 27, 5.

K. D. M. Harris, 1993. Molecular confinement. *Chemistry in Britain*, February, 132.

H. Kamimura, 1987. Graphite intercalation compounds. *Physics Today*, December, 64.

G. T. Kerr, 1989. Synthetic zeolites. *Scientific American*, July, 100.

D. Michalski, M. A. White, P. Bakshi, T. S. Cameron, and I. Swainson, 1995. Crystal structure and thermal expansion of hexakis(phenylthio)benzene and its CBr_4 clathrate. *Canadian Journal of Chemistry*, 73, 513.

M. Ogawa and K. Kuroda, 1995. Photofunctions of intercalation compounds. *Chemical Reviews*, 95, 399.

D. O'Hare, 1992. Chapter 4: Inorganic intercalation compounds. In Inorganic Materials, D. W. Bruce and D. O'Hare, Eds. John Wiley & Sons, Hoboken, NJ.

R. E. Pellenbarg and M. D. Max, 2001. Gas hydrates: From laboratory curiosity to potential global powerhouse. *Journal of Chemical Education*, 78, 896.

M. Rouhi, 1996. Container molecules. *Chemical and Engineering News*, August 5, 4.

F. Vögtle, Ed., 1991. *Supramolecular Chemistry: An Introduction* John Wiley & Sons, Hoboken, NJ.

Molecular Engineering

R. Baum, 1995. Chemists create family of "molecular squares" based on iodine or metals. *Chemical and Engineering News*, February 13, 37.

S. Borman, 2018. Material's rotors spin freely and quickly. *Chemical and Engineering News*, January 1, 8.

M. I. Fox Morone, 2018. Materials on the move. *Chemical and Engineering News*, February 26, 368.

F. Gomollón-Bel, 2017. Single molecular machine produces four chiral products. *Chemistry World*, September 21.

M. J. Snowden and B. Z. Chowdhry, 1995. Small sponges with big appetites. *Chemistry in Britain*, December, 943.

F. Stoddart, 1991. Making molecules to order. *Chemistry in Britain*, August, 714.

M. D. Ward, 2005. Directing the assembly of molecular crystals. *MRS Bulletin*, October, 705.

G. M. Whitesides, 1995. Self-assembling materials. *Scientific American*, September, 146.

Other Materials

J. Banhart and D. Weaire, 2002. On the road again: Metal foams find favor. *Physics Today*, July, 37.

M. Bloom, 1992. The physics of soft, natural materials. *Physics in Canada*, January, 7.

P.-G. de Gennes and J. Badoz, 1996. *Fragile Objects*. Springer-Verlag, New York.

N. Graham, 2001. Swell gels. *Chemistry in Britain*, April, 42.

L. D. Hansen and V. W. McCarlie, 2004. From foam rubber to volcanoes: The physical chemistry of foam formation. *Journal of Chemical Education*, 81, 1581.

M. Jacoby, 1998. Durable organic gels. *Chemical and Engineering News*, January 26, 34.

B. Kahr, J. K. Chow, and M. L. Petterson, 1994. Organic hourglass inclusions. *Journal of Chemical Education*, 71, 584.

Websites

For links to relevant websites, see PhysicalPropertiesOfMaterials.com

Part IV

Electrical and Magnetic Properties of Materials

The missing link between electricity and magnetism was found in 1820, by Hans Christian Ørsted, who noticed that a magnetic compass needle is influenced by current in a nearby conductor. His discovery literally set the wheels of modern industry in motion.

Rodney Cotterill
The Cambridge Guide to the Material World

12

Electrical Properties

12.1 Introduction

Although all the properties of materials—optical, thermal, electrical, magnetic, and mechanical—are related, it is perhaps the electrical properties that most distinguish one material from another. This distinction can be as simple as metal versus nonmetal, or it can involve more exotic properties such as superconductivity. The aim of this chapter is to expose the principles that determine electrical properties of matter.

12.2 Metals, Insulators, and Semiconductors: Band Theory

The *resistance* to flow of electric current in a material, designated R, is determined by the dimensions of the material (length L and cross-sectional area A) and the *intrinsic resistivity* (also known as *resistivity*, and represented by ρ) of the material:[*]

$$R = \rho\left(\frac{L}{A}\right) \tag{12.1}$$

where R is in units of ohms[†] (abbreviated Ω) and ρ is typically in units Ω m. The intrinsic resistivity depends not just on the specific material but also on the temperature. (We will see later how the temperature dependence of resistivity can be used to produce electronic thermometers.) Some typical resistivities are given in Table 12.1.

---

[*] The most important equations in this chapter are designated with ▌to the left of the equation.

[†] Georg Simon Ohm (1789–1854) was a German physicist. Ohm published his law ($V = IR$, where V is voltage, I is current, and R is resistance) in 1827, but it received little attention for nearly 20 years. It is now known to be one of the most fundamental principles of electronics and is called Ohm's Law.

TABLE 12.1

Electrical Resistivities, ρ, and Conductivities, σ ($= \rho^{-1}$), of Selected Materials at 25 °C

Material	ρ/Ω m	$\sigma/(\Omega^{-1}\,m^{-1})$
Metals		
Ag	1.61×10^{-8}	6.21×10^{7}
Cu	1.69×10^{-8}	5.92×10^{7}
Au	2.26×10^{-8}	4.44×10^{7}
Al	2.83×10^{-8}	3.53×10^{7}
Ni	7.24×10^{-8}	1.38×10^{7}
Hg	9.58×10^{-7}	1.04×10^{6}
Semiconductors		
Ge	0.47	2.1
Si	3×10^{3}	3×10^{-4}
Insulators		
Diamond	1×10^{14}	1×10^{-14}
Quartz	3×10^{14}	3×10^{-15}
Mica	9×10^{14}	1×10^{-15}

Source: Supramolecular Chemistry: An Introduction, F. Vögtle, John Wiley & Sons, 1991, Hoboken, NJ.

The *electrical conductivity*, σ, of a material is the reciprocal of its resistivity, ρ:

$$\sigma = \frac{1}{\rho} \tag{12.2}$$

so typical units of σ are $\Omega^{-1}m^{-1}$ ($\equiv$ S m^{-1}, where S represents Siemens* and 1 S = 1 Ω^{-1}). Electrical conductivity also can be equivalently expressed in terms of *current density, J* (units A m^{-2}, where A is amperes[†]) and electric field, ε (units V m^{-1}, where V is volts[‡]):

$$\sigma = \frac{J}{\varepsilon} \tag{12.3}$$

so, equivalently, units of σ can be expressed as A m^{-1} V^{-1} ($\equiv \Omega^{-1}m^{-1} \equiv$ S m^{-1}).

The distinguishing feature separating metals, semiconductors, and insulators is the range of σ: > 10^4 $\Omega^{-1}m^{-1}$ for metals, 10^{-3} to 10^4 $\Omega^{-1}m^{-1}$ for semiconductors, and < 10^{-3} $\Omega^{-1}m^{-1}$ for insulators (see Figure 12.1 and also Table 12.1).

---

* The Siemens unit is named after Ernst Werner von Siemens (1816–1892), a German inventor and industrialist who made significant contributions to electric power generation and electrochemistry.

† André Marie Ampère (1775–1836) was a French mathematician and physicist who taught at École Polytechnique in Paris. Ampère carried out many experiments in electromagnetism and also provided a mathematical theory for electrodynamics.

‡ Count Alessandro Volta (1745–1827) was a member of the Italian nobility who carried out many important electrical experiments. Volta was the inventor of the electrical battery and the discoverer of electrolysis of water. His home town, Como, Italy, has a museum that houses many Volta artifacts, including experimental apparatus and lab notes.

In contrast with thermal conductivities, which span a range of about 6 orders of magnitude from the poorest thermal conductor to the best (see Chapter 8), electrical conductivities (see Figure 12.1) span more than 22 orders of magnitude from the least conductive insulator to the highest-conductivity metal. (And superconductors have infinite conductivity!) The difference in ranges of thermal and electrical conductivity reflect the fact that thermal conductivity requires the collective action of phonons, whereas electrical conductivity is usually related to the motion of relatively independent electrons along a path of least resistance, giving a wider variation in electrical conductivity than thermal conductivity.

The wide variety of electrical conductivities in different materials is associated with the availability of energy levels directly above the filled energy levels in the material. The presence (and width) of a gap between the filled

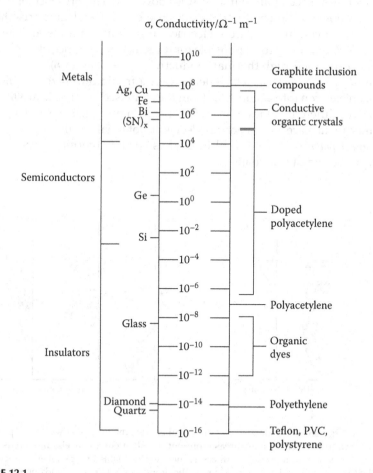

FIGURE 12.1

This diagram shows the wide range of electrical conductivity of common materials at room temperature, not including superconductors (for which $\sigma = \infty$).

electronic states and the next available electronic states leads to very different electrical properties for insulators and conductors.

12.2.1 Metals

As we have seen in Chapter 3 (especially Figures 3.2, 3.6, and 3.7), there is a continuum of electronic energy levels (called *bands*), and for a metal there are available unoccupied electronic energy levels (empty bands) immediately above the highest occupied levels. This situation is shown in Figure 12.2 for two specific metals where a realistic density of states is shown; these figures can be contrasted with the simpler representations in Figures 3.6 and 3.7.

When there are available electronic energy levels immediately above the filled levels (i.e., there is no gap), the material is a metal, and excitation of the electrons takes place at all temperatures above 0 K. The presence of excited electrons allows facile conduction of electric current and gives rise to the high electrical conductivity (low electrical resistivity) of a metal. The conduction electrons also are responsible for the shiny appearance of metals (see Chapter 3) and their high thermal conductivity (see Chapter 8).

The *work function*, Φ, for a metal is defined as the minimum energy required to remove an electron into vacuum far from the atom (see Figure 12.3a). Different metals have different values of work functions (Figure 12.3b) such that, when dissimilar metals are in contact, an electrical potential results (Figure 12.3c). This *contact potential* can be used in devices such as thermometers (see the tutorial at the end of this chapter).

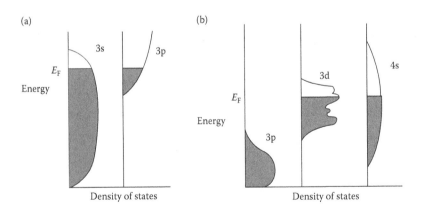

FIGURE 12.2
Electronic energy levels for metals. E_F is the Fermi energy, the energy at which the probability of occupation is ½. The shaded areas represent the filled energy levels, and bands below the valence bands are omitted. In both cases, there are available energy levels immediately above the filled bands, which give rise to metallic properties. (a) Energy bands of magnesium show the overlap in energy between the 3s band and the 3p band. (b) Energy bands of cobalt show the overlap of the energy range of the 3d and 4s bands.

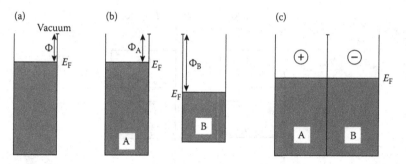

FIGURE 12.3

(a) The work function, Φ, for a metal is defined as the minimum energy required to remove an electron to vacuum far from the atom. (b) Different metals, shown here as A and B, have different work functions. (c) Dissimilar metals, when placed in contact with each other, equalize their Fermi energies, giving a contact potential, shown here as an electric field.

COMMENT: MATTHIESSEN'S RULE

Matthiessen's rule* is an empirical conclusion that the total resistivity of a crystalline metal is the sum of the resistivity due to thermal motions of the metal atoms in the lattice and the resistivity due to the presence of imperfections in the crystal. This rule allows a basic understanding of the resistivity behavior of metals as a function of temperature.

---

* Matthiessen's rule is named after Augustus Matthiessen (1831–1870), a British chemist and physicist who made significant contributions to understanding metals and alloys and opium alkaloids.

12.2.2 Semiconductors

A semiconductor has a gap (forbidden region) between the valence band and the conduction band (Figure 12.4a). At a temperature of 0 K, all the electrons of a semiconductor are in the valence band. However, for $T > 0$ K, there will be some thermally excited population of the conduction band, as shown in Figure 12.4b. It is this small population of the conduction band that allows slight electrical conductivity in semiconductors.

The structure of pure germanium, which is a semiconductor, is shown in Figure 12.5, along with a schematic representation of its electrical conductivity mechanism.

The element tin occurs as two allotropes, white tin and gray tin. White tin is the more familiar form, as it is stable above 13 °C, and we know this form to have the typical characteristics of a metal: it is shiny, ductile, and a good conductor of heat and electricity. White tin is denser than gray tin (densities are 7.2 g cm⁻³ and 5.7 g cm⁻³, respectively), and the higher density allows its electronic energy bands to overlap more, giving white tin metallic character. In contrast, gray tin has a band gap of 0.1 eV, an electrical resistivity of

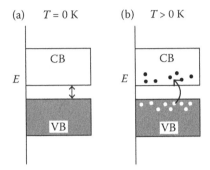

FIGURE 12.4
The energy bands of a semiconductor at (a) $T = 0$ K and (b) $T > 0$ K. Thermal energy can cause some slight population of the conduction band of a semiconductor which gives it a nonzero (but low) electrical conductivity.

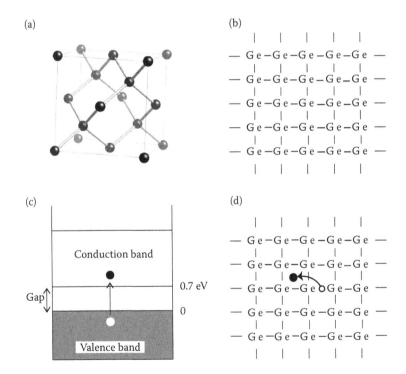

FIGURE 12.5
The semiconductor germanium: (a) The crystal structure showing tetrahedral bonding of Ge atoms. (b) A schematic two-dimensional representation of the bonding in Ge. (c) The energy level diagram of Ge, showing the formation of a "hole" (absence of electron) in the valence band and an electron in the conduction band. This electron has been promoted from the valence band to the conduction band by thermal energy. (d) An electron-hole pair in Ge.

10^{-6} Ω m, and acts as a semiconductor. Folklore has it that Napoleon lost many of his soldiers during his winter invasion of Russia due to the white tin to gray tin transformation of the tin buttons holding their uniforms closed. Since the gray form is not metallic, it is crumbly, and this is said to have led to the soldiers perishing in the cold.*

COMMENT: QUANTUM DOTS

Quantum dots are nanoscale semiconductors that have quite different electrical properties from the corresponding bulk material. The electron-hole pairs are closer together in a quantum dot than in the bulk material, making a situation referred to as quantum confinement. Because there are fewer atoms than in the bulk, the quantum dots' electronic energy levels are no longer bands but are more discrete. Although quantum dots still exhibit a band gap, it is larger than in the bulk material. Furthermore, changing the size or shape of the quantum dot by moving or adding or removing just a few atoms can substantially change the band gap. This important property allows extremely precise control of the wavelength of light emitted from a quantum dot. Furthermore, quantum dots of the same material, but with different dimensions, can emit light of different colors. These special properties make quantum dots important materials for optoelectronic applications.

COMMENT: ELECTRICAL ANISOTROPY

Many materials are electrically anisotropic, including TTF-TCNQ (below). Other examples include graphite, asbestos, and $K_2Pt(CN)_4Br_{0.3} \cdot 3H_2O$. The latter is composed of $Pt(CN)_4$ units stacked to give Pt–Pt distances of 2.89 Å, that is, nearly the same distance along these chains as in Pt metal (2.78 Å). However, the Pt–Pt distance is much greater between chains, resulting in anisotropic structure and properties. The crystals are highly reflective and electrically conducting in the direction of the chains, and transparent and electrically insulating in directions perpendicular to the chains.

---

* Although the transformation from white to gray tin is at 13 °C, as for many first-order transitions (e.g., freezing), the high-temperature phase can be supercooled. Therefore, the soldiers could have experienced temperatures slightly below 13 °C without loss of their jacket closures, but in the extreme cold of the Russian winter the tin was cooled sufficiently to ensure complete transformation of the buttons to the crumbly gray form.

12.2.3 Insulators

An insulator has a very large gap between its valence band and conduction band. Although thermal energy could promote electrons across this gap, in practice the required thermal energy is sufficient to melt the solid! This large gap essentially prevents significant electrical conductivity in insulators.

COMMENT: ONE-DIMENSIONAL CONDUCTORS

Although most molecular systems are insulators, some have high electrical conductivity; one such system is the charge-transfer complex of tetrathiafulvalene (TTF) with tetracyanoquinodimethane (TCNQ). Molecular structures for TTF and TCNQ are shown in Figure 12.6.

In the crystalline complex, TTF and TCNQ form segregated regular stacks, and there is considerable delocalization of electrons (i.e., overlap of wave functions) along these columns. TTF-TCNQ forms a charge-transfer complex; on average, each TTF molecule donates 0.59 electrons to a TCNQ. This situation allows appreciable conductivity; σ is 5×10^4 $\Omega^{-1}m^{-1}$ along the chains but 100 times less in directions perpendicular to the chains; that is, this material is highly anisotropic electrically. The unusually high metal-like electrical conductivity of TTF-TCNQ persists down to 60 K.

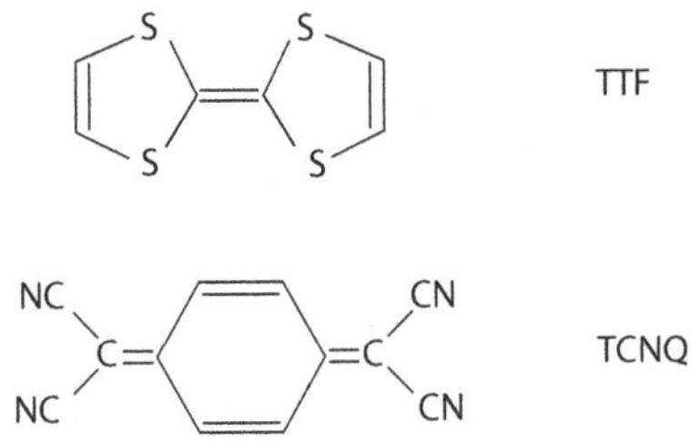

FIGURE 12.6
The molecular structures of tetrathiafulvalene (TTF) and tetracyanoquinodimethane (TCNQ).

12.3 Temperature Dependence of Electrical Conductivity

The temperature dependence of the electrical properties of a material can be used to provide important devices such as switches and thermometers. In this section we examine the principles involved in temperature dependence of electrical conductivity (or resistivity) for metals and pure (intrinsic) semiconductors.

12.3.1 Metals

The resistivity of a metal increases with temperature (see Figure 12.7), and this increase is essentially linear except at the lowest temperatures.

The increase in resistivity (decrease in electrical conductivity) with increased temperature in a metal can be directly associated with the increase in thermal motion of the atoms on their lattice sites at increased temperature. The increase in atomic motion scatters the conduction electrons and decreases their mean free path, hence decreasing their ability to carry electrical charge and increasing their resistivity.

COMMENT: WIEDEMANN–FRANZ LAW

The Debye expression for thermal conductivity (Equation 8.13) can be combined with the electronic contribution to the heat capacity of a metal (Equation 6.34) to yield

$$\kappa = \frac{1}{3}\gamma T v \lambda \qquad (12.4)$$

where γ is the electronic coefficient of the heat capacity, v is the electron's speed, and λ is the electron's mean free path. Since γ is related to the Fermi energy, E_F, for N electrons per unit volume as

$$\gamma = \pi^2 N k^2 / 2 E_F \qquad (12.5)$$

where k is Boltzmann's constant, and the electrical conductivity is given by

$$\sigma = N e^2 \lambda / m v_F \qquad (12.6)$$

where v_F is the electron speed at the Fermi energy, and m and e are the mass and charge of the electron, respectively, assuming $v \approx v_F$, leads to a ratio of

$$\frac{\kappa}{\sigma T} = \frac{\pi^2 k^2}{3 e^2} \qquad (12.7)$$

which has a value of 2.45×10^{-8} W Ω K^{-2} ($\equiv L$, the *Lorenz constant*), for all metals. Equation 12.7, which shows that the ratio of thermal conductivity to electrical conductivity at a given temperature is constant for metals, is known as the *Wiedemann–Franz law*, and it holds quite well for metals. However, in very pure metals in the temperature range where phonon scattering is important, electron–phonon interactions (which have been neglected in this derivation) lead to a breakdown of the Wiedemann–Franz law. The Wiedemann–Franz law is important in the history of our understanding of metals as it shows the electrical conductivity to be inversely proportional to temperature, in support of modeling metals as gases of electrons. Note that the Wiedemann–Franz law holds only for metals, not for semiconductors or insulators. Examination of the value of $\kappa/\sigma T$ for a material and comparison with the Lorenz constant can indicate if the material is metallic or not.

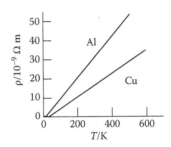

FIGURE 12.7
Resistivity as a function of temperature for two typical metals, aluminum and copper.

12.3.2 Intrinsic Semiconductors

In contrast with metals, the electrical resistivity of intrinsic (pure) semiconductors *decreases* (electrical conductivity *increases*) with increasing temperature. Results for germanium are shown as an example in Figure 12.8.

This temperature dependence can be understood as follows. At low temperatures, few electrons are able to jump the gap from the valence band to the conduction band; in fact, none has jumped at $T = 0$ K. As the temperature

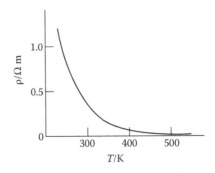

FIGURE 12.8
The electrical resistivity of pure germanium as a function of temperature.

is increased, the number of electrons that have sufficient thermal energy to jump from the valence band to the conduction band is increased. From Fermi's expression for $P(E)$, the probability of having a free electron at energy E (Equation 3.1) can be rewritten as

$$P(E) = \frac{e^{-(E-E_F)/kT}}{1+e^{-(E-E_F)/kT}} \qquad (12.8)$$

and for $|E - E_F| \ll 1$ (i.e., near the Fermi energy), Equation 12.8 can be approximated by

$$P(E) = \frac{1}{2}e^{-(E-E_F)/kT} \qquad (12.9)$$

which has the same form as the Boltzmann distribution:

$$n(E) \propto e^{-\frac{E-\bar{E}}{kT}} \qquad (12.10)$$

where $n(E)$ is the number of species in energy state E, and $\bar{E}$ is the average energy of the system. For an intrinsic semiconductor, $\bar{E}$ is E_F (since $P(E) = \frac{1}{2}$ at the Fermi energy, by definition and the distribution is symmetric* about E_F), and the most probable energy of an excited (conduction) electron will be $E_F + \frac{1}{2}E_g$ where E_g is the gap energy (i.e., most probable at the bottom of the conduction band), so Equation 12.10 becomes

$$n(E) \propto e^{-E_g/2kT} \qquad (12.11)$$

where $n(E)$ is the number of conduction electrons. The electrical conductivity, σ, is proportional to the number of conduction electrons giving

$$\sigma = \sigma_0 \, e^{-E_g/2kT} \qquad (12.12)$$

where σ_0 is a proportionality constant.

Equation 12.12 shows an interesting feature: from measurements of electrical conductivity as a function of temperature, the energy of the band gap, E_g, can be determined for a semiconductor (see Problem 12.3).

---

* Strictly speaking, the distribution is symmetric only if the density of states in the valence band and conduction band are equal, but even if they are not, the correction for asymmetry is usually small.

COMMENT: SILICON

Silicon-based semiconductors are at the heart of most of the computer devices that we use daily. Silicon (see Figure 12.9) has properties that make it very useful, and it is one of the most common elements on earth. It is hard and chemically rather inactive (mainly reacts with oxygen); it has a high melting point (1685 K), which allows for doping by solid-state diffusion at high temperatures and it has a low band gap (1.1 eV), which allows for low-power applications.

(a)   (b)

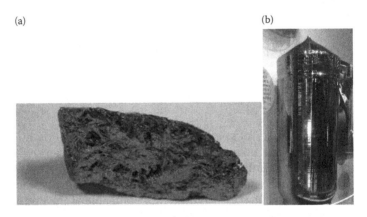

FIGURE 12.9
Photographic images of silicon, (a) at low purity (about 1 cm long) and (b) at high purity (about 1 m high), on display at Intel Corporation, Portland, Oregon.

COMMENT: ELECTRICAL PROPERTIES OF CARBON NANOTUBES

As we saw in Chapter 9, graphene (i.e., individual sheets of carbon with the graphite structure) can be rolled up to form seamless cylinders—carbon nanotubes (CNTs). The different types of CNTs can be understood in terms of their structure. An informative way to consider the structures is in terms of directional (so-called "chiral") vectors that meet up when the graphene sheet is rolled up to form the cylindrical CNT. The chiral vector is defined by integer numbers (m and n) of two unit vectors, $\hat{a}_1$ and $\hat{a}_2$, shown in Figure 12.10.

The ends of the chiral vector join to form the circumference of the CNT and the nanotube's circular cross section. Therefore, different values of m and n give rise to different CNT structures. The (5,5) CNT chiral vector and overall structure is shown in Figure 12.11, and the (9,0) CNT chiral vector and overall structure is shown in Figure 12.12.

Three categories of CNT result from the rolled graphene sheets: armchair, zigzag, and chiral CNTs. The armchair structures are formed when $m = n$ (as in (5,5) in Figure 12.11), whereas zigzag nanotubes are formed when either m or n is zero (e.g., (9,0) in Figure 12.12). All other CNTs are known as chiral.

The electronic properties of CNTs are especially interesting because of the diversity that arises despite all being formed of carbon with the same type and number of bonds. The unique electronic properties arise from the quantum confinement of electrons in the direction perpendicular to the nanotube axis. Electrons can only propagate along the nanotube axis, and the resulting one-dimensional band structure depends on the standing waves that are set up around the circumference of the nanotube. The latter depend on the structure. Calculations of the electronic band structure of CNTs show that an (m,n) carbon nanotube will be metallic when $m - n = 3I$, where I is an integer. Therefore, all armchair CNTs are metallic, as are a third of all possible *zigzag* nanotubes. The remaining possibilities are semiconducting. In a typical production of CNTs, about one third would be metallic and two thirds semiconducting. Separating CNTs by their electronic properties is one of the major challenges of materials research.

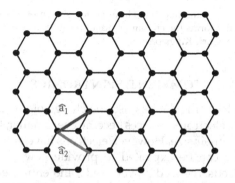

FIGURE 12.10
Portion of a graphene sheet showing the directional unit vectors $\hat{a}_1$ and $\hat{a}_2$ that define the chiral vector, generalized as $m\hat{a}_1 + n\hat{a}_2$, defining the structure of an (m,n) carbon nanotube.

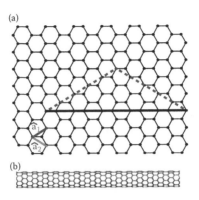

FIGURE 12.11
The so-called (5,5) carbon nanotube. The chiral vector, shown on the graphene sheet in (a), is $5\hat{a}_1 + 5\hat{a}_2$, and it forms the circumference of the (5,5) carbon nanotube shown in (b). Diagram (b) courtesy of Chris Kingston.

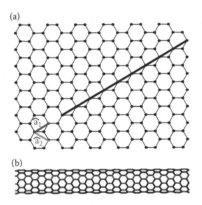

FIGURE 12.12
The so-called (9,0) carbon nanotube. The chiral vector, shown on the graphene sheet in (a), is $9\hat{a}_1 + 0\hat{a}_2$, ($= 9\hat{a}_1$), and it forms the circumference of the (9,0) carbon nanotube shown in (b). Diagram (b) courtesy of Chris Kingston.

COMMENT: SPINTRONICS

Conventional electronic devices use the charge of an electron to move information. As electronic devices become ever smaller, there inevitably will be a limit to their size. However, electrons have another property—their spin—that can be exploited to provide new electronic devices with higher speeds, reduced power needs, and enhanced functionality. Much of the research emphasis in this field of *spintronics* (i.e., spin transport electronics) concerns efficient data storage and transfer, especially for quantum computing.

COMMENT: BORON NITRIDE NANOTUBES

Boron nitride nanotubes (BNNTs) are a polymorph of boron nitride. They have cylindrical structures similar to CNTs (Figures 9.17, 12.11, and 12.12), with carbon atoms replaced by nitrogen and boron atoms. Although their structures are analogous, and both CNTs and BNNTs have similar exceptional properties such as high strength and stiffness and low density, they have very different electrical properties. Whereas CNTs are metallic or semiconducting, BNNTs have wide (~5.5 eV) band gaps, almost independent of tube chirality, and therefore are electrical insulators. The wide band gap means that BNNTs are white, whereas CNTs, with their narrow (or nonexistent) band gap are black. The boron nitride structure is much more stable thermally than the CNT structure and also resists oxidation better, therefore making BNNTs particularly attractive for high-temperature applications.

12.4 Properties of Extrinsic (Doped) Semiconductors

Defects in a semiconductor can give rise to localized electronic states as seen previously in the discussion of optical properties of semiconductors (Chapter 3). The addition of defects by doping (addition of impurities) gives rise to extrinsic semiconductors.

For example, Si (which has four valence electrons) doped with P (which has five valence electrons), results in an extra electron in the lattice and a localized energy level within the gap corresponding to this extra electron. This extra level (Figure 12.13) is called a *donor level* since P donates an extra electron. Electrons can be promoted from this level to the conduction band. Since the resultant current carriers are electrons, which are negatively charged, these are called *n-type extrinsic semiconductors*.

Holes also could carry charges and since holes are effective positive charges (they represent the absence of negative charge), these are termed *p-type extrinsic semiconductors*. An example is Si (four valence electrons) doped with Al (three valence electrons); the extra localized level in the gap due to electron deficiency in Al is an *acceptor level* since Al accepts electrons from Si, as depicted in Figure 12.14.

The importance of p-type and n-type semiconductors is realized when they are sandwiched together.

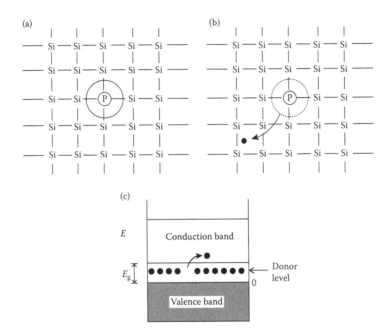

FIGURE 12.13
An n-type extrinsic semiconductor. (a) P substitutes for Si in a lattice. (b) The extra electron of P (compared with Si) can be donated to the conduction band to carry electrical charge. (c) The band model showing the donor level in the band gap.

12.5 Electrical Devices using Extrinsic (Doped) Semiconductors

Several important electronic devices can be designed using the combined properties of n-doped semiconductors and p-doped semiconductors. Some of these are described here.

12.5.1 p,n-Junction

The junction of a p-type semiconductor with an n-type semiconductor is referred to as a *p,n-junction* making a diode, with the special property called *rectification*. When an n-type semiconductor is placed next to a p-type semiconductor, the small numbers of conducting electrons in the n-type are drawn toward the interface, as are the few holes in the p-type. This electron–hole attraction is a simple consequence of unlike charges attracting. Since, for the configuration shown in Figure 12.15a, electrons are flowing left to right in the n-type semiconductor and holes (which are positive charges) are flowing right to left (implying flow of effective negative charge from left to right) in the p-type semiconductor, there is net (small) flow of electrons from left to

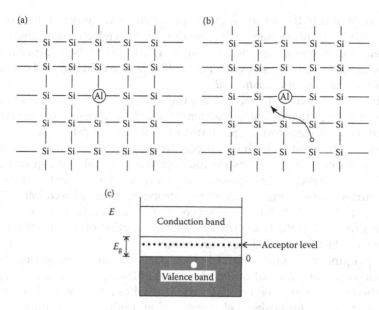

FIGURE 12.14

A p-type extrinsic semiconductor. (a) Al substitutes for Si in a lattice. (b) The missing electron of Al (compared with Si) can accept an electron from the valence band, thus leaving a hole in the valence band that can act as an electrical charge carrier. (c) The band model showing the acceptor level in the band gap and a hole in the valence band.

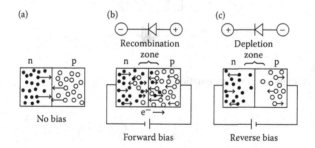

FIGURE 12.15

A p,n-junction showing rectification. Electrons shown as ● and holes as ○. (a) With no field (no bias) there is a small net flow of negative charge (left to right) due to intrinsic electrical conductivity. (b) With a forward bias, there is larger net electron flow in the material from left to right. (c) With a reverse bias, extrinsic conductivity as in (b) disappears and only a small amount of intrinsic electrical conductivity as in (a) remains.

right across the device. This current exists even in the absence of an applied electric field, and this property is referred to as *intrinsic electrical conductivity*.

Now, if an electrical field is placed across the p,n-junction such that the negative terminal is next to the n-type semiconductor and the positive terminal is next to the p-type semiconductor (an arrangement referred to as *forward bias*), the interactions of the moving charges (electrons or holes) with the field will enhance the flow of charge and there will be a larger flow of electrons left-to-right, as shown in Figure 12.15b. The field-induced current is called *extrinsic electrical conductivity*.

If the field is reversed (*reverse bias*), the electrons in the n-type semiconductor will be pulled in both directions: toward the holes on the p side and toward the positive terminal. Similarly, the holes in the p-type semiconductor will be attracted in both directions, toward the negative charge of the n-type semiconductor and toward the negative terminal (see Figure 12.15c). Therefore, the reverse bias across a p,n-junction has the effect of turning a large current into a small current; this property is called *rectification*. This situation is contrary to what is observed in normal conductors (i.e., metals), where a reversal of the field only causes the direction of the current flow to change, while its magnitude is maintained.

The p,n-junction is often referred to as a *diode*, which is an electrical device that makes use of the rectification properties of the p,n-junction. The symbol for a diode is shown in Figure 12.16; the arrow has been chosen to show the motion of current (opposite to electron motion) under forward bias.

The principle of rectification in a diode can be seen in Figure 12.17 where the total current is shown as a function of bias (forward and reverse), and the contributions to the current are explicitly shown.

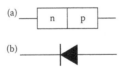

FIGURE 12.16
(a) A diode and (b) its electrical schematic symbol.

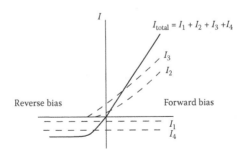

FIGURE 12.17
The total current, I_{total}, in a p,n-junction diode as a function of bias. Also shown are the contributions to the current: I_1, due to mobile electrons in the p $\rightarrow$ n direction; I_2, due to mobile electrons in the n $\rightarrow$ p direction; I_3, due to mobile holes in the p $\rightarrow$ n direction; I_4, due to mobile holes in the n $\rightarrow$ p direction.

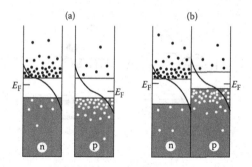

FIGURE 12.18

Electronic distributions for n-type semiconductors and p-type semiconductors, (a) when they are separated and (b) when they are in contact, showing the shift to equal Fermi energies (E_F, defined as the energy at which $P(E) = ½$). Electron occupation probabilities, ranging from zero (left side of each diagram) to one (right side), are shown by the curved line.

The properties of a p,n-junction also can be understood in terms of the electronic energy distributions. Figure 12.18 shows the distributions for an n-type semiconductor and a p-type semiconductor when they are separated and when they are placed adjacent to each other. When the two are in contact, the gaps shift so as to equalize the Fermi energies. This shift causes a contact potential (due to the difference in work function; see Section 12.2), moving electrons from the n-type semiconductor to the p-type semiconductor.

In an electric field, the effect of bias on a p,n-junction can be seen in the electronic distributions shown in Figures 12.19 and 12.20.

With the rapid change in bias of an ac source and the rectification property of a p,n-junction, a device constructed of a p,n-junction can be used to turn alternating current into direct current, for example in a wall adapter for use with a laptop computer.

A *Zener diode* is a special case of a diode, in that it has little current flow at low reverse bias as usual but more flow at large reverse bias (see

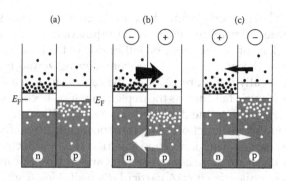

FIGURE 12.19

Electronic distributions for a p,n-junction, (a) with no bias; (b) with a forward bias (giving current carriers in both the valence band and the conduction band); and (c) with a reverse bias (giving only a few current carriers).

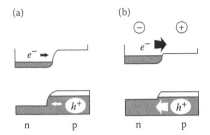

FIGURE 12.20
Electronic distributions of a p,n-junction (a) without bias (small current), and (b) with forward bias (large current).

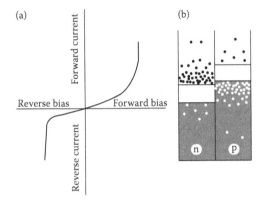

FIGURE 12.21
Zener diode: (a) showing an increase in current at large reverse bias and (b) showing the electronic bands that allow tunneling through the junction (from the conduction band of the n-type semiconductor to the valence band of the p-type semiconductor).

Figure 12.21). This property results because large reverse biases can be sufficient to cause electrons to tunnel through the junction. That is, the overlap of the wavefunctions of the electrons on either side of the junction causes some enhancement of the probability of electrons getting across the junction, as shown schematically in Figure 12.21. As the reverse bias in a Zener diode is increased, the electric field across the depletion region becomes larger. When a charge carrier gets into the depletion region, it is efficiently carried to the appropriate side of the junction. This transport can be so efficient that the motion of the charge causes ionization (creation of a free electron and a hole), which means there are more charge carriers and an amplification of current referred to as an *avalanche effect*. A particularly useful feature of Zener diodes is that slight changes in the composition of the doped semiconductors can be used to tailor-make the threshold voltage for the dramatic reverse-bias current.

COMMENT: CONDUCTING POLYMERS

Some conjugated organic polymer materials, which are normally semi-conductors with a substantial gap between the highest occupied molecular orbital (HOMO) and the lowest unoccupied molecular orbital (LUMO), can be made metallic by oxidation or reduction. The induced charge provides new energy states within the band gap. For example, polyacetylene can be converted to an electrically conducting material by oxidation or reduction. Alan Heeger,[*] Alan MacDiarmid,[†] and Hideki Shirakawa[‡] were awarded the 2000 Nobel Prize in Chemistry for their discovery and development of conductive polymers, especially notable for finding that the treatment of polyacetylene with halogen vapor increases the electrical conductivity up to 9 orders of magnitude. One of the most important properties of polymers is that they are quite robust, allowing new plastic electrical materials, including transistors. See Figure 12.22.

---

[*] Alan J. Heeger (1936–) is a professor at the University of California–Santa Barbara.
[†] Alan G. MacDiarmid (1927–2007) was a professor at the University of Pennsylvania in Philadelphia.
[‡] Hideki Shirakawa (1936–) is a professor at the University of Tsukuba in Tokyo.

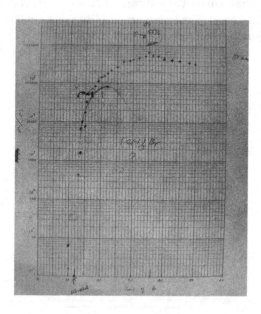

FIGURE 12.22
A photo of a page from the original laboratory notebook of Nobelist Hideki Shirakawa, on view at the Mitsubishi Museum in Yokohama, Japan, showing the great increase in electrical conductivity (y-axis, on a log scale) of polyacetylene on doping with halogen vapor (x-axis).

COMMENT: TOPOLOGICAL INSULATORS

The electrical properties of materials can be different at their surfaces, compared with their interiors. One such example is topological insulators, in which the interior of the material is an insulator, and the surface has conducting states. The band structure of the surface has special levels within the gap, allowing for metallic conduction along the surface. Topological insulators offer promise in areas such as quantum computing and low-power spintronics, as they provide access to unusual states of matter. For theoretical discoveries of topological phase transitions and topological phases of matter, David J. Thouless (1934–, University of Washington), F. Duncan M. Haldane (1951–, Princeton University), and J. Michael Kosterlitz (1943–, Brown University) were awarded the 2016 Nobel Prize in Physics.

12.5.2 Transistors

Two back-to-back p,n-junctions can work together to make a *transistor*. This device can be either a p,n,p-junction or an n,p,n-junction. Since a p,n-junction allows current to flow only in one direction, a p,n,p-junction does not allow much current to flow in either direction, unless some additional element removes or injects charge carriers into the middle. This middle element only needs to have a small current and it can control a large total current, and the resulting transistor can act as either an amplifier or a switch.

In the p,n,p-junction shown schematically in Figure 12.23, the applied electric field causes the left p,n-junction to be biased in the forward direction

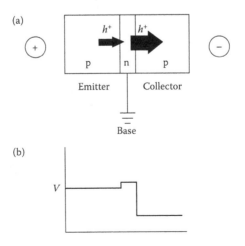

FIGURE 12.23
(a) A schematic view of a transistor. The p,n,p-device acts to amplify the current. (b) The corresponding voltage, V, across the transistor.

and the right p,n-junction to be biased in the reverse direction. Holes move from the left p-type semiconductor (the *emitter*) toward the n-type semiconductor (the *base*); if the n-layer is sufficiently thin, some of the holes will pass through the n-semiconductor to the reverse-biased junction and into the far p-layer (the *collector*). Because holes are attracted to the negative electrode at the far side of this junction, these holes greatly increase the reverse current and cause amplification. Furthermore, the number of holes reaching the reverse-bias junction increases with an increase in the potential difference between the emitter and the base, so the *gain* of the amplification (i.e., the amplification factor) can be easily controlled.

COMMENT: FIELD-EFFECT TRANSISTOR (MOSFET)

A metal oxide semiconductor *field-effect transistor*, abbreviated MOSFET, is an example of a transistor acting as a switch. The general structure of a MOSFET, shown in Figure 12.24, consists of two n-type regions in a p-type semiconductor (e.g., a silicon wafer); one n-region is the *source* and the other is the *drain*. Between the source and the drain, the wafer is covered with an insulating layer (e.g., a 10^{-7} m thick layer of silicon dioxide), that is covered by a metal that acts as the *gate*. At a positive gate voltage, electrons from the p-type semiconductor are attracted to the surface and holes are repelled, allowing a current to pass from the source to the drain in the region near the surface. Therefore, it is the field at the gate that controls the current and hence it is a field-effect transistor, acting as a switch. Since the gate does not consume much power, this low-power device can be very small (10s of nm). *Integrated circuits* consist of several such transistors made on one wafer (or chip), with the transistors connected to each other through layers of metal deposited on the chip's surface.

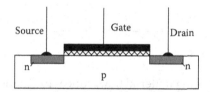

FIGURE 12.24
A field-effect transistor consists of an n-type source and n-type drain in a p-type wafer that is coated with an insulator and a metal. A field applied to the metal makes a gate that controls the current between the source and the drain, acting as a switch.

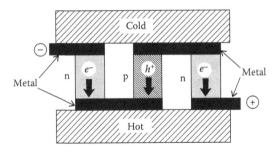

FIGURE 12.25

A schematic view of a Peltier cooling device. As current flows through the device, electrons in the n-type semiconductor and holes in the p-type semiconductor carry heat away from the upper surface, making it cooler than the lower surface. If the direction of the current is reversed, the flow of the electrons and holes is reversed, and so is the flow of heat. In a working Peltier cooler there would be many banks of n-and p-type semiconductors similar to that shown here.

COMMENT: PELTIER COOLING DEVICE

The *Peltier effect*[*] (conversion of electrical energy into a temperature gradient) is a useful *thermoelectric* property. If current is passed through a bank of n- and p-type semiconductors, as shown in Figure 12.25, heat is absorbed at one side and released at the other. (These heat effects are in addition to the Joule heating, in which the rate of energy release is expressed as I^2R, where I is the current and R is the resistance.) The Peltier effect can be used to make cooling devices using only electrical means and no moving parts. Furthermore, working in reverse, the Peltier effect can be used to convert waste heat into usable power. See also Problem 12.28 concerning thermoelectric devices.

---

[*] Jean Charles Athanase Peltier (1785–1845) was a French clockmaker. When his mother-in-law died in 1815, the inheritance allowed him to retire from his trade and take up scientific pursuits, including studies of phrenology, electricity, microscopy, and meteorology. A field trip to collect data concerning electrical charges in the atmosphere and cloud formation led to a cold from which he did not recover.

12.6 Dielectrics

A *dielectric* is an insulating material that reduces the Coulombic force between two charges. When placed between the plates of an electrical capacitor, a dielectric increases its capacitance. If C and C_0 are the capacitances with and without the substance present, respectively, then ε, the *dielectric constant* (also called the *relative permittivity*) of the material, is defined as

$$\varepsilon = \frac{C}{C_0}.$$ (12.13)

Dielectric constants of some materials are listed in Table 12.2.

Dielectric materials are polarized when they are placed in the electric field of the capacitor. That is, their center of negative charge is displaced from the center of their positive charge and there is an induced dipole moment (i.e., nonzero net vector of charge distribution in space). The dipole moment per unit volume is called the *polarization* and is given the symbol P. The dipoles align themselves in the electric field, and the aligned dipoles create an electric field that opposes the applied field, as shown schematically in Figure 12.26. Thus, the presence of the dielectric has the effect of reducing the electric field, which gives a lower voltage, V, between the capacitor plates. Since $C \propto V^{-1}$, capacitance increases when the dielectric is present.

An "ordinary" dielectric material will be unpolarized in the absence of an external electric field and can be polarized by the application of an external electric field. The sign of the polarization can be reversed by reversing the field, as shown in Figure 12.27a.

A *pyroelectric* crystal can have a nonzero electric polarization even in the absence of an applied electric field. The source of this behavior is the arrangement of the ions in the internal structure of the crystal, as shown schematically in Figure 12.28a. A material is no longer pyroelectric above a phase transition temperature (also called *critical temperature*, T_c) because the available thermal energy brings about a more symmetric structure. This high-temperature state

TABLE 12.2

Dielectric Constants for Selected Materials at $T = 300$ K, Unless Specified Otherwise[a]

Material	ε
n-Hexane	1.89
CCl_4	2.23
Benzene	2.28
Diamond	5.7
$AgNO_3$	9.0
Ge	16
Ethanol	24.3
Methanol	32.6
BaO	34
Water	78.54
Ice Ih	99 (at 243 K)
Ice VI	193 (at 243 K)
$PbZrO_3$	200 (at 400 K)

[a] ε is dimensionless.

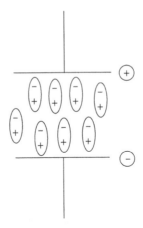

FIGURE 12.26
The dipoles of a dielectric material tend to align themselves between the plates of a capacitor. This alignment opposes the applied electric field so that the capacitance increases with the dielectric present, relative to without the dielectric.

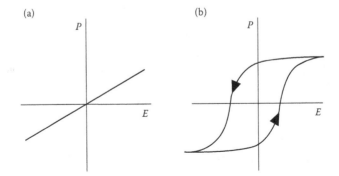

FIGURE 12.27
Polarization, P, as a function of electric field, E, for two types of materials: (a) a normal dielectric, showing zero polarization in zero field; (b) a ferroelectric material at $T < T_c$ showing spontaneous polarization even in the absence of an electric field and also showing hysteresis (i.e., the path-dependent property of different polarization depending on whether the electric field is increasing or decreasing).

is called *paraelectric*, indicating that it is not polarized but can be polarized under other circumstances, namely, lowered temperature.

A crystal is *ferroelectric* if it is pyroelectric and the spontaneous polarization can be reversed by the application of a small electric field, as shown schematically in Figure 12.28b. Ferroelectrics can be distinguished from ordinary dielectrics by their very large dielectric constants (> 1000) and by their retention of polarization even after the field is turned off. This residual polarization leads to a *hysteresis loop* (i.e., different polarizations on increasing and decreasing the electric field) as shown in Figure 12.27b. The term

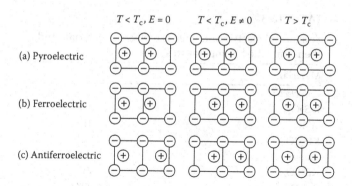

FIGURE 12.28

Schematic representations of structures of solids, all of which are paraelectric at temperatures in excess of T_c: (a) pyroelectric, showing a net spontaneous polarization even in the absence of an electric field; (b) ferroelectric, showing net spontaneous polarization that can be reversed by application of an electric field; and (c) antiferroelectric, showing microscopic ferroelectric regions that are exactly balanced by regions with the opposite polarity, so that there is no net polarization in the absence of a field but polarization can be induced by the application of an electric field.

ferroelectric is derived by analogy with the term *ferromagnetic*; in the former, electric dipoles are aligned and in the latter, magnetic moments are aligned. Again, a ferroelectric material becomes paraelectric at high temperatures.

There are other crystal structures in which there is no net polarization of a cell in the absence of an electric field, but the cell is polarized when a field is applied. These structures are *antiferroelectric*, as shown schematically in Figure 12.28c. In this case, cells (or layers) each have their own nonzero polarization, but the polarizations cancel out over the whole crystal, so there is no net polarization. Again, at higher temperatures, $T > T_c$, an antiferroelectric crystal will have no net polarization and so it will be paraelectric.

Some ferroelectric and antiferroelectric materials and their phase transition temperatures are presented in Table 12.3.

12.7 Superconductivity

The *superconducting* state, in which a material has zero direct current resistivity, was first discovered in 1911 by Gilles Hoist and Professor Kamerlingh Onnes at the prominent low-temperature laboratory in Leiden. The first material found to be superconducting was mercury, and the first evidence came from measuring its resistivity at $T = 4.2$ K, the boiling point of helium.*

---

* Heike Kamerlingh Onnes (1853–1926; Nobel Prize in Physics, 1913, for the production of liquid helium) discovered both superconductivity and the superfluid transition in liquid helium on the same day, April 8, 1911.

TABLE 12.3

Some Ferroelectric and Antiferroelectric Materials and their Transition Temperatures

Material	T_c/K
Ferroelectric	
K_2SeO_4	93
$SrTiO_3$	110
KH_2PO_4	123
Thiourea	169
KD_2PO_4	213
$BaTiO_3$	408
$NaNO_2$	436
$BaMF_4$ (M= Mn, Ni, Zn, Mg, Co)	FE at all temperatures
Antiferroelectric	
$NH_4H_2PO_4$	148
NaOD	153
KOH	227
$ND_4D_2PO_4$	242
KOD	253
$PbZrO_3$	506
WO_3	1010

The transition to the superconducting state, as seen in electrical resistivity, is shown schematically in Figure 12.29.

The property of superconductivity is particularly useful in that superconducting magnets, with their complete absence of electrical resistivity, carry current indefinitely. For example, superconducting materials are used in nuclear magnetic resonance and magnetic resonance imaging to achieve very high magnetic fields. Since they have no electrical losses (resistance along wires causes electrical losses), superconductors can have many useful applications. It is estimated that about 5% of all electricity generated is

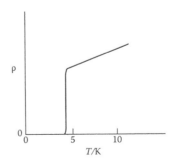

FIGURE 12.29

Schematic view of the transition to the superconducting state in mercury, as studied by immersion of mercury in liquid helium (at its boiling point of 4.2 K).

lost due to resistance during transmission. In 2014, a 1-km superconductor power-transmission cable system was installed in a commercial power grid, but the cable requires substantial cooling to allow the superconducting state. For this reason, much effort has been expended toward increasing the superconducting transition temperature with the aim of producing materials that are superconductors at room temperature. Although the maximum transition temperature has been rising over the years (the highest claimed superconducting transition temperature in mid-2018 is 203 K), the progress in increasing the superconducting temperature is unpredictable. Advancement in theoretical understanding of the superconducting state might be necessary before a room-temperature superconductor is realized.

One of the unusual properties of a superconductor is its interaction with a magnetic field. In its normal (nonsuperconducting state), the magnetic field can penetrate a material (see Figure 12.30a). Due to internal

See the video "Magnetic Levitation of Superconductor" under Student Resources at PhysicalPropertiesOfMaterials.com

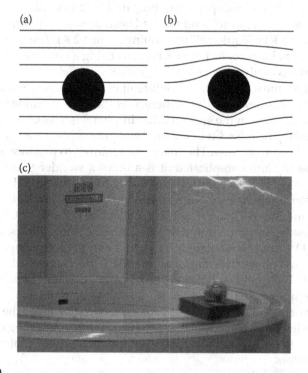

FIGURE 12.30
(a) Magnetic field can penetrate a material when it is not in its superconducting state. (b) In the superconducting state, the external magnetic field is repelled, as per the Meissner effect. (c) The Meissner effect of the cold superconducting material in the box causes levitation of the magnet above a magnetic track. (Photo taken at the Toshiba Museum, Kawasaki, Japan.)

currents, the magnetic field is repelled by a material in its superconducting state (Figure 12.30b). This factor causes a material in its superconducting state to have a strong repulsive interaction with a magnetic field, giving *levitation* of a superconductor above a magnet (Figure 12.30c), including the possibility of magnetic levitation trains (highly efficient due to low frictional losses) using superconductors. The expulsion of magnetic field by a superconductor is called the *Meissner effect*.

Since the late 1980s, there has been considerable interest in producing superconductors with very high superconducting temperatures. The main goal is to achieve a useful superconducting material in a convenient temperature range. Prior to 1986, all known superconductors were metals or alloys with very low transition temperatures. Tremendous excitement in this field arose when, in 1986, Bednorz and Müller at IBM, Zürich, found ternary perovskite-related structures with superconducting temperatures above 30 K.[*] This was followed in 1987 by the discovery, by Ching-Wu (Paul) Chu and his colleagues at the Universities of Houston and Alabama, that related compounds could be made to superconduct by cooling with liquid nitrogen (i.e., $T_c > 77$ K). The importance of this discovery was that liquid nitrogen (which boils at $T = 77$ K) is relatively easily and inexpensively prepared by condensing air. The other cryogenic (low-temperature) fluid that had been used to cool earlier superconductors ($T_c \ll 77$ K) is liquid helium (boiling point 4.2 K). The main disadvantages of liquid helium are its high cost and its being a nonrenewable resource since helium gas easily escapes Earth's gravitational field.

One of the most important high-temperature superconductors (also known as *high-T_c superconductors*, where T_c is the critical temperature below which the material is in the superconducting state) is $YBa_2Cu_3O_{8-y}$. The overall stoichiometry of Y: Ba: Cu in the final compound often leads to the abbreviation *1–2–3 superconductor*. The treatment required to produce $YBa_2Cu_3O_{8-y}$ is rather specific and complicated; it is a bit of a wonder that the material was discovered! The preparation method requires the initial reactants be ground in stoichiometric amounts, heated to 960 °C; then held for 10 hours, cooled, ground, sieved, and pressed into pellets; heated to 960 °C for 17 hours; cooled in flowing oxygen, heated to 980 °C, and held for 8 hours; then cooled slowly to 800 °C and more quickly brought down to room temperature. The Y_2O_3, BaO, CuO phase diagram, showing the 1–2–3 compound, is shown in Figure 12.31. The final result is a brittle black pellet with a superconducting transition temperature of about 120 K.

More recent developments in superconductivity include: the surprising result that MgB_2 is superconducting ($T_c = 39$ K); iron-containing compounds can be made to be superconducting; even graphene-based materials can be superconductors.

---

[*] J. Georg Bednorz (1950–) and K. Alexander Müller (1927–) are researchers at IBM, Rüschlikan, Switzerland, and corecipients of the 1987 Nobel Prize in Physics for their important breakthrough in the discovery of superconductivity in ceramic materials.

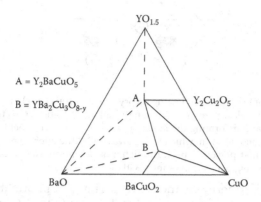

FIGURE 12.31
The $YO_{1.5}/BaO/CuO$ ternary phase diagram at 1000 °C.

One theory of superconductivity, commonly called *BCS theory*, after its initiators Bardeen, Cooper, and Schrieffer,[*] attributes the lack of electrical resistivity in the superconducting state to interactions between pairs of electrons (*Cooper pairs*), particularly at low temperatures. Although electrons normally repel each other due to their like charge, at very low temperatures they can have a special affinity for one another. This is especially so if they are paired with opposite spin so that a Cooper pair (one of each spin) travels together in the lattice. In the ionic lattice of a superconductor, one electron in the pair interacts with the cationic (i.e., positively charged) portion of the lattice by attracting it, causing a small deformation in the cationic sublattice (see Figure 12.32) corresponding to a lattice wave (phonon). This slight deviation of the atomic positions from their equilibrium positions leads to a potential gradient that makes it very favorable for the second electron in the pair to move to where the first electron just was. In this simplified view, the motion of the second electron is so facilitated that electrical resistivity is zero. The fuller picture of the superconducting state is one of a single coherent state involving many pairs of electrons.

Although it explains many features of low-temperature superconductivity, BCS theory only works if the temperature is low enough so that the available thermal energy is insufficient to break up the Cooper pairs. High-temperature superconductors (such as $YBa_2Cu_3O_{8-y}$ and its analogues) exist in a superconducting state at temperatures that exceed the range of the BCS theory; there would be too much thermal energy for the electrons

---

[*] John Bardeen (1908–1991) was an American physicist and the first person to be twice-winner of a Nobel Prize in Physics (in 1956 with William Shockley and W.H. Brattain for work leading to the invention of the transistor, and in 1972 with L.N. Cooper and J.R. Schrieffer for the superconductivity theory); Leon N. Cooper (1930–) is an American physicist who shared the 1972 Nobel Prize in Physics with Bardeen and Schrieffer for superconductivity theory work carried out while he was a research associate in Bardeen's laboratory; John Robert Schrieffer (1931–) is an American physicist who shared the 1972 Nobel Prize in Physics with Bardeen and Cooper for work on the theory of superconductivity carried out as a PhD student under Bardeen's supervision.

FIGURE 12.32
A schematic view of BCS superconducting theory. The electrons travel in Cooper pairs, one of each spin. As the first electron moves through the lattice, the cations (positively charged ions) are attracted to it, and deforming the lattice somewhat. The electric field from the deformation attracts the second electron, which then moves with little effort. An analogue to the BCS motion is that the first person to push through a busy subway station makes a path along which an accompanying person can saunter with little resistance.

to stay paired. Furthermore, there is experimental information concerning high-temperature superconductors that does not fit the BCS theory: Isotopic substitution (e.g., replacement of ^{16}O with ^{17}O) in the lattice does not change the superconducting transition temperature as the BCS theory predicts. The frequencies of the phonons would be affected by isotopic substitution, and in the BCS theory, this would change the electron–phonon interactions that give rise to the superconducting transition.

The present directions of research in superconductivity include semi-rational approaches to finding new high-T_c materials. Another area of research is increasing the current capacity and mechanical properties of high-temperature superconductors. Considerable effort is going into further experiments and theory to try to understand the mechanisms of superconductivity in these materials. In addition, new materials with interesting superconducting properties continue to be found.

12.8 Thermometry: A Tutorial

Temperature is one of the most fundamental properties required in physical measurement. Temperature can be measured many ways, from the rate of appearance of drops of sweat on a brow to the most sophisticated electronic instruments.

a. One of the most common methods of measurement of temperature in everyday application involves thermal expansion (e.g., of the fluid in a capillary tube). Can you suggest applications where this would not be a convenient method?

b. List properties that could be desirable for a thermometer.

c. If the two ends of a piece of metal wire are at different temperatures, there is an electrical potential gradient (voltage difference, ΔV) along the wire. The relationship between ΔV and the temperature difference, ΔT, is given by

$$\Delta V = S(T)\,\Delta T \qquad (12.14)$$

where $S(T)$ is the *thermoelectric power* of the wire, also known as the *Seebeck* coefficient. It is dependent on the wire's composition. Note that if electrons are the predominant electrical carriers, $S < 0$, but if holes are the predominant carriers, $S > 0$. If two wires of different material are connected as shown in Figure 12.33, then the measurement of the voltage across them, ΔV_T, is obtained:

$$\Delta V_T = \Delta V_1 + \Delta V_2 \qquad (12.15)$$

where ΔV_1 and ΔV_2 are given by Equation 12.14 for each material, 1 and 2. Here, ΔV_T is a direct measure of ΔT, as defined in Figure 12.33. This device is a thermocouple.

i.   Could both arms of a thermocouple be made of the same metal? Why or why not?

ii. A thermocouple measures temperature differences. How could one use this device to measure absolute temperatures?

iii. Would a better (more sensitive) thermocouple have a higher or lower magnitude of thermoelectric power, S?

iv. The thermoelectric power of a material depends on both its temperature and its composition. A "typical" thermoelectric power of a material used in a thermocouple is $1 \times 10^{-5}\,\text{V K}^{-1}$. If a sensitivity of $0.1\,\text{K}$ is needed in a measurement, how accurately does the

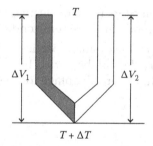

FIGURE 12.33
When dissimilar metals are joined together so that the junction is at a temperature different from the other ends, there is a voltage drop across each material as shown. This voltage arises from the different work functions of the two materials.

---

* Thomas Johann Seebeck (1770–1831) was born in East Prussia and trained as a medical doctor. However, his true interest was physics. Seebeck studied the theory of color with German scientist Johann Goethe. Seebeck's name is now mostly associated with his discovery of thermoelectricity. Although he himself thought that the magnetic field arising from a temperature gradient between junctions of dissimilar wires was fundamentally a thermomagnetic effect, we now realize that the magnetic field is due to current flow (thermoelectric effect).

thermocouple voltage need to be determined? (Note that this just takes account of one arm of a thermocouple and we should really consider $\Delta S = S_2 - S_1$ where S_1 and S_2 are the thermopowers for each of the arms.)

d. Many other electrical principles can be used to measure temperature. We have seen that the resistance of a metal will vary with temperature. How could this be used to produce a resistance thermometer? What important properties should be considered? (Hint: One of the most useful resistance thermometers is the platinum resistance thermometer.) What determines the sensitivity of a resistance thermometer?

e. The resistance of intrinsic (pure) semiconductors changes with temperature. Could this be used to produce a semiconductor resistance thermometer? Do you foresee any problems at very low temperatures?

f. Semiconductors are notoriously difficult to purify, and many of their physical properties (including electrical resistivity) depend on their purity. One way to be sure of their purity is to purposefully add impurities that overwhelm any small traces of unwanted impurities. Does this solve the low-temperature problem of using pure semiconductors as resistance thermometers? How?

g. Metal oxides that have been sintered and treated in oxidizing and/or reducing atmospheres to produce n-type semiconductors on the outside and p-type semiconductors on the inside (or vice versa) give useful thermometers called *thermistors*. A typical room-temperature resistance for a thermistor might be 10 kΩ.

   i. Would its resistance increase or decrease as the temperature is lowered? (Remember, they are semiconductors.)

   ii. Since thermistors have small particles, the electrical contact between the particles can be varied by the manufacturing process. How can this be an advantage for thermometry?

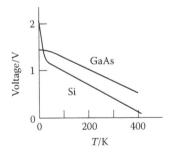

FIGURE 12.34
Forward-bias voltages of Si and GaAs diodes as functions of temperature.

iii. Thermistors can be made to be very small (less than 0.1 mm diameter). Why is this an advantage?

h. Figure 12.34 shows the temperature dependence of the forward-bias voltage of silicon and gallium arsenide diodes as functions of temperature. Could these be used to measure temperature? What voltage accuracy would be needed for each thermometer to determine the temperature at about 200 K to within 1 K?

12.9 Learning Goals

- Band theory
- Resistivity, conductivity, types of electrical materials
- Metals: band structure, work function, contact potential
- Semiconductors: band structure for pure; doping (n and p)
- Insulators: band structure
- Temperature dependence of electrical conductivity: metals, intrinsic semiconductors
- Wiedemann–Franz law for metals
- Electrical properties of CNTs
- Doped semiconductors: properties, devices (n,p-junction = diode [rectification]; p,n,p-junction = transistor [amplification, switching])
- Topological insulators
- Conducting polymers
- Peltier effect and device
- Dielectrics: pyroelectric, ferroelectric, antiferroelectric, paraelectric
- Superconductivity, Meissner effect, BCS theory
- Thermometry
- Thermoelectric devices

12.10 Problems

12.1 Consider the following electrical resistivities at 0 °C: Cu, 16 Ω nm; Ni, 69 Ω nm; Zn, 53 Ω nm; brass (Cu-Zn), ~60 Ω nm; constantan (Cu-Ni), ~500 Ω nm.

a. From these data, how does the resistivity of a solid solution (such as brass or constantan) compare with its pure components?

b. Suggest a microscopic (atomic scale) explanation for the findings of (a).

12.2 On the basis of the information in Figure 12.35, arrange the materials noted in order of decreasing electrical conductivity. Comment on how this order correlates with the position in the periodic table and the atomic size.

12.3 From Figure 12.36, which shows the intrinsic conductivity of pure germanium as a function of reciprocal temperature, estimate the band gap, E_g, for Ge.

12.4 The black-body radiation of a material can be used as a form of thermometer.

a. Would this method be more accurate at low or high temperatures? Explain.

b. What would you need to measure to determine the temperature of a black-body radiator?

12.5 Many semiconductors are known to have low-packing fractions (i.e., low densities), which could mean that they could be easily compressed. In the case of tin, a change in density could be associated

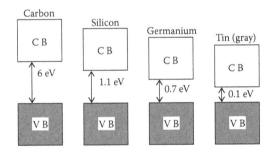

FIGURE 12.35
Various semiconductor materials, schematically showing their band gaps.

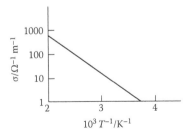

FIGURE 12.36
Intrinsic electrical conductivity of Ge as a function of reciprocal temperature.

with a phase transformation that could transform a semiconductor into a metal. In general, compression of a semiconductor will increase E_g even if there is no phase change. Can you suggest an application for this principle?

12.6 An electronic security device involves a beam of light passing across the entryway to a home. When the beam is interrupted (e.g., as an intruder blocks it), a security bell sounds. What sort of material is likely used as the beam detector? Explain your reasoning.

12.7 Boron-doped diamonds are blue (see Chapter 3). Would you expect these diamonds to be electrically conducting? Explain.

12.8 A cholesteric (also called "twisted nematic") liquid crystal has a structure as shown in Figure 5.2.

    a. If the liquid crystal molecules are polar, explain how this material can be made ferroelectric.

    b. Can the degree of polarity of the material be controlled? If so, how?

    c. Suggest an application that makes use of controlled polarity of the material.

12.9 The ice crystals in storm clouds are ferroelectric.

    a. Will the local charge of the earth just prior to an electrical storm influence the orientation of the ice crystals in the clouds? If so, how?

    b. How might this influence the optical properties of the cloud?

    c. If clouds were made of nonpolar molecules, would thunderstorms be more or less dramatic?

12.10 In Figure 12.18, the Fermi energy of an n-type semiconductor is shown as higher than that of a p-type semiconductor. Explain why. (Assume that the n- and p-semiconductors have the same dominant material, to which either n- or p-impurities have been added.)

12.11 The intralayer electronic band structure of graphite is shown schematically in Figure 12.37; the bonding σ and π states are completely full at $T = 0$ K, and since the band gap to the anti-bonding π^* molecular orbital is essentially zero (Figure 12.37), the slightest thermal energy can promote electrons. The electronic structure is quite anisotropic in graphite because the interatomic distance within the layers is 3.35 Å, compared with 1.415 Å between the layers. Use this information to explain the following:

    a. Graphite is a poor electrical conductor in the direction perpendicular to the layers and much better within the layers;

    b. Graphite is an excellent thermal conductor in the direction of the layers and a poor thermal conductor in the direction perpendicular to the layers.

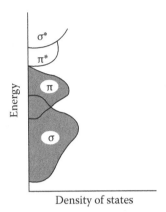

FIGURE 12.37
Band structure of graphite within the layers.

12.12 In most metals, the thermal conductivity is enhanced (over insulators) due to the transport of heat by free electrons. However, for this to happen, the electrons must be able to interact with lattice vibrations (phonons) that have thermal energy. In most superconductors, the superconducting state is due to electrons that travel together as "Cooper pairs," and these electrons, because of their long wavelengths, cannot interact with phonons. On this basis, predict whether a material would have diminished or enhanced thermal conductivity in its superconducting state.

12.13 Hydrogen, which is an insulator at ambient pressure, will become metallic at very high pressures, such as within the core of the planet Jupiter. Explain why an insulator could become metallic at high pressures.

12.14 The difference between energy levels in a system can be expressed in many different units. The band gap of a typical semiconductor is of the order of 1 eV. Calculate the equivalent to 1 eV for the following units or parameters:

   a. $J\,mol^{-1}$
   b. K (thermal energy)
   c. cm^{-1} (wavenumber)
   d. s^{-1} (frequency)
   e. nm (wavelength)

12.15 Ionic solids such as NaCl are normally considered to be insulators. However, such crystals always contain defects, for example, where an Na^+ or Cl^- ion at a particular lattice site is missing.

a. Propose a mechanism for electrical conductivity due to such defects.

b. The concentration of defects in an ionic crystal is normally very small. What does this indicate about the electrical resistivity of such a crystal?

c. The electrical conductivity of NaCl can be enhanced by doping with Ca^{2+} or Mg^{2+} ions. Explain why.

12.16 The field at which dielectric breakdown occurs is called the *dielectric strength*. Air is a dielectric with a relatively small dielectric strength. Use this information to explain how lightning occurs.

12.17 Calculate the thermal energy, kT, at room temperature in comparison with the band gap energy of a semiconductor with a band gap that corresponds to a wavelength of 1000 nm. Use this information to explain why infrared detectors using semiconductors are usually cooled to low temperatures (e.g., 77 K using liquid nitrogen) to reduce the dark current (i.e., electrical signal in the absence of an IR source).

12.18 A *bolometer* is a cooled semiconductor used to detect far infrared radiation (wavelengths as long as 5 mm). Use a density of states diagram to indicate why a bolometer's resistance changes in the presence of infrared radiation.

12.19 In Chapter 3 we saw that amorphous selenium (used in photocopiers) has a band gap of 1.8 eV. Another form of selenium, which is crystalline, has a band gap of 2.6 eV. Explain why the crystalline form has a higher band gap.

12.20 Phase diagram points with zero degrees of freedom are often used as standard points to calibrate electronic thermometers. Why?

12.21 Electrical conductivity in a metal can be changed by changing temperature, mechanical deformation, impurity concentration, or the size of crystallites. Discuss.

12.22 From Equation 12.6, and consideration of the temperature dependence of the mean free path of conduction electrons in a metal, explain the sign of the temperature dependence of the electrical conductivity of a metal.

12.23 The electrical and thermal conductivities of Al are both about 70% of that of Cu. Explain briefly why both types of conductivity should be in about the same ratio for two such materials.

12.24 Ferroelectric liquid crystals (FLCs) use liquid crystal substances that have chiral molecules in a smectic type of arrangement. The spiral nature of the structure allows microsecond switching response time which makes FLCs particularly suited to advanced displays. Why is it important that these materials be ferroelectric?

12.25 Carbon nanotubes have potential applications that arise from their special properties. However, production can lead to mixtures of metallic and semiconducting tubes. Explain why separation is important for use of carbon nanotubes to make composites with tailored electrical properties.

12.26 Carbon monoxide is a poisonous gas and it would be very useful to detect it. Fortunately, the electronic properties of semiconductors can be very sensitive to chemical reactivity, allowing design of chemical sensors. What sort of sensor materials could be used to detect CO?

12.27 The *triboelectric effect* occurs when static electricity builds up from friction between two different materials. If common fabric fibers such as cotton and wool are wound around a steel thread, the triboelectric effect is sufficient to cause charging of a capacitor (see A. Yu et al., *ACS Nano* **2017**, 11, 12764) such that the energy of everyday activities such as walking can be used to charge cell phones. Discuss factors that need to be considered to make this technology practical.

12.28 In a thermoelectric cooling device, such as the one shown in Figure 12.25, there are competing heat flows across the device. By analogy with the thermal conduction processes described in Chapter 8, the heat flux carried by the conducting electrons and holes, J_{cond}, is given by

$$J_{cond} = S^2 \sigma T \left(\frac{dT}{dz} \right) \tag{12.16}$$

where S is the thermopower (Seebeck coefficient), σ is the electrical conductivity, and dT/dz is the thermal gradient across the device. The competing backflow of heat from thermal conductivity, J_{th}, is given by

$$J_{th} = -\kappa \left(\frac{dT}{dz} \right) \tag{12.17}$$

where κ is the thermal conductivity. The ratio of the coefficients of heat flux in Equations 12.16 and 12.17 is given by

$$ZT = \frac{S^2 \sigma T}{\kappa} \tag{12.18}$$

where ZT is called the *thermoelectric figure of merit*. Note that ZT is dimensionless and represents the overall efficiency of the device.

(Strictly speaking, there is one value of ZT for the n-arm and one for the p-arm of the thermoelectric device.) Comment on what physical properties of a thermoelectric material need to be increased or decreased to improve the efficiency of the device. Discuss both electrical and thermal properties.

12.29 Thermoelectric materials also can be used to make devices that scavenge waste heat and turn it into power. Comment on some of the potential applications of such devices.

See the video "Thermoelectric fan" under Student Resources at PhysicalPropertiesOfMaterials.com

12.30 The speed of sound in a gas is another way to determine temperature, via acoustic thermometry. Give the equation that relates the average speed of a gas molecule to the temperature.

12.31 Composites made with carbon nanotubes (CNT) as filler can have interesting properties. The electrical and thermal conductivities of pure individual metallic carbon nanotubes are $\sim 10^8$ S m^{-1} and 2000 W m^{-1} K^{-1}, respectively. A typical matrix for a CNT composite would be epoxy with electrical and thermal conductivities of 10^{-13} S m^{-1} and 0.2 W m^{-1} K^{-1}, respectively.

   a. Based on the rule of mixtures (Equation 11.5), calculate the expected electrical and thermal conductivities of a CNT composite with 1% CNT loading.

   b. Typical observed electrical and thermal conductivities of a CNT composite with 1% CNT loading are 10^{-6} S m^{-1} and 0.25 W m^{-1} K^{-1}. Compare these with the calculated values from (a).

   c. CNTs with electrical conductivity of 10^8 S m^{-1} are metallic, whereas epoxy, with an electrical conductivity of 10^{-13} S m^{-1}, is an insulator and a 1% CNT composite with an electrical conductivity of 10^{-6} S m^{-1} is a semiconductor. Explain why there is such a large difference in electrical conductivity with such a small loading of CNTs.

   d. CNTs with thermal conductivity of 2000 W m^{-1}K^{-1} are excellent heat conductors, while epoxy, with a thermal conductivity of 0.2 W m^{-1}K^{-1}, is a thermal insulator, as is a 1% CNT composite with a thermal conductivity of 0.25 W m^{-1}K^{-1}. Explain why there not such a large difference in thermal conductivity (as there was for electrical conductivity) with this loading of CNTs.

   e. Suggest an application for materials with high electrical conductivity and low thermal conductivity.

Further Reading

General References

Advances in silicon carbide electronics. Special issue. J. C. Zolper and M. Skowronski, Eds. *MRS Bulletin*, April 2005, 273–311.

Molecular transport junctions. Special issue. C. R. Kagan and M. A. Ratner, Eds. *MRS Bulletin*, June 2004, 376–410.

Organic spintronics. Special issue. J. S. Moodera, B. Koopmans, and P. M. Oppeneer, Eds. *MRS Bulletin*, July 2014, 578–620.

M. Ya Axbel, I. M. Lifshitz, and M. I. Kganov, 1973. Conduction electrons in metals. *Scientific American*, January, 88.

P. Ball, 1994. *Designing the Molecular World*. Princeton University Press, Princeton, NJ.

A. Barton, 1997. *States of Matter: States of Mind*. Institute of Physics Publishing, Bristol.

R. J. Borg and G. J. Dienes, 1992. *The Physical Chemistry of Solids*. Academic Press, Oxford.

R. A. Butera and D. H. Waldeck, 1997. The dependence of resistance on temperature for metals, semiconductors, and superconductors. *Journal of Chemical Education*, 74, 1090.

W. D. Callister, Jr. and D. G. Rethwisch, 2013. *Materials Science and Engineering: An Introduction*, 9th ed. John Wiley & Sons, Hoboken, NJ.

B. S. Chandrasekhar, 1998. *Why Things Are the Way They Are*. Cambridge University Press, Cambridge.

P. Chaudhari, 1986. Electronic and magnetic materials. *Scientific American*, October, 136.

A. K. Cheetham and P. Day, Eds. 1992. *Solid State Chemistry Compounds*. Oxford University Press, Oxford.

P. G. Collins and P. Avouris, 2000. Nanotubes for electronics. *Scientific American*, December, 62.

A. H. Cottrell, 1967. The nature of metals. *Scientific American*, September, 90.

M. de Podesta, 2002. *Understanding the Properties of Matter*, 2nd ed. CRC Press, Boca Raton, FL.

R. A. Dunlap, 1988. *Experimental Physics*. Oxford University Press, New York.

H. Ehrenreich, 1967. The electrical properties of materials. *Scientific American*, September, 195.

A. B. Ellis, M. J. Geselbracht, B. J. Johnson, G. C. Lisensky, and W. R. Robinson, 1993. *Teaching General Chemistry: A Materials Science Companion*. American Chemical Society, Washington, DC.

G. G. Hall, 1991. *Molecular Solid State Physics*. Springer-Verlag, New York.

R. Hoffmann, 1987. How Chemistry and Physics Meet in the Solid-State. *Angewandte Chemie – International Edition in English*, 26, 846.

T. S. Hutchison and D. C. Baird, 1968. *The Physics of Engineering Solids*. John Wiley & Sons, Hoboken, NJ.

C. Kittel, 2004. *Introduction to Solid State Physics*, 8th ed. John Wiley & Sons, Hoboken, NJ.

P. G. Nelson, 1997. Quantifying electrical character. *Journal of Chemical Education*, 74, 1084.

C. N. R. Rao and J. Gopalakrishnan, 1997. *New Directions in Solid State Chemistry.* Cambridge University Press, Cambridge.

J. F. Shackelford, 2004. *Introduction to Materials Science for Engineers,* 6th ed. Prentice Hall, Upper Saddle River, NJ.

J. Singleton, 2001. *Band Theory and Electronic Properties of Materials.* Oxford University Press, New York.

L. E. Smart and E. A. Moore, 2012. *Solid State Chemistry,* 4th ed. CRC Press, Boca Raton, FL.

G. Stix, 1999. Bad connections. *Scientific American,* December, 50.

L. H. Van Vlack, 1989. *Elements of Materials Science and Engineering,* 6th ed. Addison-Wesley, Reading, MA.

F. Vögtle, Ed., 1991. Chapter 10: Organic semiconductors, conductors and superconductors and Chapter 11: Molecular wires, molecular rectifiers and molecular transistors. In *Supramolecular Chemistry: An Introduction.* John Wiley & Sons, Hoboken, NJ.

Electrochromic Materials

P. M. S. Monk, R. J. Mortimer, and D. R. Rosseinsky, 1995. Through a glass darkly. *Chemistry in Britain,* May, 380.

Electronic Devices

Advanced flat-panel displays and materials. Special issue. R. M. Wallace and G. Wilk, Eds. *MRS Bulletin,* November 2002, 186–229.

Alternative gate dielectrics for microelectronics. Special issue. *MRS Bulletin,* March 2002.

Fabrication of sub-45-nm device structures. Special issue. J. J. Watkins and D. J. Bishop, Eds. *MRS Bulletin,* December 2005, 937–982.GaN and related materials for device applications. Special issue. S. J. Pearton and C. Kuo, Eds. *MRS Bulletin,* February 1997, 7–57.

Harvesting energy through thermoelectrics: Power generation and cooling. Special issue. T. M. Tritt and M. A. Subramanian, Eds. *MRS Bulletin,* March 2006, 188–229.

Macroelectronics. Special issue. R. H. Reuss, D. G. Hopper, and J.-G. Park, Eds. *MRS Bulletin,* June 2006, 447–485.

Organic-based photovoltaics. Special issue. S. E. Shaheen, D. S. Ginley, and G. E. Jabbour, Eds. *MRS Bulletin,* January 2005, 10–52.

Perovskite photovoltaics. Special Issue. M. K. Nazeeruddin and H. Snaith. *MRS Bulletin,* August 2015, 641–686.Polymeric and organic electronic materials and applications. Special issue. A. J. Epstein and Y. Yang, Eds. *MRS Bulletin,* June 1997, 13–56.

Solid-state century. Special issue. *Scientific American Presents,* 1997.

Transparent conducting oxides. Special issue. B. G. Lewis and D. C. Paine, Eds. *MRS Bulletin,* August 2000, 22–65.

O. Auciello, J. F. Scott, and R. Ramesh, 1998. The physics of ferroelectric memories. *Physics Today,* July, 22.

S. Bauer, R. Gerhard-Multhaupt, and G. M. Sessler, 2004. Ferroelectrets: Soft electroactive foams for transducers. *Physics Today*, February, 37.

D. Bradley, 2001. What memories are made of. *Chemistry in Britain*, March, 28.

K. Bourzac, 2018. 'Atomristors' made from 2-D materials. *Chemical and Engineering News*, January 15, 6.

S. M. Condren, G. S. Lisensky, K. R. Nordell, T. F. Kuech, and S. A. Stockman, 2001. LEDs: New lamps for old and a paradigm for ongoing curriculum modernization. *Journal of Chemical Education*, 78, 1033.

R. Dagani, 2001. Here comes paper 2.0. *Chemical and Engineering News*, January 15.

R. Dagani, 2001. Nanotube single-electron transistor works at room temperature. *Chemical and Engineering News*, July 9, 10.

R. Dagani, 2001. Polymer transistors: Do it by printing. *Chemical and Engineering News*, January 1, 26.

R. Dagani, 2001. Semiconductor nanowires light up a nano-LED. *Chemical and Engineering News*, January 8, 7.

R. Dagani, 2001. Single molecule wired into circuit. *Chemical and Engineering News*, October 22, 14.

F. Gomollón, 2017. World's smallest diode warms-up for real-life applications. *Chemistry World*, May 12.

M. Gross, 2001. From e-ink to e-paper. *Chemistry in Britain*, July, 22.

B. Gross Levy, 2001. New printing technologies raise hopes for cheap plastics electronics. *Physics Today*, February, 20.

N. Holonyak Jr. 2005. From transistors to lasers and light-emitting diodes. *MRS Bulletin*, July, 509.

W. E. Howard, 2004. Better displays with organic films. *Scientific American*, February, 76.

M. Jacoby, 2000. Data storage: New materials push the limits. *Chemical and Engineering News*, June 12, 37.

M. Jacoby, 2000. Semiconductors meet organics. *Chemical and Engineering News*, April 17, 32.

M. Jacoby, 2001. Carbon nanotube computer circuits. *Chemical and Engineering News*, September 3, 9.

M. Jacoby, 2001. The ol' switcheroo comes in a new size. *Chemical and Engineering News*, June 25.

B. E. Kane, 2005. Can we build a large-scale quantum computer using semiconductor materials? *MRS Bulletin*, February, 105.

J. R. Minkel, 2002. Falling in line. *Scientific American*, April, 30.

J. R. Minkel, 2005. Shrinking memory. *Scientific American*, February, 33.

R. E. Newnham, 1983. Structure-property relations in ceramic capacitors. *Journal of Materials Education*, 5, 941.

I. M. Ross, 1997. The foundations of the silicon age. *Physics Today*, December, 34.

P. E. Ross, 2006. Viral nano electronics. *Scientific American*, October, 52.

H. Sevian, S. Müller, H. Rudman, and M. F. Rubner, 2004. Using organic light-emitting electrochemical thin film devices to teach materials science. *Journal of Chemical Education*, 81, 1620.

A. G. Smart, 2017. Polymer-based transistors bring fully stretchable devices within reach. *Physics Today*, March, 14.

S. H. Voldman, 2002. Lightning rods for nanoelectronics. *Scientific American*, October, 90.

P. Yang, 2005. The chemistry and physics of semiconductor nanowires. *MRS Bulletin*, February, 85.

Ice

B. Vonnegut, 1965. Orientation of ice crystals in the electric field of a thunderstorm. *Weather*, 20, 310.

Metals

J. Emsley, 1996. By Jove, metallic hydrogen! *Chemistry in Britain*, June, 14.
W. A. Harrison, 1969. Electrons in metals. *Physics Today*, October, 23.
A. R. Mackintosh, 1963. The Fermi surface of metals. *Scientific American*, July, 110.
E. Wilson, 1998. Deuterium goes metallic. *Chemical and Engineering News*, August 24, 11.

Molecular Materials and Plastics

Commercialization of organic electronics. Special issue. F. So, J. Kido, and P. Burrows, Eds. *MRS Bulletin*, July 2008, 663–705.
Electrifying plastics. *Chemical and Engineering News*, October 16, 2000.
Electroactive organic materials. Special issue. Z. Bao, V. Bulovic, and A. B. Holmes, Eds. *MRS Bulletin*, June 2002, 441–464.
Electronic properties of organic-based interfaces. Special issue. L. Kronik and N. Koch, Eds. *MRS Bulletin*, June 2010, 417–465.
Hybrid organic-inorganic materials. Special issue. D. A. Loy, Ed. *MRS Bulletin*, May 2001, 364–401.
Materials for stretchable electronics. Special issue. S. Wagner and S. Bauer, Eds. *MRS Bulletin*, March 2012, 207–260.
Organic single crystals. Special issue. V. Podzorov, Ed. *MRS Bulletin*, January 2013, 15–71.
Stretchable and ultraflexible organic electronics. Special issue. D. J. Lipomi and Z. Bao, Eds. *MRS Bulletin*, February 2017, 93–142.
G. P. Collins, 2004. Next stretch for plastic electronics. *Scientific American*, August, 74.
G. P. Collins, 2005. A future in plastics. *Scientific American*, December, 55.
R. Dagani, 2003. Softer, pliable electronics. *Chemical and Engineering News*, January 6, 25.
S. R. Forest, 2004. The path to ubiquitous and low-cost organic electronic appliances on plastic. *Nature*, 428, 911.
M. Freemantle, 2001. Superconducting organic polymer. *Chemical and Engineering News*, March 12, 14.
M. Freemantle, 2003. Solutions for OLED displays. *Chemical and Engineering News*, February 24, 6.
B. Gross Levi, 2000. Nobel Prize in chemistry salutes the discovery of conducing polymers. *Physics Today*, December, 19.
J. R. Heath and M. A. Ratner, 2003. Molecular electronics. *Physics Today*, May, 43.
M. Jacoby, 2001. Nanotube conduction. *Chemical and Engineering News*, April 30.
R. B. Kaner and A. G. MacDiamid, 1988. Plastics that conduct electricity. *Scientific American*, February, 105.
G. Malliaras and R. Friend, 2005. An organic electronics primer. *Physics Today*, May, 53.
M. Najdoski, L. Pejov, and V. M. Petruševski, 1999. Pyroelectric effect of a sucrose monocrystal. *Journal of Chemical Education*, 76, 360.

J.-F. Tremblay, 2016. The rise of OLED displays. *Chemical and Engineering News,* July 11, 30.

M. D. Ward, 2001. Chemistry and molecular electronics: new molecules as wires, switches and logic gates. *Journal of Chemical Education,* 78, 321.

B. Wessling, 2001. Polymers show their metal. *Chemistry in Britain,* March, 40.

Nanomaterials

Quantum dot light-emitting devices. Special issue. D. Talapin and J. Steckel, Eds. *MRS Bulletin,* September 2013, 685–742.

S. K. Blau, 2017. Conduction electrons flow like honey. *Physics Today,* November, 22.

M. Gross, 2002. The smallest revolution. *Chemistry in Britain,* May, 36.

C. M. Leiber, 2001. The incredible shrinking circuit. *Scientific American,* September, 58.

M. Ouyang, J.-L. Huang, and C. M. Lieber, 2002. Fundamental electronic properties and applications of single-walled carbon nanotubes. *Accounts of Chemical Research,* 35, 1018.

H. I. Smith and H. G. Craighead, 1990. Nanofabrication. *Physics Today,* February, 24.

Y.-W. Son, M. L. Cohen, and S. G. Louie, 2006. Half-metallic graphene nanoribbons. *Nature,* 444, 347.

G. M. Whitesides and J. C. Love, 2001. The art of building small. *Scientific American,* September, 38.

R. M. Wilson, 2017. The carbon nanotube integrated circuit goes three-dimensional. *Physics Today,* September, 14.

Paraelectrics

R. E. Cohen, 2006. Relaxors go critical. *Nature,* 441, 941.

S. B. Lang, 2005. Pyroelectricity: From ancient curiosity to modern imaging tool. *Physics Today,* August, 31.

Semiconductors

Semiconductor quantum dots. Special issue. A. Zunger, Ed. *MRS Bulletin,* February 1998, 15–53.

G. H. Döhler, 1983. Solid-state superlattices. *Scientific American,* November, 144.

M. W. Geis and J. C. Angus, 1992. Diamond film semiconductors. *Scientific American,* October, 84.

G. C. Lisensky, R. Penn, M. L. Geselbracht, and A. B. Ellis, 1992. Periodic properties in a family of common semiconductors. *Journal of Chemical Education,* 69, 151.

J. R. Minkel, 2002. Charging up diamonds. *Scientific American,* November, 36.

S. O'Brien, 1996. The chemistry of the semiconductor industry. *Chemical Society Reviews,* 25, 393.

A. Pisanty, 1991. The electronic structure of graphite. *Journal of Chemical Education,* 68, 804.

E. Sandre, A. LeBlanc, and M. Danot, 1991. Giant molecules in solid state chemistry. *Journal of Chemical Education,* 68, 809.

Superconductors

C_{60} made to superconduct at 117 K. *Chemical and Engineering News*, September 3, 2001, 34.

High performance YBCO-coated superconductor wires. Special issue. Y. P. Paranthaman and T. Izumi, Eds. *MRS Bulletin*, August 2004, 533–589.

High-temperature superconductivity. *Physics Today*, June 1991 (special issue with several articles on this topic).

F. J. Adrain and D. O. Cowan, 1992. The new superconductors. *Chemical and Engineering News*, December 21, 24.

J. Bardeen, 1990. Superconductivity and other macroscopic quantum phenomena. *Physics Today*, December, 25.

P. C. Canfield and S. L. Bud'ko, 2005. Low-temperature superconductivity is warming up. *Scientific American*, April, 80.

P. C. Canfield and G. W. Crabtree, 2003. Magnesium diboride: Better late than never. *Physics Today*, March, 34.

R. J. Cava, 1990. Superconductors beyond 1–2–3. *Scientific American*, August, 42.

L. L. Chang and L. Esaki, 1992. Superconductor quantum heterostructures. *Physics Today*, October, 36.

P. C. W. Chu, 1995. High-temperature superconductors. *Scientific American*, September, 162.

A. Chubukov and P. J. Hirschfeld, 2015. Iron-based superconductors, seven years later. *Physics Today*, June, 46.

C. D. Cogdell, D. G. Wayment, and D. J. Casadonte Jr., 1995. A convenient, one-step synthesis of $YBa_2Cu_3O_{7-x}$ superconductors. *Journal of Chemical Education*, 72, 840.

G. P. Collins, 2004. High-temp knockout. *Scientific American*, May, 28.

D. L. Cox and M. B. Maple, 1995. Electronic pairing in exotic superconductors. *Physics Today*, February, 32.

R. Dagani, 2001. Superconductor stuns physicists. *Chemical and Engineering News*, March 2, 13.

S. Dann, 2001. No resistance. *Chemistry in Britain*, December, 18.

P. Day, Ed., 2006. *Molecules into Materials*. World Scientific, Singapore.

A. de Lozanne, 2006. Hot vibes. *Nature*, 442, 522.

J. de Nobel, 1996. The discovery of superconductivity. *Physics Today*, September, 40.

P. I. Djurovich and R. J. Watts, 1993. A simple and reliable chemical preparation of $YBa2Cu3O_{7-x}$ superconductors. *Journal of Chemical Education*, 70, 497.

E. A. Ekimov, V. A. Sidorov, E. D. Bauer, N. N. Mel'nik, N. J. Curro, J. D. Thompson, and S. M. Stishov, 2004. Superconductivity in diamond. *Nature*, 428, 542.

R. J. Fitzgerald, 2014. Better superconducting wires. *Physics Today*, May, 17.

M. Freemantle, 2000. Holes raise C_{60} superconductivity temperature. *Chemical and Engineering News*, December 4, 12.

T. H. Geballe, 1993. Superconductivity: From physics to technology. *Physics Today*, October, 52.

A. M. Goldman and N. Markovič, 1998. Superconductor-insulator transitions in the two-dimensional limit. *Physics Today*, November, 39.

B. Gross Levi, 2000. Learning about high-T_c superconductors from their imperfections. *Physics Today*, March, 17.

B. Gross Levi, 2004. New experiments highlight universal behavior in copper oxide superconductors. *Physics Today*, September, 24.

R. M. Hazen, 1988. Perovskites. *Scientific American*, June, 74.

L. Hoddeson, 1999. John Bardeen and the BCS theory of superconductivity. *MRS Bulletin*, January, 50.

K. D. Irwin, 2006. Seeing with superconductors. *Scientific American*, November, 86.

J. R. Kirtley and C. C. Tsuei, 1996. Probing high-temperature superconductivity. *Scientific American*, August, 68.

Y. Maeno, T. M. Rice, and M. Sigrist, 2001. The intriguing superconductivity of strontium ruthenate. *Physics Today*, January, 42.

A. P. Malozemoff, J. Mannhart, and D. Scalapino, 2005. High-temperature cuprate superconductors get to work. *Physics Today*, April.

J. Mannhart and P. Chaudhari, 2001. High-T_c bicrystal grain boundaries. *Physics Today*, November, 48.

A. McCook, 2001. A warmer superconductor? *Scientific American*, October, 22.

B. Raveau, 1992. Defects and superconductivity in layered cuprates. *Physics Today*, October, 53.

D. van Delft and P. Kes, 2010. The discovery of superconductivity. *Physics Today*, September, 38.

D. G. Walmsley and X.-H. Zheng, 2017. Sulfur hydride and superconductivity theory. *Physics Today*, July, 14.

E. Wilson, 2001. Boron gives up its resistance. *Chemical and Engineering News*, July 16, 7.

E. Wilson, 2001. Superconducting nanotubes. *Chemical and Engineering News*, July 2, 8.

Thermoelectrics

Materials for energy harvesting. Special issue. T. Mori and S. Priya, Eds. *MRS Bulletin*, March 2018, 176–219.

Thermoelectric materials and applications. Special issue. T. M. Tritt and M. A. Subramanian, Eds. *MRS Bulletin*, March 2006, 188–229.

P. Ball, 2014. Harvesting heat. *Chemistry World*, November, 52.

F. J. DiSalvo, 1999. Thermoelectric cooling and power generation. *Science*, 285, 703.

D. M. Rowe, Ed., 2005. *Thermoelectrics Handbook*. CRC Press, Boca Raton, FL.

B. C. Sales, 1998. Electron crystals and phonon glasses: A new path to improved thermoelectric materials. *MRS Bulletin*, January, 15.

A. Shakouri, 2011. Recent developments in semiconductor thermoelectric physics and materials. *Annual Review of Materials Research*, 41, 399.

J. R. Sootsman, D. Y. Chung and M. G. Kanatzidis, 2009. New and old concepts in thermoelectric materials. *Angewandte Chemie International Edition*, 48, 8616.

T. M. Tritt, 2011. Thermoelectric phenomena, materials and applications. *Annual Review of Materials Research*, 41, 433.

E. J. Winder, A. B. Ellis, and G. Lisensky, 1996. Thermoelectric devices: Solid-state refrigerators and electrical generators in the classroom. *Journal of Chemical Education*, 73, 940.

Thermometry

S. Carlson, 1999. A homemade high-precision thermometer. *Scientific American*, March, 102.

S. Uchiyama, A. P. De Silva, and K. Iwai, 2006. Luminescent molecular thermometers. *Journal of Chemical Education*, 83, 720.

Topological Insulators

Topological insulators. Special Issue. C. Felser and X.-L. Qi, Eds. *MRS Bulletin*, October 2014, 843–879.

M. Z. Hasen and C. L. Kane, 2010. Colloquium: Topological insulators. *Reviews of Modern Physics*, 82, 3045.

M. Z. Hasen and J. E. Moore, 2011. Three-dimensional topological insulators. *Annual Review of Condensed Matter Physics*, 2, 55.

Websites

For links to relevant websites, see PhysicalPropertiesOfMaterials.com

13

Magnetic Properties

13.1 Introduction

Magnetic materials have revolutionized our lives, from the recording and playing of our favorite music to readable magnetic strips on our bank cards to magnetic materials storing information in our computers. The great advantage of magnetic materials is that they can store so much information in such a small space.* In this chapter, we explore the principles behind magnetic materials.

13.2 Origins of Magnetic Behavior

The magnetic properties of a material can be understood in microscopic terms when we consider the special properties of moving electrons. It is for this reason that electrical and magnetic properties are related and some of the terms are parallel in their use. It is the motion of a charge (i.e., electrons; note that the motion of the nuclei contributes only about 0.1% as much as the electrons to magnetic properties) that creates a magnetic field. Before we can proceed to consider magnetism in detail, it is useful to define some terms.

A *magnetic field* is a region surrounding a magnetic body. This region can induce magnetic fields in other bodies. Magnetic materials, such as iron filings, can align themselves to represent *field lines* (also called *flux lines*), as shown in Figure 13.1.

A magnet can be considered to be formed from two poles. We are familiar with this in a bar magnet. However, no matter how finely a bar magnet is divided, both poles are always present. The unit pole (or magnetic monopole) concept is a useful theoretical idea to describe quantitative aspects of

---

* Magnetic materials contain dramatically increasing amounts of information. Magnetic tape could store a few kilobytes in the 1940s. Four decades later, the 1-GB-per-square-inch storage density barrier was broken using a magnetoresistive head. Compare that with the data storage capacity of today's computers!

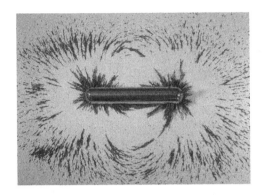

FIGURE 13.1
Magnetic field patterns, as indicated by iron flings aligned by the magnetic field of a bar magnet.

magnetism. A *unit pole* is defined as a body that will repel an equal unit pole placed 1 cm away in a vacuum, with a force of 10^{-5} N.

The *magnetic field strength*, $\bar{H}$, is a vector that measures the force acting on a unit pole placed at a fixed point in a magnetic field (in vacuum). $\bar{H}$ also measures the magnitude and direction of the magnetic field. The SI units of the magnitude of the magnetic field strength, H, are A m^{-1}.

The lines of force of a magnet describe the free path traced by an imaginary magnetic monopole in a magnetic field. For the space surrounding the source of the magnetic field, there is an *induction* whose magnitude, B, is the *flux density*. The induction and the magnetic field strength are related in vacuum by

$$\bar{B} = \mu_0 \bar{H} \tag{13.1}$$

where μ_0 is the *permeability of vacuum*. If there is a material in the magnetic field, the relation is similar*:

$$\bar{B} = \mu \bar{H} \tag{13.2}$$

where μ is the *permeability* of that material.

Equation 13.2 is analogous to the electrical relation

$$J = \sigma \frac{V}{l} \tag{13.3}$$

where J is the electrical current density, σ is the electrical conductivity, and V/l is the voltage gradient. Comparison of Equations 13.2 and 13.3 shows that

---

* The most important equations in this chapter are designated with ▌ to the left of the equation.

magnetic induction (B) is analogous to electric current density (J); magnetic field strength (H) is analogous to voltage gradient (V/l); and permeability (μ) is analogous to electrical conductivity (σ).

Equation 13.2 can be rewritten to explicitly show the contribution of the material to the induction:

$$\bar{B} = \mu\,\bar{H} = \mu_0\left(\bar{H} + \bar{M}\right) \tag{13.4}$$

where $\bar{M}$ is the *magnetization* of the material, and the term $\mu_0\bar{H}$ represents the additional magnetic induction field associated with the material. The SI units for induction, B, are tesla* (abbreviated as T) and for magnetization, M, are A m^{-1}. The *relative permeability*, μ_r, of a material is defined as

$$\mu_r = \frac{\mu}{\mu_0} \tag{13.5}$$

and the *magnetic susceptibility*, χ, is defined as

$$\chi = \frac{\bar{M}}{\bar{H}} \tag{13.6}$$

where both μ_r and χ are unitless.

The sign and magnitude of χ indicate the magnetic class of a material. If χ is negative and independent of H, the material is expelled from a magnetic field, and this material is *diamagnetic*. If χ is small and positive (10^{-5} to 10^{-2}) and independent of H, the material is slightly attracted to a magnetic field; this material is *paramagnetic*. If χ is large and positive (10^{-2} to 10^6) and dependent on H, the material is very attracted to a magnetic field; this material is *ferromagnetic*. The influence of the material on the magnetic field lines for a diamagnet (flux density decreases), a paramagnet (flux density increases), and a superconductor (magnetic field repelled by the Meissner effect) are shown in Figure 13.2.

The origins of diamagnetism, paramagnetism, and ferromagnetism are all related to the electrons in the materials, and understanding this relationship draws heavily on quantum mechanics. Briefly, there are two important ways in which electrons contribute to magnetism: *orbital magnetism* (due to the motion and resulting orbital angular momentum of the electrons in "orbits" about the nucleus) and *spin magnetism* (due to the spin-up and spin-down properties of an electron).

---

* Nikola Tesla (1856–1943) was born in Croatia and made his major contributions as an electrical engineer in the United States. This Serbian-American is best known as the inventor of the ac induction motor, an idea that he sold to George Westinghouse for commercialization. Tesla was an eccentric, brilliant, reclusive inventor; he carried out many dangerous electrical experiments in his private laboratory, often putting out the power in his region.

(a) (b) (c)

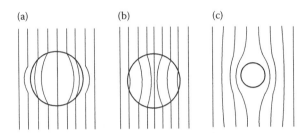

FIGURE 13.2
Magnetic flux density contours in (a) a diamagnetic material; (b) a paramagnetic material, and (c) a superconducting material.

COMMENT: SUPERCONDUCTORS AS THERMAL SWITCHES

One of the properties of the superconducting state is that electrical resistivity is exactly zero. Another property is that its normal resistivity is restored in the presence of magnetic fields in excess of a *critical magnetic field*, B_c. The critical field depends on the critical temperature of the material, T_c, approximately as

$$B_c = B_0\left(1 - (T / T_c)^2\right) \tag{13.7}$$

where B_0 is the value of B_c as $T \to 0$ K.

Although conduction electrons allow metals to conduct heat more efficiently than in insulators, the paired electrons (Cooper pairs) in a superconductor do not interact with thermal phonons and therefore a superconductor does not conduct heat as well as a metal. This property can allow a superconductor to be used as a thermal switch.

For example, a link of tin or lead can provide a connection between a low-temperature sample and its surroundings. In its metallic state, it is an efficient thermal conductor and can be used as such (e.g., during cooling processes). Further cooling puts the link in its superconducting state and thermally isolates the system (e.g., to carry out thermal measurements). Similarly, the thermal link can be activated or deactivated by turning a magnetic field on and off. Such a thermal link avoids problems such as vibration and heating associated with mechanical heat switches at very low (*cryogenic*) temperatures.

Diamagnetic behavior, which is characterized experimentally by the (usually weak) repulsion of the material from a magnetic field, arises due to the (orbital) motion of the electrons being influenced by the presence of a magnetic field. In a diamagnetic material, all the spins are paired (represented as ↑↓, i.e., spin up and spin down, in equal amounts) so the atoms have no resultant magnetic moment, and an applied magnetic field has little influence except induction associated with the motion of the electrons (current flows in a direction to produce a magnetic field that opposes the inducing field and repulsion results). All materials have this diamagnetic effect, but in materials with unpaired spins it can be overwhelmed by more powerful attraction of the material into the magnetic field, e.g., from paramagnetism and ferromagnetism.

If there are unpaired electrons, as in many transition metal species and also some other molecules such as O_2, the total electronic spin must be nonzero, and paramagnetism results. We have already seen that spin can contribute to the magnetic properties of the system. In *ordinary paramagnets* (often referred to as *Pauli* paramagnets) the magnetic properties are defined by the properties of the individual atoms, and spin is one of two important considerations. The other is orbital magnetism, which can arise for certain electronic states of some atoms. Orbital magnetism can give additional attraction of this system to a magnetic field, similar to the magnetic effect associated with an electric current flowing in a closed-loop of wire.

In contrast with diamagnets and ordinary paramagnets, a ferromagnet relies on more than the single-atom properties to derive its interaction with a magnetic field. The origin of this strong, attractive force is cooperative interactions among the magnetic moments of individual atoms arranged on a lattice. Iron is the archetypal example from which this class of materials derives its name. There are two competing interactions involving magnetic spins to keep in mind: *Exchange interactions* tend to keep spins aligned (↑↑) due to electron–electron and electron–nuclear interactions and *magnetic dipole interactions* tend to align spins antiparallel (↑↓) due to long-range interactions that favor this configuration.

In a ferromagnetic metal, the importance of many-atom effects in determining magnetism is directly related to the electronic band structure. A high density of states in the bands makes it possible to have reduced electron repulsion (i.e., reduced energy) by having electrons with parallel spins singly occupying energy levels near the Fermi energy. This situation requires a high density of unoccupied states near the Fermi level. In the absence of a magnetic field, averaged over the whole sample there would be approximately equal numbers of spins "up" and "down," as shown in Figure 13.3a. An external magnetic field would lower the energy of electrons aligned with the field and raise the energy of those electrons that are aligned opposing the applied magnetic field (Figure 13.3b). To compensate for this, some electrons realign their spins into the lower-energy levels, as shown in Figure 13.3c, leading to net magnetization. For metals with nearly full, narrow 3d bands (e.g., iron, nickel, cobalt), the density of states near the Fermi level is particularly high, and the cost of promoting electrons to

---

* Wolfgang Pauli (1900–1958) was an Austrian theoretical physicist and winner of the 1945 Nobel Prize in Physics for the enunciation of what we now call the Pauli exclusion principle. Although a brilliant theoretician, his experimental contemporaries considered Pauli to be "jinxed," and he was regarded as bad luck in a laboratory. One story has it that an important experiment went disastrously wrong, and this was blamed on Pauli passing through that city on the night train.

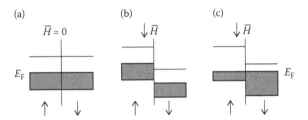

FIGURE 13.3
Electronic energy levels of a ferromagnetic material. (a) In the absence of a magnetic field, there are equal numbers of electrons with both spins. (b) At the instant when a magnetic field is applied as shown, the aligned electrons have lower energies, and the electrons with opposing alignment have increased energy. (c) To reduce the system energy in the magnetic field, some electrons realign their spins into lower energy levels, producing net magnetism.

higher levels is so small that it is very favorable energetically to have large numbers of unpaired electrons. As a result, these materials are ferromagnetic. There is a particularly favorable situation for ferromagnetism in the middle of the first row of the transition elements; the electronic band structures of some metals are shown in Figure 13.4.

> See the video "Floating Magnet" under Student Resources at PhysicalPropertiesOfMaterials.com

In the absence of a magnetic field, a piece of iron (or other ferromagnet) might not be magnetized, depending on the sample's history. The reason is that the spins are aligned in a given direction only within small volumes within the crystal (about 10^{-14} m^3 each), called *domains*. Different domains have different magnetic orientations such that the overall crystal is not magnetized. With only a few spins, the preferred orientation is aligned, but as more spins are added, magnetic dipolar interactions tend to favor antiparallel arrangements. The result is domains of different spin orientations, as shown in Figure 13.5.

Diamagnetic, paramagnetic, and ferromagnetic effects are summarized in Table 13.1.

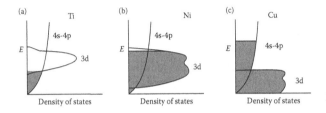

FIGURE 13.4
Electronic band structure diagrams for (a) titanium which is diamagnetic as most valence electrons are in the diffuse 4s-4p band; (b) nickel which is ferromagnetic since the Fermi level is in the region of high density of states that makes exchange interactions energetically feasible, and (c) copper which is diamagnetic since the Fermi level is in the diffuse 4s-4p band. The occupied levels are shaded.

COMMENT: MOLECULAR MAGNETS

Chemists have been designing new magnetic materials with molecular building blocks to have control over their properties. For example, molecular materials generally have low density and are amenable to property tuning by minor changes in the synthesis. The first-discovered molecular magnet, $[Fe(C_5(CH_3)_5)_2]^{\bullet+}[(NC)_2C=C(CN)_2]^{\bullet-}$ is ferromagnetic below 4.8 K. Studies of molecular magnets have shown that spin coupling is not sufficient to cause ferromagnetism; intermolecular interactions also are required. Depending on the spin interaction, molecular magnets can be either ferromagnetic or antiferromagnetic.

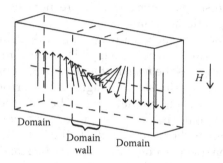

FIGURE 13.5
Domains in a ferromagnetic material. The region between domains is called the *domain wall*.

TABLE 13.1

Diamagnetic, Paramagnetic, and Ferromagnetic Effects

Effect	Field	χ	Dominant Origin	Magnitude of Magnetic Effect
Diamagnetism	Weak repulsion	<0, Independent of field strength	Induction associated with orbital motion; due to paired electrons; always present	Small
Pauli paramagnetism	Weak attraction	10^{-5}–10^{-2}, Independent of field strength	Unpaired spins of individual atoms + possibly orbital magnetism	Small
Ferromagnetism	Strong attraction	10^{-2}–10^{6}, Depends on field strength	Cooperative interactions; exchange ($\uparrow\uparrow$) and magnetic dipole ($\uparrow\downarrow$) compete	Large

13.3 Magnetic Induction as a Function of Field Strength

Equations 13.1 and 13.2 show the relations between induction, B, and magnetic field strength, H, for free space (vacuum) and a material, respectively. For a paramagnetic material, $\mu > \mu_0$ (typically $\mu = 1.01\ \mu_0$), which leads to the induction-field strength relation shown in Figure 13.6. For a diamagnetic material, $\mu < \mu_0$ (typically, $\mu = 0.99995\ \mu_0$); this induction-field strength relation is also shown in Figure 13.6.

Ferromagnetic materials show much more dramatic induction with increasing field strength, as shown in Figure 13.7 (see also Table 13.2). This relationship is associated with their very large relative permeabilities that arise, as mentioned already, due to cooperative effects among the magnetic moments within the material.

Figure 13.7 shows that the induction in a ferromagnetic material has some further interesting properties that are not observed for diamagnetic and paramagnetic materials. When the magnetic field strength is increased, the induction reaches a maximum value called the *saturation induction* (B_s). Furthermore, when the field is reduced to zero, there is still some induction, known as the *remanent induction, B_r.* (This is a well-known phenomenon; when a previously "unmagnetized" ferromagnetic material is brought up to a magnet, magnetism can be induced and can remain even after the first magnet has been removed.) If the field is reversed, at a certain value known as the *coercive field, H_c,* the induction finally becomes zero. The induction–field strength relation on increasing the field is not the same as on decreasing the field, and this gives rise to a hysteresis loop (Figure 13.7).

It is the cooperative effects of the magnetic moments on adjacent atoms that give such large magnetic effects in the case of a ferromagnet. This situation is shown schematically in Figure 13.8, where a trip along the hysteresis loop,

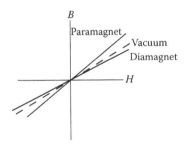

FIGURE 13.6
Induction (B) as a function of magnetic field strength (H) for free space (vacuum), a typical paramagnetic material, and a typical diamagnetic material. The differences between the slopes in vacuum (slope $= \mu_0$) and the paramagnetic material (slope $\approx 1.01\ \mu_0$) and diamagnetic material (slope $\approx 0.99995\ \mu_0$) are exaggerated here for clarity.

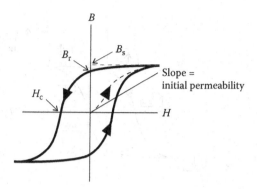

FIGURE 13.7

Relationship between induction (B) and magnetic field strength (H) for a ferromagnet. The broken line is the initial magnetization, and the solid line is the dependence after the first magnetization, where the arrow heads indicate whether H is increasing or decreasing. The value B_s is the saturation induction, and B_r is the remanent induction (induction remaining when the field is removed). The value H_c is the coercive field required to reach zero induction. The hysteresis loop shows that the induction on increasing field is not the same as the induction on decreasing field.

TABLE 13.2

Properties of Selected Soft Magnets

Material	Initial Relative Permeability (μ_r at $B \approx 0$)	Hysteresis Loss/ (J m^{-3} per Cycle)	Saturation Induction, B_s/T
Commercial iron	250	500	2.16
Fe-4% Si, random	500	50–150	1.95
Fe-3%Si, oriented	15,000	35–140	2.0
45 Permalloy ($Ni_{45}Fe_{55}$)	2700	120	1.6
Mumetal ($Ni_{75}Cu_5Cr_2Fe_{18}$)	30,000	20	0.8
Supermalloy ($Ni_{79}Fe_{15}Mo_5$)	100,000	2	0.79

starting at zero induction, first shows saturation to be a result of maximum spin alignment with the applied field. When the field is removed, there are still some residual aligned domains. As the field is reversed, more domains align in the reverse direction until finally they are fully aligned (saturation in the reverse direction).

The hysteresis loop of a ferromagnet can allow applications of these materials. The hysteresis loop can be very broad (*hard magnets* because the domain walls are difficult to move) or narrow (*soft magnets* because the domain walls are easier to move); these properties are illustrated in Figure 13.9.

One major use of metallic ferromagnets is in the core of power transformers. Soft magnets are used there because of the small width of their hysteresis loop and consequent higher energy efficiency (the area of the loop

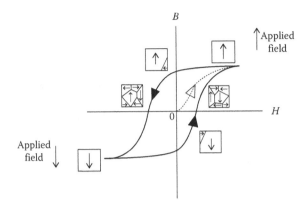

FIGURE 13.8
An induction-magnetic field hysteresis loop for a ferromagnetic material, showing schematic domain structures at various points in relation to the applied magnetic field.

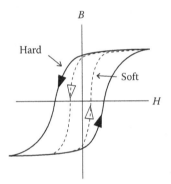

FIGURE 13.9
Comparison of typical hysteresis loops for a hard and a soft ferromagnet.

is proportional to the energy loss in one pass through the loop). The ease of magnetization and demagnetization makes soft magnets particularly useful in alternating current applications. It is also important that a magnet used in this application has sufficiently high saturation induction to minimize the size of the transformer core. Properties of some soft magnetic materials are given in Table 13.2.

Hard magnets are useful as so-called *permanent magnets* because their large hysteresis loop leads to large residual magnetization values. Alloys such as samarium–cobalt, platinum–cobalt, and alnico (an alloy of aluminum, nickel, and cobalt) are useful hard magnets.

Experimentally, a soft magnet can be distinguished from a hard magnet by the value of the coercive field, H_c, necessary to return the induction to zero

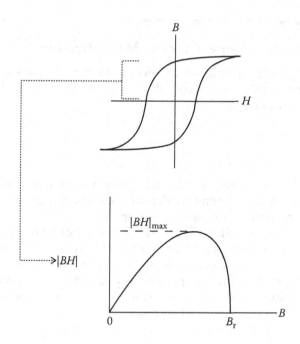

FIGURE 13.10

Plotting of the product $|BH|$ from the demagnetization quadrant as a function of B defines a maximum, designated $|BH|_{max}$, for a magnet, where $|BH|_{max}$ is a measure of the power of a permanent magnet.

following magnetization, with $|H_c| > 1000$ A m^{-1} as the barrier beyond which a magnet is considered to be hard (permanent magnet). Another criterion is $|BH|_{max}$, as defined in Figure 13.10. This value is indicative of the energy required for demagnetization and is a measure of the magnetic strength of a material. Typical values of $|BH|_{max}$, B_r, and H_c are given in Table 13.3.

TABLE 13.3

Properties of Selected Hard Magnets

Material	Remanence, $B_r/(V \text{ s m}^{-2})$	Coercive Field, $H_c/(kA \text{ m}^{-1})$	Maximum Demagnetizing Product, $|BH|_{max}/(kJ \text{ m}^{-3})$
Carbon steel	1.0	−4	1
Alnico V	1.2	−55	34
Ferroxdur (BaFe$_{12}$O$_{19}$)	0.4	−150	20
Rare earth-cobalt	1.0	−700	200
Nd$_2$Fe$_{14}$B	—	−1600	—

13.4 Temperature Dependence of Magnetization

For a normal (Pauli) paramagnetic material in which the paramagnetic species act independently, the susceptibility obeys the *Curie*[*] law as the temperature, T (in kelvin), changes:

$$\chi = \frac{C}{T} \tag{13.8}$$

where C is the *Curie constant*. This relationship is shown schematically in Figure 13.11, and the temperature dependence reflects the increase in thermal randomization with greater thermal energy.

When there is a possibility of cooperative magnetic behavior, the temperature dependence of the susceptibility is different. At very high temperatures, there is too much thermal energy to allow cooperative magnetism; below a threshold temperature such cooperation is possible. For a ferromagnet, this critical temperature is the *Curie temperature, T_C*, and the temperature dependence is described by the *Curie–Weiss*[†] *law*

$$\chi = \frac{C}{T - T_C}, \qquad T > T_C. \tag{13.9}$$

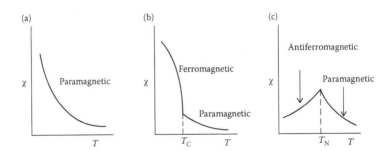

FIGURE 13.11
The temperature dependence of magnetic susceptibility for (a) a paramagnet; (b) a ferromagnet (showing the ferromagnetic–paramagnetic transition at T_C); and (c) an antiferromagnet (showing the antiferromagnetic–paramagnetic transition at T_N).

---

[*] Pierre Curie (1859–1906) was a French physicist and codiscoverer (with his brother Jacques-Paul Curie) of piezoelectricity in 1877. His famous PhD thesis concerned important work on magnetism; in 1903, he shared the Nobel Prize in Physics with his wife, Marie Skłodowska Curie, and Henri Becquerel for their work on radioactivity. His life ended prematurely when he was hit by a vehicle while crossing Rue Dauphine in Paris.
[†] Pierre-Ernest Weiss (1865–1940) was a French physicist who introduced the concept of magnetic domains.

Below T_C, a ferromagnet has aligned magnetic moments that give rise to spontaneous magnetization even in the absence of a field. The moments (aside from domain effects) would be totally aligned at $T = 0$ K. The Curie temperatures of selected ferromagnets are given in Table 13.4.

For some materials, there is a different temperature dependence to the susceptibility, expressed as

$$\chi = \frac{C}{T + T_N}, \qquad T > T_N \tag{13.10}$$

where T_N is the *Néel* temperature. These materials are *antiferromagnetic*; their magnetic spins are aligned in layers, but alternate layers are of opposite magnetic orientation such that there is no net alignment of spins in the system. An example of an antiferromagnetic structure is NiO; its magnetic structure is shown in Figure 13.12a. Antiferromagnetism arises in this case due to the *superexchange interaction* driven by the covalent Ni-O-Ni bonding as follows. The interaction of the d orbitals of Ni with the oxygen 2p orbital gives some covalent bonding, with partial transfer of an electron from the oxygen to the metal orbitals. As shown in Figure 13.12b, if one Ni has spin up, only a spin-down electron from oxygen can be transferred to this orbital (Pauli exclusion principle). Therefore, bonding to an adjacent Ni must involve the spin-up electron of oxygen, which can be transferred to Ni only if its unpaired electron is in the spin-down state; this situation leads to an antiferromagnetic structure.

TABLE 13.4

Curie Temperatures (T_C, Ferromagnet) and Néel Temperatures (T_N, Antiferromagnet) for Selected Materials

Ferromagnets	T_C/K
Co	1400
Fe	1043
Ni	631
CrO_2	392
Gd	292
EuO	69
Antiferromagnets	T_N/K
NiO	530
CoO	292
FeO	198
MnO	122
NiF_2	83

---

[*] Louis Eugène Néel (1904–2000) was a French physicist and winner of the 1970 Nobel Prize in Physics for work on antiferromagnetism and ferrimagnetism.

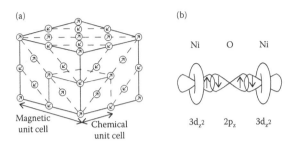

FIGURE 13.12
Antiferromagnetism in NiO. (a) In the antiferromagnetic structure of this material, alternate layers of aligned Ni spins are oriented in opposing directions such that there is no net magnetic moment. (b) Antiferromagnetism in NiO arises from the Ni-O-Ni bonding arrangement which dictates that alternate nickel atoms have opposing spins; this is an example of superexchange.

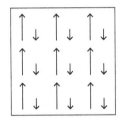

FIGURE 13.13
A schematic view of a ferrimagnetic structure. For Fe_3O_4, a ferrimagnetic material, unbalanced magnetic moments arise from different numbers of unpaired electrons on Fe^{3+} and Fe^{2+}.

Another type of magnetism, *ferrimagnetism*, sometimes known as unbalanced antiferromagnetism, is illustrated in Figure 13.13; in ferrimagnetism, there are two spin alignments, but they are not balanced out. *Ferrites* are ferrimagnets with the general formula Fe_2O_3MO, where M is a metal cation (e.g., Zn, Cd, Fe, Ni, Cu, Co, Mg, etc.). Magnetite, Fe_3O_4, is an example of a ferrite that has been known and used (e.g., as a compass) since ancient times. More recently, ferrites have been used in recording tapes and transformer cores. Ferrites are among the most important materials in magnetic applications.

In practice, paramagnets, ferromagnets, and antiferromagnets can be distinguished by the form of χ^{-1} as a function of temperature, as shown in Figure 13.14. Equations 13.8–13.10 can be generalized as

$$\chi = \frac{C}{T - \Theta} \tag{13.11}$$

where Θ, called the *Weiss constant*, is zero for a paramagnet, positive for a ferromagnet, and negative for an antiferromagnet.

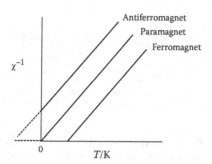

FIGURE 13.14

Variation of χ^{-1} as a function of temperature (extrapolated from high temperature) for a paramagnet, ferromagnet, and antiferromagnet. In each case, the $\chi^{-1}=0$ intercept value allows the calculation of the Weiss constant, Θ, from Equation 13.11.

COMMENT: MAGNETIC RESONANCE

Magnetic fields can influence the energy states of atoms and ions, and this situation can be used to probe the structure and dynamics of materials.

For a paramagnetic ion of spin ½, with a doubly degenerate ground state in the absence of a magnetic field, the energy levels will split as a function of magnetic field B (see Figure 13.15a), with a gap of $g\beta B$, where g is the *Landé splitting factor* (related to the spin of the system in question), and β is the *Bohr magneton* (a constant related to the magnetic moment of an electron). If at a given magnetic field B, the system is subject to radio frequency ν (energy $h\nu$) such that $h\nu = g\beta B$, the radio frequency will be absorbed and g can be determined. The value of g (and its anisotropy) can indicate the magnetic environment of a given nucleus. In the slightly more complex case of a zero-field splitting of the ground state (Figure 13.15b), measurements at more than one field can be used to determine g.

Resonance experiments at microwave frequencies ($\approx 10^{10}$ s^{-1}) can be used to investigate electronic states of paramagnetic ions by *electron spin resonance* (ESR), also called *electron paramagnetic resonance* (EPR).

Similar principles can be used to investigate states associated with the nucleus of an atom with nonzero spin by *nuclear magnetic resonance* (NMR) spectroscopy. NMR is arguably the most important structural technique in all of chemistry today, as it allows one to determine the atomic bonding arrangement (i.e., structure of molecules). NMR is a very sensitive technique; it is said that magnetic resonance imaging of brain processes during exposure to music can distinguish persons who have had musical training from those who have not.

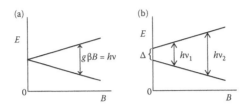

FIGURE 13.15
(a) A paramagnetic ion with a doubly degenerate ground state has its ground state degeneracy lifted in a magnetic field. (b) The zero-field splitting of a ground state can be determined by energy absorptions at two different magnetic fields: $h\nu_1 = \Delta + g\beta B_1$ and $h\nu_2 = \Delta + g\beta B_2$.

COMMENT: MAGNETORESISTANCE

Magnetoresistance is the change in electrical conductivity of a conductor when it is placed in a magnetic field. Usually the effect is small, of the order of a few per cent. In 1988, two research groups independently discovered that some layered materials showed large magnetoresistance, now called *giant magnetoresistance* (GMR). Peter Grünberg and Albert Fert shared the 2007 Nobel Prize in Physics for their discoveries. In both their investigations, GMR resulted from magnetic multilayers, nanometer-thick layers of ferromagnetic (Fe), and paramagnetic (Cr or Cu) materials.

Only the electrons near the Fermi level participate in electrical conduction. From Figure 13.3, it is clear that the properties of the electrons at the Fermi energy, including conduction, are different for spin-up and spin-down electrons in a ferromagnet, whereas for a paramagnetic material, spin-up and spin-down electrons contribute equally to electrical conduction. In the absence of a magnetic field, the system can be modeled as parallel resistors, one for the spin-up electrons and one for the spin-down electrons, as shown in Figure 13.16a, where $R_\uparrow$ and $R_\downarrow$ represent the resistances of the different spins. The overall resistance in the absence of the field, R_0, is given by

$$R_0 = \frac{R_\uparrow + R_\downarrow}{2}. \tag{13.12}$$

In the presence of a magnetic field, the situation changes as shown in Figure 13.16b, and the resistance, R_H, is given by

$$R_H = \frac{2R_\uparrow R_\downarrow}{R_\uparrow + R_\downarrow} \tag{13.13}$$

which gives a change in resistance due to the magnetic field,

$$\Delta R = R_H - R_0 = \frac{-\left(R_\uparrow + R_\downarrow\right)^2}{2\left(R_\uparrow + R_\downarrow\right)}. \tag{13.14}$$

It is clear from Equation 13.14 that the giant magnetoresistance originates with $R_\uparrow \neq R_\downarrow$.

Even larger magnetoresistive effects, called *colossal magnetoresistance*, have been found in bulk materials such as manganese perovskites. However, they require much larger applied fields, which limits their technological applications.

Materials with large magnetoresistance effects could be used for injection of spins in spintronic materials and rapid magnetic switching. GMR sandwich-structured materials have led to a great increase in the data storage of hard-disk drives, using GMR materials in the read heads.

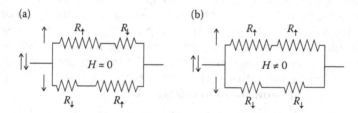

FIGURE 13.16
The resistance analog of the ferromagnet–paramagnet–ferromagnet multilayer system exhibiting giant magnetoresistance, (a) with no field and (b) in a magnetic field. The resistance of the paramagnetic layer is assumed to be negligible. The spin-up (↑) and spin-down (↓) electrons have different resistances.

13.5 Magnetic Devices: A Tutorial

a. Although much of the discussion in this chapter has concerned pure materials, one of the beauties of materials science is that it does not limit itself only to pure materials, and in purposefully added impurities, new properties can arise. Magnetic materials are no exception. Steel (which can be considered to be impure iron) is a much harder magnet than very pure iron. Why? Consider domain growth in a magnetic field. Will domains generally change the magnetic properties more for hard or soft magnetic materials?

b. Many of the first-row transition metals can have very useful ferromagnetic properties; however, ferromagnetism is not observed in the second- and third-row transition metals. Given that the 4d and 5d bands are more diffuse than the 3d bands, can you reconcile this finding? Consider the cost of promotion of electrons (exchange interaction).

c. The first-row transition elements can be combined with lanthanides to give among the most powerful magnets known. Examples are $SmCo_5$ and $Nd_2Fe_{14}B$. Although they could potentially contribute directly to magnetism, the f electrons associated with the lanthanide atoms are too localized to exhibit band structure. In pure ferromagnetic lanthanide metals, the source of the ferromagnetism is delocalized d electrons; these d electrons interact with the localized f electrons to cause alignment of the d and f electrons (via exchange interactions) to reduce electron repulsion. Suggest qualitatively how such powerful magnetism arises in the transition metal–lanthanide alloys.

d. Chromium dioxide, CrO_2, was commonly used for audio cassettes, where it was chosen for its magnetic properties. CrO_2 is a ferromagnet with a Curie temperature of 392 K. Its 3d orbitals form a very narrow band, allowing ferromagnetism. Later transition metal oxides (e.g., MnO_2) have localized 3d electrons and are insulators or semiconductors. The earlier transition metal oxide, VO_2, is a semiconductor at room temperature but becomes metallic above 340 K; although it is paramagnetic for $T > 340$ K, the spins are localized and not ordered cooperatively. This shows that CrO_2 has a special place among the first-row transition metal oxides.

　i. The position of CrO_2 among the first-row transition metal oxides is somewhat similar to that of Fe, Co, and Ni in the pure transition metals: it is at the balance point between wide bands of delocalized electrons (earlier elements) and localized electrons (later elements). Why does this balance point occur at atomic number 24 for the oxides and somewhat later (atomic numbers 26, 27, and 28) for the pure elements? Consider the metal–metal separation for the oxides compared to the pure metals and how the metal–metal separation influences the overlap of the 3d orbitals and the width of the band.

　ii. CrO_2 is used in magnetic recording tapes. Once used commonly for audio recording, magnetic tape is still used in favor of disk recording when dealing with large amounts of data, due to the low cost per bit. A recording tape consists of a polyester tape impregnated with needle-like crystals of CrO_2. In audio recording, the sound to be recorded activates a diaphragm in a microphone, and the vibrations cause fluctuating electric current in a coil of wire wrapped around an iron core in the recorder head, over which the tape passes. The varying current causes a varying magnetic field in the iron core, which in turn magnetizes the CrO_2 particles on the tape. The direction and magnetization of the CrO_2 particles is a record of the electrical impulses from the microphone. To play back a tape, the process is reversed. Sketch diagrams of the recording and playback mechanisms showing the magnet and the tape. Should the iron core be a hard ferromagnet or a soft ferromagnet? What about CrO_2? Are there any other considerations that are important for a good magnetic tape material? (You might consider the value of the Curie temperature for CrO_2 among other factors.)

e. For a computer memory device, it is most useful to have a magnetic material that switches cleanly between two different states (full induction in one direction or the other direction) as the magnetic field changes. Sketch the ideal $B–H$ hysteresis loop for such a material. Examples of nearly ideal behavior are $(CoFe)O$ and Fe_2O_3.

13.6 Learning Goals

- Magnetic field strength, magnetic flux density
- Magnetic susceptibility
- Diamagnetic, paramagnetic, ferromagnetic, antiferromagnetic, ferrimagnetic
- Exchange interactions, magnetic dipole interactions
- Domains, domain walls
- Relationships between magnetic induction and magnetic field strength: saturation induction, remanent induction, coercive field
- Hysteresis in magnets
- Hard versus soft magnets; permanent magnets
- Temperature dependence of magnetization; Curie temperature for ferromagnets; Néel temperature for antiferromagnets
- Magnetoresistance

13.7 Problems

13.1 Although manganese is diamagnetic, some alloys containing manganese, such as Cu_2MnAl, are ferromagnetic. The Mn–Mn distance in these alloys is greater than in pure manganese metal. Suggest why this situation could lead to ferromagnetism.

13.2 The sizes and orientations of magnetic domains in a material can be altered by mechanical means. For example, domain orientations for two magnetic materials are shown in Figure 13.17 where the "textured" (oriented) structure is a result of cold rolling. Sketch the initial portion (starting at zero field) of the magnetization (B-H) plot for the random and for the textured materials.

13.3 Is there such a thing as a "nonmagnetic" material? Explain.

13.4 Suggest possible uses for the Meissner effect exhibited by superconducting magnetic materials.

13.5 Are magnetic effects in crystalline noncubic solids inherently isotropic or anisotropic? Explain.

13.6 From Equation 13.11, derive equations for the three lines in Figure 13.14.

13.7 The so-called permanent magnets maintain their magnetization even in the absence of magnetic fields. This situation is achieved by using fine-grained alloys with "nonmagnetic" inclusions that inhibit domain wall movement. In addition, rod-shaped particles aligned along the direction of magnetization will hinder demagnetization. Which would make a better permanent magnet: Fe with a coercive force of 50 A m^{-1} and remanence of 1.2 T or $SmCo_5$ with a coercive force of 6×10^5 A m^{-1} and remanence of 0.9 T? Explain.

13.8 Arrange the transition metal oxides listed in Table 13.4 in order of increasing covalency on the basis of trends in their Néel temperatures.

13.9 Ferroelectric materials also can be described as "hard" or "soft", by analogy with ferromagnetic materials. Sketch a diagram (analogous to Figure 13.9) that distinguishes a hard ferroelectric material from a soft ferroelectric material.

13.10 Rare earth magnets are especially important for capture and use of renewable energy. For example, a 1 MW wind turbine can use up to 1 tonne of rare earth permanent magnets containing neodymium,

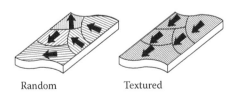

Random   Textured

FIGURE 13.17
The initial magnetic structure is random, but cold rolling leads to a textured structure with magnetic alignment in the direction of rolling.

dysprosium, and terbium. Comment on the availability of these elements. Can other elements that are not rare earths be substituted?

13.11 In a magnetic system, the phases can change with T, p and magnetic field. For a single magnetic component at constant pressure, what is maximum number of phases that can coexist? Show your reasoning.

13.12 If a ferromagnetic material is cooled in the absence of a magnetic field, its entropy changes as a function of temperature as shown in the upper curve of Figure 13.18.

a. If a magnetic field is turned on while the temperature is held constant, what happens to the entropy, and what is the microscopic origin of the change in entropy?

b. If the field is then turned off, the entropy stays the same. What happens to the temperature of a sample that is thermally isolated from its surroundings?

c. Suggest an application for this *magnetocaloric effect*.

13.13 In 1879, Edwin H. Hall found that when charge passes through a material in a magnetic field, as shown in Figure 13.19, a voltage builds up across the material. We now call this the *Hall effect*, and recognize that the transverse Hall voltage, V_H, is due to the transverse (Lorentz) force on the charge carriers. For a current, I, and a magnetic field, B, the Hall voltage is given by

$$V_H = \frac{IB}{ned} \qquad (13.15)$$

where n is the density of negative charge carriers, e is the electron charge, and d is the thickness of the material. Explain how the Hall effect can provide useful information concerning semiconductors with very low dopant levels.

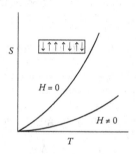

FIGURE 13.18
The influence of magnetic field and temperature on the entropy of a ferromagnetic material.

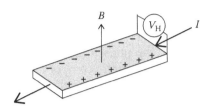

FIGURE 13.19
A current (I) passing through a material that is in a magnetic field (B) will experience a transverse Hall voltage, V_H.

Further Reading

General References

High-performance emerging solid-state memory technologies. Special issue. H. Goronkin and Y. Yang, Eds. *MRS Bulletin*, November 2004, 805–851.

Magnetic nanoparticles. Special issue. S. A. Majetich, T. Wen, and O. T. Mefford, Eds. *MRS Bulletin*, November 2013, 899–944.

Materials for magnetic data storage. Special issue. H. Coufal, L. Dhar, and C. D. Mee, Eds. *MRS Bulletin*, May 2006, 374–418. New materials for spintronics. Special issue. S. A. Chambers and Y. K. Yoo, Eds. *MRS Bulletin*, 2003, 706–748.

D. D. Awschalom, M. E. Flatté, and N. Samarth, 2002. Spintronics. *Scientific American*, June, 66.

P. Berger, N. B. Adelman, K. J. Beckman, D. J. Campbell, A. B. Ellis, and G. C. Lisensky, 1999. Preparation and properties of an aqueous ferrofluid. *Journal of Chemical Education*, 76, 943.

W. D. Callister, Jr. and D. G. Rethwisch, 2013. *Materials Science and Engineering: An Introduction*, 9th ed. John Wiley & Sons, Hoboken, NJ.

A. K. Cheetham and P. Day, Eds., 1992. *Solid State Chemistry. Compounds*. Oxford University Press, Oxford.

G. P. Collins, 2003. Getting warmer. *Scientific American*, March, 30.

G. P. Collins, 2004. Magnetic soot. *Scientific American*, July, 26.

P. Day, Ed., 2006. *Molecules into Materials*. World Scientific, Singapore.

M. de Podesta, 2002. *Understanding the Properties of Matter*, 2nd ed. Taylor & Francis, Washington, DC.

A. Geim, 1998. Everyone's magnetism. *Physics Today*, September, 36.

G. G. Hall, 1991. *Molecular Solid State Physics*. Springer-Verlag, New York.

R. E. Hummel, 2004. *Understanding Materials Science*, 2nd ed. Springer, New York.

M. Jacoby, 2006. Putting a spin on electronics. *Chemical and Engineering News*, August 28, 30.

M. Jacoby, 2013. Powerful pull to new magnets. *Chemical and Engineering News*, January 7, 23.

D. Jiles, 1998. *Introduction to Magnetism and Magnetic Materials*, 2nd ed. Taylor & Francis, Washington, DC.

J. R. Minkel, 2004. A pulse for magnetic memory. *Scientific American*, January, 31.

A. P. Ramirez, 2005. Geometrically frustrated matter–magnets to molecules. *MRS Bulletin,* June, 447.

C. N. R. Rao and J. Gopalakrishnan, 1997. *New Directions in Solid State Chemistry.* Cambridge University Press, Cambridge.

J. F. Shackelford, 2004. *Introduction to Materials Science for Engineers,* 6th ed. Prentice Hall, Upper Saddle River, NJ.

L. E. Smart and E. A. Moore, 2012. *Solid State Chemistry,* 4th ed., CRC Press, Boca Raton, FL.

S. A. Solin, 2004. Magnetic field nanosensors. *Scientific American,* July 2004, 71.

N. A. Spaldin, 2003. *Magnetic Materials: Fundamentals and Device Applications.* Cambridge University Press, Cambridge.

S. Thompson, 2001. Magnets made to order. *Chemistry in Britain,* March, 34.

L. Van Vlack, 1989. *Elements of Materials Science and Engineering,* 6th ed. Addison-Wesley Publishing Company, Reading, MA.

M. Wegener, M. Kadic and C. Kern, 2017. Hall-effect metamaterials and "anti-Hall bars". *Physics Today,* October, 14.

M. Wilson, 2006. Scanning tunneling microscope measures the spin-excitation spectrum of atomic-scale magnets. *Physics Today,* July, 13.

Disordered Systems

D. S. Fisher, G. M. Grinstein, and A. Khurana, 1988. Theory of random magnets. *Physics Today,* December, 56.

D. L. Stein, 1989. Spin glasses. *Scientific American,* July, 52.

Magnetic Devices and Processes

Materials for heat-assisted magnetic recording. Special Issue. M. T. Kief and R. H. Victoria, Eds. *MRS Bulletin,* February 2018, 87–124.

Materials for magnetic storage. Special issue. H. Coufal, L. Dhar and C. D. Mee, Eds. *MRS Bulletin,* May 2006, 374–418.

M. H. Kryder, 1987. Data-storage technologies for advanced computing. *Scientific American,* October, 117.

S. Langenbach and A. Hüller, 1991. Adiabatic demagnetization of antiferromagnetic systems: A computer simulation of a plane-rotor model. *Physical Review B,* 44, 4431.

R. M. White, 1980. Disk-storage technology. *Scientific American,* August, 138.

Molecular Materials

Molecule-based magnets. Special issue. J. S. Miller and A. J. Epstein, Eds. *MRS Bulletin,* November 2000, 21–71.

Polymer is magnetic below 10 K, 2001. *Chemical and Engineering News,* November 19, 62.

A. J. Epstein, 2003. Organic-based magnets: Opportunities in photoinduced magnetism, spintronics. Fractal magnetism, and beyond. *MRS Bulletin,* July, 492.

J. S. Miller and A. J. Epstein, 1994. Forthcoming attractions. *Chemistry in Britain,* June, 477.

P. F. Schewe, 2002. Light-activated plastic magnet. *Physics Today*, April, 9.

J.-F. Tremblay, 2000. Ferromagnetic, conductive molecular magnet created. *Chemical and Engineering News*, November 27, 5.

Other Magnetic Systems

R. Dagani, 1992. New material is optically transparent, magnetic at room temperature. *Chemical and Engineering News*, July 20, 20.

L. M. Falicov, 1992. Metallic magnetic superlattices. *Physics Today*, October, 46.

I. Gilbert, C. Nisoli and P. Schiffer, 2016. Frustration by design. *Physics Today*, July, 54.

J. L. Miller, 2017. Ferromagnetism found in two-dimensional materials. *Physics Today*, July, 16.

Websites

For links to relevant websites, see PhysicalPropertiesOfMaterials.com.

Part V

Mechanical Properties of Materials

The power of any Spring is in the same proportion with the Tension thereof: That is, if one power stretch or bind it one space, two will bend it two, and three will bend it three, and so forward. And this is the Rule or Law of Nature, upon which all manner of Restituent or Springing motion doth proceed.

Robert Hooke, 1676

14

Mechanical Properties

14.1 Introduction

In many cases, mechanical properties are the most important factor in determining potential applications of a material. Stiffness, tensile strength, and elastic properties are important in material applications as seemingly diverse as sound production from piano strings to the strength of dental porcelain to the protection of a bulletproof vest. Some high-temperature superconductors have very useful electrical and magnetic properties but are limited in their applications due to their mechanical properties. The aim of this chapter is to provide a basis for consideration of mechanical properties, based on a microscopic picture that has developed from experimental observation. We begin with relevant definitions.

Many of the mechanical properties of a material take into account its behavior under certain forces. If a force F is applied to a material of cross section A, it develops a *stress*, σ:[*]

$$\sigma = \frac{F}{A} \tag{14.1}$$

where the units of σ (e.g., N m$^{-2} \equiv$ Pa) show that stress and pressure are similar concepts.

Under this force, there also will be a *strain*, ε, due to a change from the original length, l_0, by an amount Δl:

$$\varepsilon = \frac{\Delta l}{l_0}. \tag{14.2}$$

It is apparent from Equation 14.2 that strain is unitless.

At low stress and low strain, stress is proportional to strain, that is,

$$\sigma \propto \varepsilon \tag{14.3}$$

---

[*] The most important equations in this chapter are designated with ▮ to the left of the equation.

so, at low σ and low ε,

$$\frac{\sigma}{\varepsilon} = E \tag{14.4}$$

where E, which is constant for a material at low ε, is called *Young's modulus* (also known as *elastic modulus*) for the material. Equation 14.4, which gives a linear relationship between force and extension, is a form of Hooke's law (see also Chapter 7). The proportionality constant, E, is a measure of the *stiffness* (i.e., resistance to strain) of a material, but it is quite distinct from the strength. A small value of E means that a small stress gives a large extension (e.g., as in rubber); a large value of E indicates that the material is very stiff. For example, E for diamond is ~10^3 MPa, whereas E for rubber is about 7 MPa.

A related concept is *Poisson*[*] *ratio*, ν, defined as

$$\nu = -\frac{\left(\dfrac{\Delta l_2}{l_2}\right)}{\left(\dfrac{\Delta l_1}{l_1}\right)} \tag{14.5}$$

where l_1 and l_2 are the original dimensions, respectively, in the direction of and perpendicular to the extension force, and Δl_1 and Δl_2 are the respective changes in length.

Equation 14.4 describes the mechanical behavior of a material at low stress; a more complete picture is given in Figure 14.1. From this figure, the proportionality of stress to strain at low stress is shown, as well as the deviation from linearity at higher stress. The stress at which elongation is no longer reversible is called the *elastic limit*. Below the elastic limit, that is, in the region of elastic behavior, strain is reversible: It disappears after the stress is removed. For metallic and ceramic materials, the elastic region shows a linear relationship between stress and strain, but for polymers such as rubber, the relationship can be nonlinear.

The *yield strength* is defined as the value of the stress when the strain is 0.2% more than the elastic region would allow; that is, the strain is 1.002 times the extrapolated elastic strain.

Plastic deformation (*plastic strain*) is permanent strain and it occurs beyond the yield strength, as removal of the stress leaves the material deformed.

The *tensile strength* is the maximum stress experienced by a material during a test in which it is being pulled in tension, and *ductility* is the strain at failure (usually expressed as percent).

---

[*] Siméon-Denis Poisson (1781–1840) was a French mathematician, physicist, astronomer, and engineer. He made important contributions in the areas of mathematics (Poisson distribution), celestial mechanics, electricity, magnetism, and mechanics.

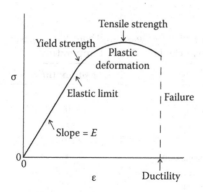

FIGURE 14.1

The stress (σ)–strain (ϵ) relationship for increasing strain on a ductile material. At low stress, the material is elastic and obeys Hooke's law with a proportionality of E, Young's modulus. The yield strength is defined as the stress at which the strain is 0.2% greater than predicted by Hooke's law. Plastic deformation, in which the sample remains deformed even after the stress is removed, occurs beyond the yield strength. The tensile strength is the maximum value of the stress when the material is in tension (i.e., the load pulls the material). Ductility is defined as % elongation at failure = 100% × $\epsilon_{failure}$.

Toughness is a measure of the energy required to break a material. It is different from strength (a measure of the stress required to break or deform a material): tensile strength is the maximum stress value on a stress–strain curve, and toughness is related to the area under the stress–strain curve up to failure. Toughness varies with the material: it is 0.75 MPa m$^{1/2}$ for silica glass, 4 MPa m$^{1/2}$ for diamond, and up to 100 MPa m$^{1/2}$ for steel.

Hardness is another mechanical property. It is the resistance of a material to penetration of its surface. For example, the *Brinell* hardness number (BHN) is a hardness index based on the area of penetration of a very hard ball under standardized load.

Typical values of mechanical properties for some materials are given in Tables 14.1 and 14.2 and Figure 14.2.

Figure 14.1 presents a summary of experimental findings for a *ductile* material (i.e., one that undergoes plastic deformation). The low stress data (i.e., below the elastic limit) can be understood in terms of the interatomic potential energy that is responsible for holding the solid together. Such a potential, V, is shown in Figure 14.3 as a function of atomic separation, r. The corresponding force, F, written for the simplified one-dimensional case (see Equation 5.2), is given by

$$F = -\frac{dV}{dr} \tag{14.6}$$

---

* Johan August Brinell (1849–1925) was a Swedish metallurgist. His apparatus for testing hardness was first shown at the Paris Exhibition of 1900.

TABLE 14.1

Typical Values of Tensile Strength and Young's Modulus (*E*)
for Various Materials at Room Temperature

Material	Tensile Strength/MPa	E/MPa
Diamond	~10^3	1.2×10^6
Carbon nanotubes	~10^5	1×10^6
Kevlar	4000	1.8×10^5
High-strength carbon fiber	4500	2×10^5
High-modulus carbon fiber	2500	2×10^5
High-tensile steel	2000	2×10^5
Superalloy	1300	2×10^5
Titanium	1200	1.2×10^5
Spider webs (drag line)	1000	1×10^4
Aluminum	570	7×10^4
Bone	200	2×10^4
Bamboo	100	1×10^4
Nylon	100	3×10^3
Rubber	100	≈ 7
Lexan	86	2.4×10^3

TABLE 14.2

Poisson Ratio for Various Materials at Room Temperature

Material	ν
Fe	0.17
TiC	0.19
Borosilicate glass	0.2
Al_2O_3	0.26
W	0.28
Carbon steel	0.3
Al	0.33
Cu	0.36
Pb	0.4
Nylon 66	0.41

and $F(r)$ is also shown in Figure 14.3. The corresponding stress–strain relationship, as shown in Figure 14.1, is similar to the force–separation relationship, $F(r)$, of Figure 14.3 (with stress comparable to force [but with opposite sign]) in the elastic region. Young's modulus, E, is related to the interatomic potential as

$$E = \frac{1}{r_0}\left(\frac{\partial^2 V}{\partial r^2}\right)_{r=r_0}, \tag{14.7}$$

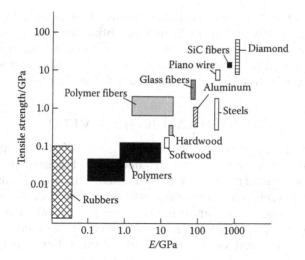

FIGURE 14.2
Tensile strength and Young's modulus ranges for a number of types of materials.

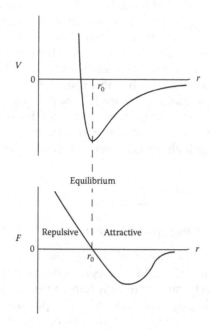

FIGURE 14.3
An intermolecular potential, $V(r)$, as a function of interatomic distance, r, leads to a force–separation relationship, $F(r)$, which can be used to deduce the elastic portion of the stress–strain relationship for a material (i.e., below its elastic limit). The intermolecular separation marked r_0 corresponds to the equilibrium position ($F = 0$).

which shows that E depends on the curvature of the potential well at $r = r_0$. Note that the pairwise intermolecular potential does not directly explain stress–strain curves beyond the elastic limit because after this point the shapes of these curves are determined by progressive destruction of the crystalline lattice, which cannot be understood in terms of two-atom interactions.

COMMENT: AMORPHOUS METALS

Even metals can become amorphous or glassy when they are cooled so rapidly that the atoms cannot arrange themselves in the more stable crystalline form. The first amorphous metal discovered was a gold-based alloy, prepared by *quenching* from the melt at about 10^6 K s^{-1}. The resulting *metallic glass* is amorphous. Its production can be assisted by the presence of many types of atoms of very different atomic radii in the alloy, which reduces the probability of nucleation and growth of crystals. Alloys based on zirconium and palladium are very good glass formers, but amorphous metals based on titanium, copper, magnesium, iron, and other metals have been made. Rapid quenching leads to thin samples of amorphous metals, but more recently, techniques have been developed to produce bulk metallic glass. Because amorphous metals do not have the defects that limit the strength of their crystalline counterparts, they have higher fracture strengths, higher elastic limits, and they are harder. One commercial amorphous metal has almost twice the tensile strength of high-grade titanium. However, the lack of defects also means that amorphous metals are not very ductile, so they can easily shatter in tension. One of the most useful properties of amorphous metals is their processability: Since they soften and flow, they can be shaped into objects by injection molding. This feature allows them to be used in applications where high strength and unusual shapes are required, such as golf club heads and cell phone cases.

14.2 Elasticity and Related Properties

Many materials with low values of Young's modulus are also very elastic, that is, reversibly deformed to high stress. As with many other properties of materials, this property can be strongly temperature dependent.

An example of a highly elastic material is a polymer. Its elasticity stems from the coiling of long molecular chains; extension can take place by uncoiling without much change in interatomic distance, so it takes relatively little energy (see Figure 14.4). However, polymers with a high degree of structural organization (either a high degree of crystallinity or extensive crosslinking) are not very elastic, so more energy is required to deform their structures,

Coiled   Stretched

FIGURE 14.4
Uncoiling of an elastic polymer takes relatively little energy because the interatomic distances are not much changed.

and the deformation is not reversible. It is found that polymer properties such as the glass transition temperature (T_g; whether the temperature is less than or greater than T_g reflects the rigidity of the polymer; see Chapter 6) can be related to mechanical properties such as Young's modulus, strength, and toughness. For example, Lexan (a very strong polycarbonate with such diverse uses as bubbles on space helmets and impact-resistant bumpers) has a very high glass transition temperature ($T_g = 149$ °C, compared with 45 °C for nylon and −45 °C for polyethylene), which means that Lexan is in its rigid glassy state at room temperature. This high T_g correlates with Lexan's good strength and toughness characteristics at room temperature; polymers with lower T_g values would require lower temperatures for similar mechanical properties.

Young's modulus, E, is related to the speed of sound in a material, v, by

$$v = \left(\frac{E}{\rho}\right)^{1/2} \tag{14.14}$$

where ρ is the density of the material. The speed of sound can be important in determining the desired properties for a material. For example, low-density (0.4–0.5 g cm^{-3}) pine with high Young's modulus (11–18 GPa) can be used to produce materials with high sound speeds (4800–6000 m s^{-1}) used for construction of violins.

Young's modulus is a measure of the elasticity of a material as it is extended or compressed. Considering compression further, by analogy with Equation 14.4, we can write

$$p = Kc \tag{14.15}$$

where p is the hydrostatic pressure, c is the compression of the material (change in dimension/original dimension), and the proportionality constant in this version of Hooke's law, K, is the *isothermal bulk modulus*, defined as

$$K = -V\left(\frac{\partial p}{\partial V}\right)_T, \tag{14.16}$$

that is, K is the reciprocal of the isothermal compressibility (β_T) as discussed in Chapter 6.

COMMENT: STRETCHING A POLYMER

As a polymer is stretched, the polymer itself becomes more ordered due to the alignment of the previously disorganized chains. Thus, if the polymer is the thermodynamic system, entropy considerations for uncoiling indicate

$$\Delta S_{system} < 0. \tag{14.8}$$

However, from the second law of thermodynamics, any process that can happen must have

$$\Delta S_{total} > 0. \tag{14.9}$$

and since

$$\Delta S_{total} = \Delta S_{system} + \Delta S_{surrounding} \tag{14.10}$$

it follows that

$$\Delta S_{surroundings} > 0. \tag{14.11}$$

From the definition of reversible entropy change for the surroundings,

$$\Delta S_{surroundings} = \frac{q_{srroundings}}{T} \tag{14.12}$$

and since $T > 0$, it follows that, on extension of a piece of elastic polymer,

$$q_{srroundings} > 0 \text{ so } q_{elastic} < 0. \tag{14.13}$$

In other words, extension of the elastic polymer is an ordering process, which must therefore be exothermic, as can be verified by extending an elastic band while holding it next to your cheek or lip (very sensitive thermometers!). You will find that the polymer gives off heat when stretched.

COMMENT: SEEING STRESS WITH POLARIZED LIGHT

When an isotropic material is placed under a nonuniform mechanical stress, it can become anisotropic in some regions. If the material is transparent, then this stress can be revealed by observations through polarization filters. This situation is shown in Figure 14.5.

The principle involved is the production of birefringence within the stressed sample (see also Chapter 5). When linearly polarized light hits the sample, optical anisotropy leads to the production of two rays of light within the sample. When these rays emerge from the sample and reach the analyzer (another piece of polarizing film), light of only selected wavelengths will pass through, and the hue of the observed light will depend on the stress. This *stress birefringence* process is shown schematically in Figure 14.6.

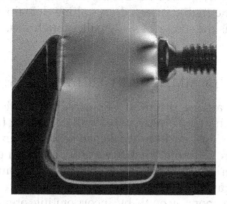

FIGURE 14.5
A bar of Plexiglass (also known as poly(methyl methacrylate), or PMMA) under stress, viewed through crossed polarizers, shows stress polarization.

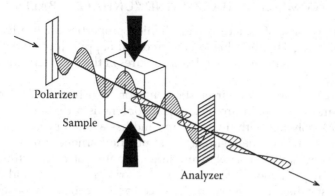

FIGURE 14.6
A schematic view of stress birefringence. Light is first polarized and then passed through an anisotropic material (e.g., a material under stress) leading to two exiting polarizations that can be separated by an analyzer. Each color (wavelength) of light will behave differently, and the observed color will depend on the stress. The color patterns on the front cover of this book were produced by polarized light passing through a piece of plexiglass that was under stress.

COMMENT: STRENGTH OF SPIDER SILK

Spider silk has amazing mechanical properties, and we now understand these properties at a molecular level. For example, ^{2}H NMR studies of the dragline silk produced by deuterium-fed spiders shows that the crystalline fraction of the silk consists of two types of alanine-rich regions: one highly oriented and the other less densely packed and poorly oriented. The weakly oriented sheets account for the high compressive strength of spider silk, and the coupling of the poorly oriented crystallites to the highly oriented crystalline domains and the amorphous regions, is the source of spider silk's toughness.

COMMENT: ARTIFICIAL BONES

Our bones are among the most important mechanical parts in our bodies. Although many broken bones will eventually heal themselves, there are conditions, such as weakened load-carrying bones, where replacements are required. Hydroxyapatite, the main part of the rigid structural material in natural bones, has been combined in the laboratory with organic polymers to produce strong artificial bones that promote natural bone growth by incorporating organic matter. The complex organoapatite has stiffness comparable with apatite; it also has more than twice the tensile strength.

COMMENT: "HAPPY" AND "UNHAPPY" BALLS

The importance of structure to mechanical properties can be illustrated by bouncing "happy" [neoprene] and "unhappy" [polynorbornene] balls, as shown in Figure 14.7. The structures of the polymers are shown in Figure 14.8.

The differences in their elastic behavior stem from their compositions and their chain architectures: neoprene is a three-dimensional network polymer with linear chains and has a small portion of cross-links that prevent chain slippage and polynorbornene contains linked cyclic units and has a relatively large quantity of nonvolatile liquid such as naphthenic oil. The rigidity of the ring makes polynorbornene less elastic than neoprene (which can coil and uncoil in response to different stresses). Furthermore, the liquid in polynorbornene can absorb the energy imparted to the "unhappy ball" when it is dropped. Thus, two materials that look and feel very similar can have remarkably different mechanical properties.

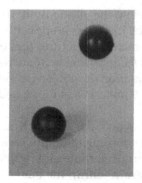

FIGURE 14.7
Although they look very similar, the neoprene ball (on the right) bounces, while the polynorbornene ball (on the left) does not. (Balls from Arbor Scientific, Ann Arbor, MI.)

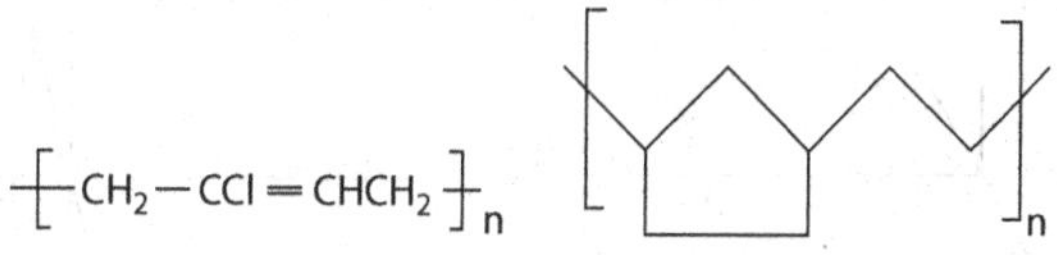

$$-\left[CH_2-CCl=CHCH_2\right]_n$$

Neoprene   Polynorbornene

FIGURE 14.8
Molecular structures of the repeat units of neoprene and polynorbornene. Both materials are polymers.

COMMENT: HUMIDITY PLAYS A ROLE IN BASEBALL

Humidity can change the elastic properties of polymers, especially those in which hydrogen bonds play an important role. For example, due to its polyamide (wool) core, a baseball can lose as much as 12 g of water on heating at 100 °C for 12 h, and such dehydration leads to a harder ball that is less deformable, causing the collisions to be more elastic. The change in properties with degree of hydration explains why it is easier to hit a home run on a cool, dry evening than on a hot, humid night.

14.3 Beyond the Elastic Limit

In the region of reversible stress–strain relations, i.e., the elastic region, a strained material will return to its original dimensions once the stress is removed. Furthermore, below the elastic limit, the strain observed is independent of the rate at which the stress is applied.

At larger stress, a *brittle* material fractures (i.e., the reversible stress–strain region ends abruptly with *failure*), whereas a ductile material is permanently deformed (i.e., the deformation is no longer reversible). The distinction between brittle and ductile materials is illustrated in the stress–strain relations of Figure 14.9.

When a large stress (beyond the elastic limit) is applied to a ductile material, on removal and subsequent reapplication of the stress, the strain path does not retrace its original route and hysteresis results (Figure 14.10); that is, there is residual strain within the material even when the stress is

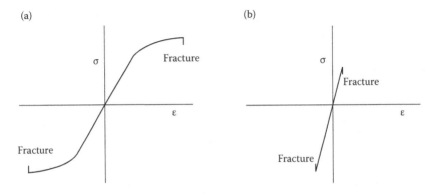

FIGURE 14.9
Stress–strain relations, including compression ($\varepsilon < 0$) and tension ($\varepsilon > 0$), for (a) a ductile material and (b) a brittle material. The brittle material shown is stronger in compression than in tension.

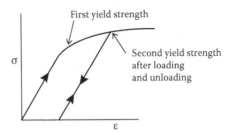

FIGURE 14.10
The application of stress beyond the elastic limit of a ductile material can lead to permanent deformation, as shown here by the hysteresis on removal of a high stress. The increase in yield strength after loading and unloading is due to work hardening.

removed. The area within the hysteresis loop is proportional to the energy given to the system. For example, in the case of a fly landing in a spider web, the web extends due to the force of the fly landing, and then, when the stress is over, the kinetic energy of the fly has been transferred to the silk of the web, and it is manifest as a slight increase in both the temperature and the extension of the silk. The low Young's modulus and high tensile strength of the spider silk ensure that the web does not break when the fly contacts it.

High stress can lead to inelastic deformation; in addition, at high stress the rate of application of the stress becomes important. For example, a bullet shot into ice will penetrate to about the same depth as one shot into water since at this high stress the H_2O molecules do not have time to rearrange themselves in either ice or water. In a less extreme example, polymers beyond their elastic limit and subjected to stress may *creep*, that is, continue to deform with time.

The temperature of a material can greatly influence the way in which it behaves at very high stress. For a polymer below its glass transition temperature, brittle fracture will occur; that is, there will be fracture with little deformation. Amorphous materials tend to be brittle at relatively low temperature (often, room temperature is low enough) because the energy required to deform the material by moving atoms or molecules past one another is much greater than the energy required to propagate cracks (see Section 14.5). For a polymer above its glass transition temperature, the fracture mode will be ductile (i.e., with considerable plastic deformation) and dependent on the strain rate, with slow strain rates allowing time for rearrangement of the polymer chains, leading to greater deformation.

Far beyond the elastic limit, pure metals or *alloys* (mixtures of several metals) often can be manipulated without fracture. For example, they can be rolled, extruded into wires, etc. The high ductility of metals can be attributed to the atomic packing, that is, the rather nondirectional arrangement of the atoms, which provides mobility of *dislocations* (packing irregularities; also see Section 14.4). For example, if a piece of metal is bent, the atoms rearrange their bonds with other atoms by sending clouds of dislocations flying around but with little overall disruption of the packing. On the other hand, if a sheet of ice is bent, the more directional bonds in ice mean that much more reorganization of the bonding is required than for the piece of metal. Therefore, dislocations, which could allow bending, are less mobile in a solid such as ice because of the directional nature of its bonds. It might seem therefore that ice will simply continue to deform elastically, i.e., show a linear relationship between σ and ε, but in fact it fails in an entirely different way, by propagation of a crack. This failure happens at rather low stresses. As a result, the nonductile (brittle) ice will dissipate the energy by failure at low stress, whereas metal will bend.

COMMENT: CONTROLLED STRESS

Prince Rupert drops, named after a 17th-century prince who saw them at a Bavarian glassworks, are formed when drops of molten glass are rapidly cooled by drop-

See the video "Prince Rupert Drop" under Student Resources at PhysicalPropertiesOfMaterials.com

ping them into cold water. The exterior cools more quickly than the interior, so the outer surface of the drops is in compression. This stress, which makes the drops very strong, can be seen by viewing the glass through crossed polarizers (as in the stressed Plexiglass in Figure 14.5). If the tail of the droplet is broken, the stress is released explosively, and the glass shatters into many very small pieces. Since the glass breaks into so many pieces, these drops are much safer to handle than glass that breaks into sharp shards.

Glass that has been *tempered* (i.e., toughened by having its outer surface in compression), either by quick cooling or by chemical treatment, has many applications, including eyeglasses, windshields, and cookware. An important example is the Corning product, Gorilla® Glass, a thin, light, damage-resistant glass used for portable electronic devices. This alkali-aluminosilicate glass obtains its surface strength and crack resistance by ion exchange. During its processing, it is immersion in a hot potassium salt ion exchange bath, trading the Na^+ for large K^+. The bulkier K^+ ions give residual compressive stress at the surface, increasing toughness, similar to the thermally induced stress in Prince Rupert drops.

COMMENT: BULLETPROOF VESTS

Kevlar, a form of poly (*p*-phenyleneterephthalamide) (see Figure 14.11) was invented at DuPont.* Kevlar has an ultra-high Young's modulus (1.8×10^5 MPa), as well as high tensile strength and high thermal stability, and low extension at the break point. The strength of Kevlar is related to its structure: fibrils are interconnected by small tie bundles in a pleated sheet with a 500-nm repeat distance. The polymer is highly crystalline, and it would take a very large amount of energy to disturb the conformation of the chains. Conversely, the chains can absorb significant amounts of energy (buckling from *trans* to *cis* conformations) without much damage, allowing Kevlar to stop bullets in bulletproof vests. Kevlar is an example of a material in which ease of microscopic deformation leads to macroscopic strength.

---

* Kevlar was invented by DuPont chemist Stephanie Kwolek (1923–2014). She was inducted into the National Inventor's Hall of Fame in 1995 for this important invention.

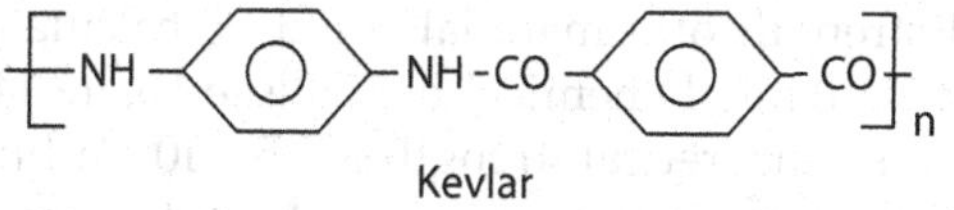

Kevlar

FIGURE 14.11
Molecular structure of Kevlar, showing the repeat unit of this important polymer.

14.4 Microstructure

The microstructure, i.e. structure at the sub-mm length scale, can play a very important role in mechanical properties of materials. Even for a pure material, the influence of the presence of grain boundaries makes the mechanical properties different for a single crystal compared with an agglomerated powder of the same material. For a multicomponent system, the mechanical properties will depend on whether the material is a single phase, as in a solid solution or alloy, or contains multiple phases.

The microstructure of a material can be controlled by processing. The cooling rate of a material from its melt can greatly influence the size of its grains, with slower cooling leading to larger grains, due to the high energy of the corresponding grain surfaces. Similarly, annealing (heating to a temperature just below the melting point) will cause grains to grow. *Sintering*, the process by which powders are compacted into a solid mass by use of heat and/or pressure, can have a similar influence on grain sizes. The reduction in numbers of grain boundaries can change the ductility of a metal. (See Cymbals: A Tutorial for an example.)

The addition of other components also can change the microstructure of a material, by introduction of secondary phases. Such defects and dislocations can greatly influence the mechanical properties, as we now examine in detail.

14.5 Defects and Dislocations

No material is perfect, and it is often the type and location of imperfections that determine the mechanical properties of a material. Examples of *point defects* (atoms missing or in irregular places in the lattice, including *lattice vacancies, substitutional impurities, interstitial impurities,* and *proper* [or *self*] *interstitials*) are shown in Figure 14.12. Examples of *linear defects* (groups of atoms in irregular positions, including *screw dislocations* and *edge dislocations*) are shown in Figures 14.13 and 14.14. An indication of the concentration of imperfections is that, if point defects and linear imperfections in 1 kg of aluminum were placed side by side, the defects would cover the equator more than a thousand times!

The theoretical strength of a material could be calculated from the force required to break individual chemical bonds (see Figure 14.3), per unit area. For copper, this gives a theoretical strength of 2.5×10^9 Pa, but copper is known to deform and break at much lower stresses. In fact, many pure metals start deforming at less than 1% of their theoretical strength. The reason for the difference is the ease of motion of dislocations in the lattice, allowing the material to deform, for example, through the breaking and reforming of a single row of atomic bonds along the dislocation line, as shown in the edge dislocation

COMMENT: HUME-ROTHERY RULES

Empirical rules established by William Hume-Rothery (1899–1968) describe the conditions under which an element can dissolve into a crystalline metallic solid forming a solid solution. For substitutional solid solutions, the following are required: (1) the atomic radii of the solute and solvent should not differ by more than 15%, (2) the crystal structures of the solvent and solute must match, (3) the solvent and solute should have similar electronegativities, and (4) the solute and solvent should have similar valences for maximum solubility. For interstitial solid solutions, the rules are (1) the solute atoms should be smaller than the vacancies formed in the parent structure, and (2) the solute and solvent should have similar electronegativities.

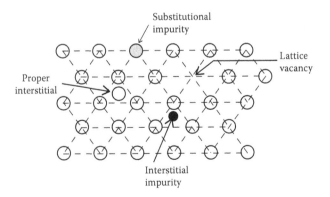

FIGURE 14.12
Point defects in an otherwise periodic monatomic crystal. A *lattice vacancy* is the absence of an atom at a lattice site; a *substitutional impurity* indicates the presence of an impurity atom at a lattice site; an *interstitial impurity* indicates the presence of an impurity atom at a location other than a lattice site; a bulk atom at a position other than at a lattice site is called a *proper interstitial*.

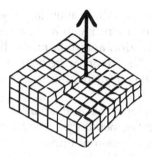

FIGURE 14.13

A *screw dislocation* in a crystal. This defect arises when atoms spiral around an axis, shown here with an arrow. Note that at the front of the crystal there is a step, but there is no step at the back edge.

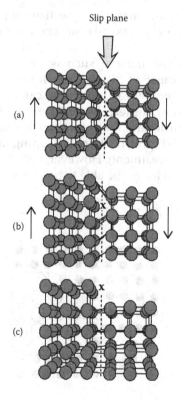

FIGURE 14.14

The motion of an *edge dislocation* (a layer of extra atoms within the crystal structure), indicated by x, from (a) start, to (b) intermediate time, to (c) finish. In the process of moving, bonds are broken and re-formed along the *slip plane* (the plane along which the slippage takes place, shown here as a dashed line). Arrows indicate that one side of the crystal is moving with respect to the other. This process is the basic way in which ductile materials rearrange bonds between atoms, allowing for ductility.

of Figure 14.14. The motion of a dislocation in a crystal can be considered to be similar to the motion of a wrinkle in a rug when the wrinkle is pushed. In some cases, such as tin, zinc, and indium, the high speed of dislocation-assisted motion leads to a "cry" that can be heard when rods of these materials are bent (see Section 14.6 for further discussion of the role of defects in the strength of a material).

Imperfections can also play a role in the hardness of a material. For example, a major practical aspect of metallurgy is development of harder materials, and one method involves addition of impurities that reduce the mobility of dislocations (see Figure 14.15). Hardness can be enhanced if there are small aggregates of impurities (e.g., gold alloy with 10% copper is 10 times harder than pure gold). Thermal history can play an important role because heating induces impurities to clump together, reducing the hardness of the material. Dislocations are immobilized at *grain boundaries* (the interfaces between homogeneous regions of the material), so decreasing the grain size can help harden a material. In a *work-hardened* material, a mechanical process has introduced more dislocations, reducing the overall dislocation mobility, making the material harder.

To summarize: A ductile material, such as a metal, bends well below its theoretical strength because of dislocations, for example those that allow planes to slip past one another easily. A metal can be hardened by decreasing the number of dislocations (e.g., by heat treatment that anneals out dislocations), or pinning dislocations by work hardening (deformation causes dislocation tangles), and precipitation hardening or grain refining (by mechanical and/or heat treatment). However, if dislocations are too immobilized, the metal will no longer be able to bend and it will snap in a brittle fashion under load.

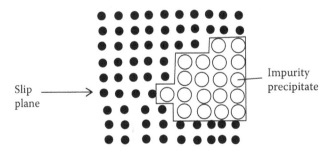

FIGURE 14.15
A dislocation associated with a slip plane can proceed no further when it encounters the grain boundary at an impurity precipitate, just as a wrinkle in a rug cannot move further along the rug when it encounters a piece of heavy furniture. The presence of the precipitate hardens the material. *Annealing* the material, that is, heating it to a temperature high enough to mobilize impurity precipitates allowing them to become larger but fewer in number, will allow dislocations to move more easily. Therefore, annealing decreases the hardness of the material.

COMMENT: HEAT TREATMENT OF ALUMINUM

The binary Al–Cu phase diagram (Figure 14.16) allows an illustration of the influence of thermal treatment on mechanical properties. From the phase diagram, cooling 96 mass% Al/4 mass% Cu (this is so-called *Duralumin*, or 2000 series aluminum alloy), will give, at room temperature, a two-phase mixture. One phase is $CuAl_2$ (note that this is an approximate, nonstoichiometric composition) and it is dispersed in an aluminum-rich phase. If this mixture is cooled slowly from the melt, precipitates of $CuAl_2$ are large and far apart; dislocations can rather easily avoid the precipitates in this material, and the material is soft. If cooled quickly, the dispersion of $CuAl_2$ in the Al-rich phase is much finer (smaller grains), and the high concentration of (small) precipitates ensures more interaction of dislocations, resulting in a harder material.

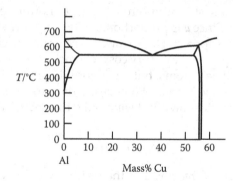

FIGURE 14.16
A portion of the Al–Cu phase diagram. An intermetallic compound of approximate composition $CuAl_2$ is formed at ~55 mass % Cu.

14.6 Crack Propagation

One important factor concerning the mechanical strength of a material is the response of its structure to stress. Many materials will crack when they are stressed, and in this section, we examine this in terms of both structure and energetics.

Cracks begin at flaws in materials, often at the surface. The propagation (growth) of a crack in a material both *releases* energy (the strain energy) and *costs* energy (due to increased surface area [recall that surfaces are higher in energy than layers within the bulk materials] and energy to reorganize the material in the region of the crack). The competition between the energy cost and the energy gain can lead, in certain circumstances, to crack propagation.

FIGURE 14.17
Schematic view of a crack of depth d in a material. Stress, σ, is released primarily in the shaded areas, which have a total area $\approx d^2$.

First, let us consider the energy that is released when a stressed brittle material is cracked. In the schematic two-dimensional crack shown in Figure 14.17, a crack of depth d can release stress on both sides of the crack such that a total area of about d^2 has its stress released. Therefore, the internal energy gained by the surroundings in this cracking process, $\Delta_{stress}U$, is

$$\Delta_{stress}U = -ad^2 \tag{14.17}$$

where a is a proportionality constant ($\approx \sigma^2/E$), and the negative sign indicates that the energy of the system is lowered by $\Delta_{stress}U$. Lowering energy is always a favorable process. Since a is proportional to σ^2, the more stress one puts on a body, the easier it is for a crack to grow, as is commonly experienced.

However, there are also the energy costs of crack growth. One is the kinetic energy of the mobilized atoms, but we ignore this factor here as it is small compared with the energy costs of increasing the surface area, $\Delta_{surf}U$. For the two-dimensional crack shown in Figure 14.17, the energy cost of increasing the surface is

$$\Delta_{surf}U = 2Wd \tag{14.18}$$

where W is the *work of fracture*, and the sign of $\Delta_{surf}U$ now shows that this costs energy. The work of fracture includes two contributions: a contribution for the energy to create a new surface and a contribution to reorganize the material in the region of the crack. These two components mean that more than just the surface energy of the material is important; one must also take into account how far into the material the damage extends. The work of fracture, W, can be expressed in terms of the *surface energy* of a material, G_s, where G_s is typically 1 J m^{-2} (see Chapter 10). For a ductile material (for example, a metal, that is, a material that can undergo permanent strain before it fractures), W is 10^4–10^6 times G_s since the crack motion is accompanied by defects[*] well below the surface of the material.[†] (Note that the equations here

---

[*] The defects are clouds of dislocations emerging from the crack tip.

[†] We are familiar with the fact that defects decrease strength, in that metal can be broken easily by repeated bending. When a piece of metal is bent, the dislocation concentration at the bend site increases considerably. They tangle around one another, and this eventually leads to the material becoming brittle, so W is decreased and the material breaks. A piece of metal that has been bent is quite stiff (have you ever tried to bend a piece of copper tubing that has already been bent? It is very difficult!), but it can be returned to its original flexible state by heating it to anneal out the defects.

are for a two-dimensional simplification of a three-dimensional effect.) On the other hand, W for a brittle material such as glass is only about 10 G_s since the influence of stress does not reach very far below its surface. Such a low value of W means that the material could be susceptible to brittle failure. This conclusion can be very important in practice: evidence indicates that components of the Titanic were brittle due to impurities in the steel, lowering W relative to better steel and leading to catastrophic failure (brittle fracture) on collision with an iceberg.

At the balance point between the release of stress energy and the cost of surface energy, the crack has a particular length, l_G, called the *Griffith length*,* such that, with stress σ and Young's modulus E,

$$l_G = \frac{2WE}{\pi\sigma^2} \ . \tag{14.19}$$

(Equation 14.19 holds for a three-dimensional crack; for an edge crack, $l_G = WE/\sigma^2$.) At crack lengths less than the Griffith length, the cost in surface energy and internal rearrangement energy exceeds the energy gain from release of stress, and the crack does not grow. However, if the stress on the material is increased, more stress energy is released, and the crack will grow. For cracks longer than the Griffith length, stress energy releases more energy than the energetic costs of increased surface area and internal rearrangement, and the crack grows spontaneously, provided that the stress remains.

Typical values of l_G vary considerably from one material to another. For example, a 1-m crack in the hull of a ship can be "safe" (less than the Griffith length under normal seagoing stresses; it would cost more energy to grow further, so it does not propagate spontaneously). On the other hand, a microscopic crack in a glass material can grow spontaneously when stressed. Anyone who has used a file to start a crack in a piece of glass tubing will be familiar with this; the crack will propagate across the tube with relatively little stress. This situation is also seen in automobile windshields: A small crack can exist for some time, but when the stress is sufficient (e.g., from hitting a pothole, or from having the increased interior air pressure in a sun-warmed car), the crack can grow rapidly and spontaneously, causing the windshield to shatter.

The state of the surface of a crack also is important in determining the rate at which it will grow. For example, it is common to see a glassblower wet the scratch on a glass rod before breaking it. The water wets the surface in the crack, lowering the surface energy of the glass, thus requiring less stress to achieve crack propagation. Indeed, moisture content can be one of the important variables in crack propagation rates.

---

* Alan A. Griffith (1893–1963) was a British researcher who studied strengths of materials while a researcher at the Royal Aircraft Establishment, Farnborough, England.

COMMENT: ADDED STRENGTH IN FIBERS

In 1920, Griffith suggested that the strengths of materials are limited by microscopic cracks on their surfaces. Griffith based his conclusion on studies of glass rods and fibers, and his finding of remarkably high strength when the diameter was less than about 10^{-6} m. He concluded that these fibers were relatively free of surface cracks and therefore much stronger than glass rods. We now understand that the increased strength arises from the decreased probability of finding a strength-limiting flaw in a fiber compared with a bulk material.

An additional factor adding to the strength of crystalline fibers is the reduction in the number of dislocations and/or their mobilities in fibers, compared with bulk materials, especially since dislocations reduce the strength of the material.

Today, fibers are commonly used for strength; for example, sporting equipment such as fishing rods, bicycles, and tennis rackets often contains carbon fibers for added strength. Graphite/metal composite fibers with values of Young's modulus over 600 GPa have been produced. Carbon nanotubes can provide fibers with strengths far in excess of that of steel (see Table 14.1).

COMMENT: IMPURITIES IMPORTANT IN FAILURE

Impurities tend to be concentrated at grain boundaries where they can be 10^5 times more abundant than in the bulk of a sample. This factor can make the material harder because dislocation motion is more difficult, but the material will be more susceptible to brittle failure because the grain boundaries cannot deform with the impurities there, and cracks will tend to begin at the grain boundaries. For example, phosphorous segregates at grain boundaries in steel, and, to maintain its mechanical properties, steel must have less than 0.05% phosphorous.

COMMENT: POLYMER CLASSIFICATIONS

Synthetic polymers can be categorized on the basis of their thermal behavior as *thermoplastics* or *thermosets*.

Thermoplastics can be repeatedly softened when heated and hardened when cooled. Examples include polystyrene, acrylics, polyethylenes, vinyls, nylons, and polyfluorocarbons. Thermoplastics consist of linear or branched polymer molecules that are not connected to the adjacent molecules; as they are heated, the thermal motion can over-come the van der Waals attraction, allowing softening.

Thermoset polymers, as their name implies, become hard on heating. Furthermore, this process is irreversible as it happens due to the onset of new chemical bonds (*crosslinks*) between adjacent polymer molecules. Many of the network polymers, such as vulca-nized rubbers, epoxies, and phenolic and polyester resins, are ther-mosets, and they can be prepared in suitable shapes by the use of a mold. Thermosets are usually stiffer, harder, and more brittle than thermoplastics.

Super-hard polymers have been produced by irradiation with high-energy ions, hardening the polymer from a tensile strength of about 0.1 GPa (unhardened) to 22 GPa (hardened), exceeding the tensile strength of steel (2 GPa). The irradiation process breaks chemical bonds that can re-form with crosslinkage. The new bonds can also cause further delocalization of the π-bonds and therefore increase both the strength and electrical conductivity of the polymers after irradiation.

COMMENT: GLASS CERAMICS

Small crystallites in a glass have been found to inhibit crack propaga-tion. This property is utilized in the production of glass ceramics such as Corningware (developed at Corning Glass Works by S. Donald Stookey, 1915–2014). In that case, very fine particles of titanium oxide act as highly dispersed nuclei (10^{12} nuclei per mm^3 of molten glass) on which SiO_2 crystallites can form. In production, the object is first made by blowing or molding in order to achieve the appropriate shape. Then, controlled heat treatment is used to *devitrify* (that is, to crystal-lize) up to 90% of the material. Because the small grain sizes scatter visible light, the material loses its transparency and becomes white and opaque following devitrification. It is thought that the small crys-tallites create many grain boundaries, and these immobilize disloca-tions (making the material harder) and also prevent crack propagation (making the material more resistant to failure). Furthermore, many glass ceramics have low thermal expansion coefficients (see Chapter 7) so that thermally induced stress is reduced. These two factors—high tensile strength and low thermal expansion—make it possible to place a cold Corningware dish in a hot oven without catastrophic failure. Glass ceramics were discovered in 1957 after the accidental overheat-ing of a furnace.

COMMENT: CERMETS

A *cermet*, as the name implies, is a composite composed of a mixture of a *ceramic* and a *metal*, with properties different from either the ceramic or the metal. Cermets can be formed by mixing the powdered components and compacting them under pressure, followed by sintering. Cermets also can be formed by internal oxidation of dilute solutions of one metal in another. The metal solute oxidizes, leaving a metal oxide in the main metal. Cermets can have properties different from either of their components, including increased strength and hardness, better high-temperature resistance, improved wear resistance, and better resistance to corrosion. Cermets are especially useful for cutting and drilling tools.

14.7 Adhesion

Archaeologists have shown that adhesives were used thousands of years ago: Pitch and natural resins held spearheads to shafts. Glue, one of the oldest adhesives known, is derived from animal skins and bones.

Adhesion of two materials results when the atoms or molecules are in such intimate contact that weak intermolecular forces (such as van der Waals interactions, which fall off as $1/r^6$, where r is the intermolecular distance) can collectively strengthen the contact. Even when two pieces of ceramic are broken and can be held back in place, the distances between all but a few contact points are too large to allow adhesion to take place on its own. An adhesive can flow into all the crevices and make intimate contact between both parts; when the adhesive hardens (due to evaporation of the solvent, or polymerization processes, for example), the interactions allow adhesion.

To adhere, an adhesive must first *wet* the substrate; that is, it must spread across its surface. For this reason, the surface tension of the liquid adhesive is an important consideration. A familiar experience is that all liquids bead up on Teflon® surfaces, and virtually nothing will adhere to the surface.

A second criterion for a good adhesive is the strength of the interactions, both between the adhesive and substrate (*adhesive strength*) and within the adhesive (*cohesive strength*). For example, increased crosslinkage in polymers enhances cohesive strength.

Flexibility can be an important factor in a successful adhesive; for example, pure epoxy resins are brittle and subject to cohesive failure. However, the addition of a toughening agent, such as a liquid rubber, leads to rubber-rich domains dispersed in the cured epoxy matrix. This situation leads to a tougher adhesive as fracture energy can be dissipated in the rubber particles.

Of course, there must also be good adhesion between the rubber particles and the epoxy.

Although strong adhesives have such diverse applications as bookbindings from the last century to holding on the wings of modern airplanes, there are also applications where only mild adhesion is required. The most notable example of this is the development of Post-it® notes. Although the *low-tack* (i.e., weak) adhesive was developed by 3M scientist Spence Silver, it was another 3M scientist, Art Fry, who, in seeking a way to temporarily mark his pages in a music book, found its first use. In this case, the adhesion is rather weak so that the bond can be easily broken.

COMMENT: UNUSUAL VISCOELASTIC BEHAVIORS

Some polymers behave elastically at low stress (or slow application of stress), yet behave like viscous liquids at high stress (or fast application of stress). This situation allows these materials to be liquid-like while under stress, yet solid-like when left alone. An example of this so-called non-Newtonian behavior is "no-spill" paint, which is solid-like in the paint can (exhibiting elastic behavior at low stress), and yet can be applied with a brush (viscous behavior under application of stress). The origin of this unusual behavior is the arrangement of the chains of the polymer (Figure 14.18): at low (or slow) stress, the polymer molecules cluster together and the material is "solid-like," while under shear stress the molecules align and the network can be broken, allowing for flow as a viscous "liquid."

Related principles explain why a mixture of corn starch and water thickens on rapid stirring (the molecules do not have time to align) but flows when left alone. Similarly, Silly Putty™ bounces (i.e., acts elastically at high stress) and yet can be slowly extruded into threads (given sufficient time, molecules align). In these two cases, due to the molecular structures, the structure responds elastically to rapid shearing forces but inelastically with slower application of stress.

Shear

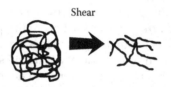

FIGURE 14.18
The arrangement of polymer chains changes under shear stress. Without stress, there is a rigid network. Under shear stress, the polymer chains can be aligned, allowing easier flow, for example, in no-spill paint.

COMMENT: THE IMPORTANCE OF OTHER
PHASES AND MICROSTRUCTURE

When a material is composed of two or more components, some of its properties are additive and can be determined by suitably weighted values of the properties of the phases present, that is, obey a rule of mixtures (Equation 11.5). Color, density, heat capacity, and sometimes thermal and electrical conductivity are additive, at least to a first approximation. However, mechanical properties are highly interactive, that is, different from the weighted average of the constituent phases, and this finding can be used to great advantage in designing materials with particular mechanical properties.

For example, materials can be strengthened by the addition of fillers, such as carbon with rubber, sand with clay, sand with tar or asphalt, or plastic with wood flour. The additives increase the resistance of the material to deformation or flow. Similarly, the additions of ferrite and carbide to steel increase the hardness and the tensile strength.

The coarseness of the *microstructure* also plays an important role in mechanical properties. Very fine sand, when added to asphalt, produces a more viscous mixture than an equal amount of gravel. Similarly, steel with a very fine microstructure of ferrite and carbide will be much harder and stronger than steel with the same carbon content but a much coarser microstructure. The microstructure of a material can be controlled by its preparation and by heat treatment.

14.8 Electromechanical Properties: The Piezoelectric Effect

One very interesting effect that relates mechanical and electrical properties is the *piezoelectric effect*, in which an electric field across a crystal causes a strain (i.e., change in dimensions), and vice versa. This effect was first discovered by Pierre and Jacques-Paul Curie in 1870. It is responsible for such seemingly diverse phenomena as Wint-O-green lifesavers™ lighting up when crunched, and some children's shoes illuminating when the shoe moves.

The requirement of the piezoelectric effect is an ionic crystal, such as quartz, that has a *non-centrosymmetric* structure (in quartz this is due to the tetrahedral arrangement of four oxygen atoms around each silicon atom). The difference between an ionic centrosymmetric structure (positive ions symmetrically surrounded by negative ions; no net dipole or electric polarization of the crystal) and an ionic non-centrosymmetric crystal (positive ions not symmetrically surrounded by negative ions; nonzero net dipole and electric polarization of the crystal) is shown schematically in Figure 14.19. As we have already seen in Chapter 12, a polar non-centrosymmetric ionic crystal can be ferroelectric, meaning it has a nonzero electric dipole. (Recall that the term *ferroelectric* was coined to be analogous to the term *ferromagnetic*, which refers to a collective net magnetic dipole moment in the structure and originally referred to iron-based permanent magnets.)

Stress applied to a centrosymmetric ionic crystal does not change the (zero) electric polarization of the crystal (see Figure 14.19a). However, if the structure is non-centrosymmetric, the stress would change the electric polarization (Figure 14.19b). In a non-centrosymmetric (piezoelectric) ionic crystal, stress has the effect of changing the electric field across the crystal. If this crystal happens to be part of an electrical circuit, it can have induced electrical changes. For example, the gravitational force of a 0.5-kg mass applied across a 0.5 mm thick quartz plate of dimensions 10 cm × 2 m induces opposing charges of 2×10^{-9} C on the faces. The electric field from a piezoelectric device can amount to more than 10^4 V, and the resulting sparks can be used for ignition, such as in piezoelectric propane barbecue lighters. The circuit for shoes that make use of the piezoelectric effect is shown schematically in Figure 14.20.

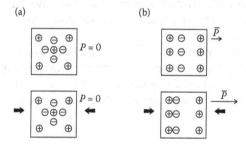

FIGURE 14.19
A schematic representation of (a) a centrosymmetric ionic crystal with and without an applied stress, showing no change in electric polarization, P, that is, no piezoelectric effect and (b) a non-centrosymmetric ionic crystal with and without an applied stress, showing a change in electric polarization, P, on application of stress, that is, the piezoelectric effect.

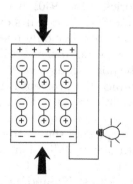

FIGURE 14.20
Schematic representation of a piezoelectric crystal in use in some running shoes. Stress on the shoe hitting the floor induces an electric field, which causes an electrical potential and a momentary burst of current and illumination of the light in the circuit.

The converse also is true: If an electric field is applied across a piezoelectric crystal, the crystal will have an induced strain. This relationship can be used to control very fine manipulations. For example, very fine motions (on the order of nanometers) can be controlled using a piezoelectric crystal in the circuit: A small change in the electric field across the crystal can stress the crystal, changing the crystal's dimensions. This can be coupled to a mechanical part (e.g., the tip of an atomic force microscope) to control its motion. The head of an inkjet printer contains hundreds of piezoelectric inkwells, and each can be electrically induced to contract and squirt its ink.

If the applied electric field is alternating, the thickness of the crystal oscillates with the same frequency, and the amplitude of the change can be considerable if it is the frequency of a normal lattice mode. For quartz, a 1.5-V silver oxide battery can be used to give 10^6 oscillations per sec. The natural frequency in one particular direction in quartz is especially independent of temperature such that frequency stabilities of 0.2 parts in 10^9, corresponding to an accuracy of 1 s per 150 years in a quartz timepiece, can be achieved. A few decades ago, wristwatches were based on mechanical springs and required adjustment at least once a week. Now, quartz-crystal watches require virtually no adjustment.

The conversion of mechanical energy in a piezoelectric material into electrical energy (and vice versa) is the basis for many *transducers*. One example is the piezoelectric crystals in our ears, which turn sound waves (mechanical motion) into electrical impulses that are carried along auditory nerves to our brain.

COMMENT: MODELING FRACTURE

There is a very successful theory for fracture that describes when a crack will begin to move in response to complicated applied stresses. It was created by George Irwin (1907–1998) who, as a researcher at the Naval Research Laboratories in the 1940s, was called upon to find out why a large fraction of the first US missiles exploded during launch. The theory, called *fracture mechanics*, only needs as input the energy necessary to move a crack a small distance. However, calculating this energy requirement at the microscopic level is still a work in progress.

There is one case in which the problem is beginning to be understood: crack motion in a brittle crystal at low temperatures. Cracks can move in two ways under these circumstances. At moderate velocities, up to some fraction of the speed of sound, the crack *cleaves* the crystal (i.e., breaks it along a natural line of division), leaving only phonons in its wake. However, when there is too much energy in the crack tip, the motion of the crack becomes unstable, and the crack begins to generate increasing amounts of subsurface damage and fissures in more than just the cleavage direction, leading to destruction.

COMMENT: SMART MATERIALS

Materials that respond to their environment are known as *smart materials*. For example, buildings could respond to the strains induced by an earthquake, bandages could contract and apply pressure to a wound in response to bleeding, and other devices could change when under chemical stress or in magnetic fields. Such devices, which use composite materials, sensors, piezoelectric devices, and shape-memory alloys and which rely on phase transformations for their various "states" are already a reality.

COMMENT: SOUNDING OUT IMPROVED MATERIALS

Researchers have developed an inexpensive method to investigate the mechanical properties of new composite materials or film coatings. A test sample is submerged in water, and an acoustic wave transducer sends a pulsed sound wave through it. The speed of the reflected wave enables determination of Young's modulus, and the direction of the reflected wave provides information concerning crystal planes or defects in the material. The instrument makes use of a curved transducer (for a lensing effect) made of an inexpensive piezoelectric plastic film.

See the video "Dental Drills"
under Student Resources at
PhysicalPropertiesOfMaterials.com

14.9 Shape-Memory Alloys: A Tutorial

It is possible that a given material can exist in different polymorphs with very different mechanical properties. One example is carbon, in its graphite and diamond phases.

a. Give another example of a material with different properties in different solid phases.

Another material that can change its mechanical properties with a change in phase is an alloy of nickel and titanium, called *Nitinol* (after *Ni*ckel *Ti*tanium *N*aval *O*rdnance *L*aboratory, where it was discovered in 1965). This material has two phases, a high-temperature *austenite* phase that is very difficult to bend, and a low-temperature *martensite* phase that is very flexible. The high-temperature phase can be achieved by immersing a piece of Nitinol in boiling water. On cooling to room temperature, the austenite phase can remain, but it is metastable with respect to the martensite phase. This metastability leads to a hysteresis loop in the physical properties.

b. Sketch Young's modulus, E, for Nitinol as a function of temperature (heating and cooling paths), from −50 °C to 100 °C.

The structure of Nitinol in the austenite phase is more symmetric than in the martensite phase, as shown in Figure 14.21. There are several variations in the way that martensite cells can be oriented: two ways are shown schematically in Figure 14.22, in contrast with the single arrangement of the austenite form.

c. Based on Figure 14.22, explain why the martensite phase is more flexible than the austenite phase.

d. Nitinol can be bent easily into a new shape when it is in the martensite phase. What do you expect will happen to the shape of a Nitinol wire when it is heated into the austenite phase?

e. Can you suggest applications for this interesting property of Nitinol (see Figure 14.23)?

f. It would be useful to be able to shape Nitinol wire in the high-temperature form, so that when it is heated, it returns to this particular shape (i.e., not just a straight wire). This can be accomplished by annealing. Annealing takes place at high temperature, when there is sufficient thermal energy to allow easier mobility of the atoms, although the temperature is not hot enough to melt the material. For example, intricate

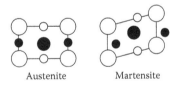

Austenite   Martensite

FIGURE 14.21
Nitinol in its austenite and martensite phases. ○ is Ni in the higher layer; o is Ni in the lower layer; ● is Ti in the higher layer; • is Ti in the lower layer.

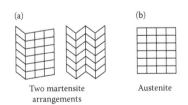

(a)   (b)

Two martensite   Austenite
arrangements

FIGURE 14.22
Packing arrangements in (a) martensite and (b) austenite structures of Nitinol.

(a)   (b)   (c)

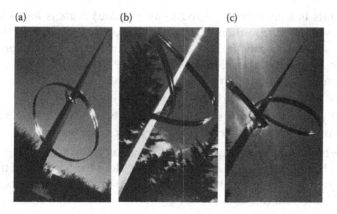

FIGURE 14.23
A sculpture made of Nitinol, "Le Totem du Futur" by the late French sculptor Jean-Marc Philipe. The photos are of the sculpture at (a) 5 °C; (b) 20 °C; (c) 35 °C. (Photos by J.-M. Philipe. With permission.)

laboratory glassware is annealed to remove stress after the glass is blown and before it is used. Why?

g. Explain what happens to a Nitinol wire when it is heated and shaped to a "V" in a flame (austenite phase). Use a diagram to show a schematic view of the internal structure. Also show the structure in the martensite phase. (These materials are also known as *shape-memory alloys* or *shape-memory metals*.)

h. Although many applications of Nitinol can be imagined, these would be limited if the transition temperature were fixed at 70 °C, as it is for the alloy $Ni_{0.46}Ti_{0.54}$. Can you suggest a way to modify the transition temperature?

> See the video "Memory Metal" under Student Resources at PhysicalPropertiesOfMaterials.com

i. Ni-Ti is not the only *memory metal* (martensitic) system; others include Cu-Zn-Al and Cu-Al-Ni. Ni-Ti has an advantage in that it is more corrosion resistant. However, Ni-Ti is more difficult to machine. Discuss how corrosion resistance and machining ability could be important in applications.

14.10 Cymbals: A Tutorial

The best cymbals are made from tin-rich bronze. *Bronze* is a solid solution, mainly copper, with added tin and possibly other elements. (A solid solution of all metals is sometimes referred to as an *alloy*.) *Brass* is also sometimes used for cymbals, and, although it has inferior properties, it is cheaper. Brass is a solid solution of zinc (minor component) in copper (major component), sometimes also with other elements.

The atomic diameter of Sn is about 15% greater than that of Cu, whereas the atomic diameter of Zn is about 4% greater than that of Cu.

a. Would you expect higher solid-state solubility of Sn in Cu, or Zn in Cu? Explain, considering the Hume-Rothery rules (see Section 14.4).

b. The presence of Sn in Cu (giving bronze) or Zn in Cu (giving brass) gives a harder material than pure Cu. Why does the addition of the second element harden Cu? Which element, Sn or Zn, would be expected to harden Cu more?

c. Which would you expect to have a higher speed of sound: Cu-Sn bronzes or Cu-Zn brasses?

The Cu-rich portion of the Cu–Sn phase diagram is shown in Figure 14.24. The phase labeled "α" is a solid solution of Sn in Cu, rich in Cu, commonly known as α bronze.

d. If Cu-Sn with less than 10 mass % Sn is first heated to 700 °C to make it homogeneous and then cooled slowly to room temperature, how many phases will exist at room temperature? What if it is cooled quickly?

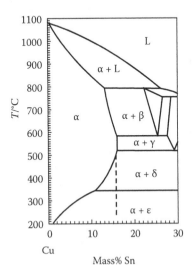

FIGURE 14.24
Copper–tin phase diagram in the copper-rich region. The solid lines indicate the equilibrium phase diagram and the dashed line indicates the metastable boundary most often encountered in industrial preparations of bronze.

e. The material resulting from fast-cooling in (d) is stiff but brittle. Why is a brittle material not suitable for a cymbal?

f. For samples richer in Sn ($\approx$ 20 mass %) and $T < 800$ °C, a second phase will occur. This can be the body-centered cubic δ-phase, which is very brittle. However, the body-centered cubic β-phase is less brittle. How does the type of second phase depend on the cooling?

g. Which second phase would be more desirable for a good cymbal, δ or β?

The first step in making a cymbal is *casting* (i.e., pouring the liquid metal into a mold). The appropriate quantities, for example, $\approx$ 20% Sn and $\approx$ 80% Cu, of high-purity Cu and high-purity Sn are combined with a small amount of silver. The metals are heated together to 1200 °C. This is considerably above the liquid temperature (see Figure 14.24).

h. Why is the temperature so high when 20 mass% Sn in Cu melts at about 900 °C?

i. Oxides are a problem in the manufacture of bronze, so high-grade carbon is added as a "flux." What does the carbon do?

The molten alloy is poured into a rotating, centrifugal mold. The castings are left to cool and age at room temperature for up to a week. This aging step seems to lead to a better product, possibly due to formation of very small and finely distributed precipitates.

j. What would the presence of very small and finely distributed precipitates do to the hardness of the material?

The next step is *rolling*. The casting is heated to about 800 °C and then subjected to several rolling steps, resulting in about 50% reduction in thickness. Each rolling step is in a different direction. This gives repeated recrystallization and a fine crystal structure with minimal alignment of crystallites, making the cymbal tougher with an increased speed of sound. Each casting is reheated and rolled from 8 to 15 times, depending on the size of the cymbal. The degree of heating (time and temperature) and variations in the rolling procedure are two sources of differences in acoustical properties of otherwise "identical" cymbals.

k. Traditionally, cymbals are made in open-pit fires, but now temperature-controlled ovens are used. Does the change of heating method increase the reproducibility of the cymbal process?

At this point, the casting is heated again and the cup (central indentation of the cymbal) is stamped with a hot press. This process is called *cupping*. (In some simpler processes, the cup is shaped in the original casting.)

In the *tempering* step, the cymbal is toughened and strengthened by heating to 800 °C, followed by quick cooling in an aqueous saline solution. The purpose of the saline solution is to provide a rapid and reproducible means of quenching. Furthermore, the salt helps remove any residual oxides that form in the grain boundaries and decrease ductility and workability.

l. Explain why oxides at grain boundaries would decrease ductility.

The tempering process replaces the brittle δ-phase with the quenched β-phase. While this technique is used in other aspects of metallurgy to make the material harder, here the purpose is to make the material less brittle.

In the next process, called *pressing*, the alloy is forced into a tapered concave cymbal shape, immediately after tempering and before crystals have a chance to grow. This step ensures that slight residual tensile stresses occur in the metal; these are necessary to create the sound. For some cymbals, the pressing process is preceded by a step called back-bending, in which the stack of cymbals is bent backward by about 90° a few times, in different positions and directions, in order to reduce the elasticity of the metal.

m. The pressing process is an example of work hardening. How does this step change the yield strength of a material?

While the prior steps ensure that the phase of the bronze is appropriate (hard, high speed of sound), the *hammering* process controls the shape of the cymbal, which is important to introduce its acoustical properties. A traditional hand-hammered cymbal could take 90 min to hammer, but the first cymbal of the batch could take considerably longer to adjust for variations. Although traditionally done by hand, hammering is now also done by machine.

In the first of a three-step *lathing* process, the cutting tool is used to make one continuous cut on the backside from the central cup to the edge. The top is cut in two steps. The first stiffens the cymbal and puts it into its final shape by using a templated backer. The final cut of the top side introduces the tonal grooves, which are essential for the proper acoustics and the final taper. The final sound of the cymbal depends on a number of factors during the lathing process, including the speed, pressure, and shape of the tool, as each can influence the dissipation of heat.

n. Why is it important to consider heat dissipation in the lathing process?

One of the most important variations in the making of a cymbal is the material itself. In cheaper alloys, such as B8 bronze (8 mass% Sn), the α-phase is dominant (see Figure 14.24), but the material will not have a high speed of sound. This gives B8 cymbals a narrower frequency spectrum than the preferred B20s, and thus a "harder" sound. Still cheaper brass cymbals have a more compact sound with shorter sustain and less brightness. Variations in tempering, pressing, hammering, and lathing also can influence the sound.

o. How could post-manufacture conditions influence the sound of a cymbal?

14.11 Learning Goals

- Stress, strain, Young's modulus, Poisson ratio
- Tension, compression
- Elastic limit, yield strength, plastic deformation, tensile strength, compressive strength, toughness, ductility, hardness
- Stiffness and relationship to interatomic potential
- Ductile versus brittle
- Stress birefringence
- Material failure
- Microstructure
- Defects and dislocations: point defects (lattice vacancies, substitutional impurities, interstitial impurities, and proper [or self] interstitials); line defects (screw dislocations and edge dislocations); slip plane and relationship to ductility
- Hume-Rothery rules
- Hardness: defects and dislocations, impurities, annealing, work-hardening
- Crack propagation: surface energy, stress energy, work of fracture, Griffith length
- Adhesion
- Piezoelectric effect
- Smart materials, including shape-memory alloys
- Cymbals as an example of materials processing influencing structure and properties

14.12 Problems

14.1 Many polymers change their mechanical properties as a function of temperature. With the same temperature (x) axis, sketch two curves: (a) the DSC trace of a polymer passing through its glass transition temperature, T_g, and (b) its Young's modulus over the same temperature range.

14.2 Two model bridges have been built and they are identical in every way, except that one bridge is 1/2 scale in each dimension with respect to the other bridge. Which bridge would support a larger load? Explain.

14.3 Based on your answer to Problem 14.2, which is better to contain a gas at very high pressure: a small-diameter tube or a large-diameter tube?

14.4 Two materials, a and b, differ in their values of Young's modulus, E, with $E_a \gg E_b$. The two materials are polymers of the same molecular structure, and the only difference between them is that one material is crystalline and the other is amorphous.

    a. Which material is crystalline?

    b. What are their relative values of

        i. Debye temperatures?

        ii. Speed of sound?

        iii. Heat capacities at room temperature?

        iv. Thermal conductivities at room temperature?

14.5 a. Explain how the mechanical properties of a material can influence its applicability as a toner for use in the photocopying process.

    b. What role does temperature play in consideration of the properties of a toner?

14.6 The Leaning Tower of Pisa was restored. During the work, the ground was frozen with liquid nitrogen ($T = 77$ K) in order to "prevent dangerous vibrations during the next phase of the salvage project." What is the influence of temperature on the mechanical properties (Young's modulus, speed of sound, Debye temperature) of the ground?

14.7 Most materials expand when heated. However, stressed rubber contracts when heated. Explain this phenomenon. *Hint*: Consider whether entropy increases or decreases on contraction of rubber.

14.8 A tub is to be made of metal, and in its use, it will be under some (calculable) maximum stress. It does not matter for this application if there is a crack in the tub, but it is important that the material not fail; that is, any cracks must not propagate. Explain how knowledge of the Griffith length for several potential tub materials will aid in the selection of an appropriate metal for this application.

14.9 The resin of fiber-reinforced composite materials (e.g., FibreGlass®) can be damaged by water. For example, the glass transition temperature, T_g, can be reduced by swelling of the resin caused by uptake of water. Explain how lowering T_g can adversely influence the mechanical properties of this material.

14.10 One of the most common construction materials is *concrete*, usually made by mixing Portland cement with sand, crushed rock, and water. The resulting material is very complex, resulting primarily from the reaction of Portland cement with water, as the adhesion results from hydrates. The setting process of concrete is exothermic. In some situations, such as outdoor use in winter, this heat source is useful. In other applications, such as large concrete structures, the exothermicity of the reaction can lead to stress. What is the source of the stress? Discuss factors that could be considered to lower the stress during concrete setting.

14.11 An advance in materials for turbine blades is the production of a *superalloy* that can form as single crystals, that is, without grain boundaries. What properties will be affected by the absence of grain boundaries?

14.12 Glass can be strengthened chemically by ion exchange to give a surface of lower thermal expansion coefficient than the underlying (bulk) glass. How does this treatment strengthen the glass?

14.13 Glass can be strengthened thermally by prestressing it so that all components are under compression at high temperatures. This step is followed by cooling.

a. Would this technique work best for low-thermal-expansion glass or for high-thermal-expansion glass?

b. Explain how the shape and size of the glass item may play important roles in the success of this tempering method.

14.14 Window glass that has been *tempered* (i.e., heat-treated for strengthening) will show patterns when viewed through a polarizer. This can be observed for side and rear car windows, for example. (Windshields are usually laminated [layered] structures and do not show this effect.) Explain the origin of the polarization patterns.

14.15 *Piezomagnetism* is the induction of a magnetic moment by stress. For example, some materials become magnetic after they are struck with another object. By analogy to the term *piezoelectricity*, explain the origins of piezomagnetism. Include a discussion of any necessary conditions.

14.16 *Piezoresistivity* is the variation in electrical resistivity of a material produced by applied mechanical stress. It is said that this property is observed to some extent in all crystals, but it is maximal in semiconductors. Why is it largest in semiconductors?

14.17 Explain how stress on a crystal can change its degree of birefringence. (This is a *piezooptic effect*.)

14.18 The addition of a second metal to a pure metal usually leads to an alloy with higher yield strength, tensile strength, and hardness. However, the addition of the impurity can change other properties. Would you recommend adding a second component to copper wires used for transmission of electrical power? (The alloy could have greater strength than pure copper.) Explain.

14.19 When a rubber band is immersed in liquid nitrogen, it loses its elastic properties and can be shattered easily with a hammer. Why?

14.20 a. Diamond is a very hard material. However, if a diamond is struck with a hammer, it will shatter. Explain why the diamond, and not the hammer, breaks.

b. If a diamond is squeezed between two metal surfaces, the diamond will become embedded in the metal. Why do the diamond and metal not shatter as in (a)?

c. In designing diamond drills, it is important to prevent fracture of the diamonds. Explain why polycrystalline diamond is often used for these applications. Consider the force per unit area on the diamond surfaces.

14.21 Copper can be drawn into wires that are strengthened by work hardening due to increased concentration and entanglement of dislocations.

a. How would work-hardening a copper wire influence its electrical resistivity? Why?

b. How would the electrical resistivity of a drawn copper wire change after annealing and then cooling to room temperature? Why?

14.22 Polyethylene bags are more difficult to stretch (i.e., plastically deform) along the polymer backbone direction than in perpendicular directions (between polymers). Explain.

14.23 Polymers are inherently weak materials. They can be strengthened by various measures, for example, cross-linking (an example is the S-S bonds formed in vulcanization of rubber), chain stiffening (for example, addition of large groups such as C_6H_5 that introduce steric hindrance), and/or chain crystallization (by suitable preparation and/or heat treatment). Discuss why each of these methods leads to stronger polymers.

14.24 Give the relative values of Young's modulus for a clean hair and a hair coated with (dried) hair gel.

14.25 Draw stress–strain diagrams for two materials, one very tough and one not so tough.

14.26 Explain how the temperature dependence of ductility could make a ship unsafe in cold waters and yet seaworthy in warm waters.

14.27 Pure gold (100% Au) is often referred to as 24 karat (or 24 k). This purity of gold is not commonly used for jewelry because it is too soft. Other metals are usually added to the gold, lowering the karat number (e.g., to 18 k or 10 k), making the material harder.

a. Explain how the impurities increase the hardness of the material.

b. If the "gold" was to be used for electrical contacts, explain how the impurities would be expected to change the electrical properties of the material.

c. Would you expect the impurities to change the thermal conductivity of the material? Explain.

14.28 A typical laptop computer display uses a very small fluorescent lamp for the backlight.

a. These lights do not produce much heat. How is that a particular advantage for this application?

b. These lamps are very thin, but not very strong. Why is the latter property a concern?

14.29 Would the thermal conductivity of an alloy in its amorphous metal form be expected to be higher or lower than that of the corresponding crystalline form of the alloy? Explain.

14.30 Most materials have a positive Poisson ratio, and thin laterally when extended (Figure 14.25). However, some materials have the unusual property of lateral expansion when stretched, leading to a negative Poisson

See the video "Auxetic Structure" under Student Resources at PhysicalPropertiesOfMaterials.com

ratio. These materials are known as *auxetic* (from the Greek *auxesis*, meaning "increase"). An example of the type of structure that can lead to auxetic behavior is shown in Figure 14.26. Suggest applications for auxetic materials.

14.31 For ionic crystals, two special cases of point defects arise. One is from the loss of a cation–anion pair, leaving behind a vacancy on two adjacent sites (one cationic, the other anionic), leaving a *Schottky defect*, as shown in Figure 14.27a. The other is a *Frenkel defect*, formed when a lattice ion migrates into an interstitial position, leaving a vacancy behind, as shown in Figure 14.27b. The vacancy–interstitial pair is referred to as a Frenkel defect. Discuss ways in which the presence of Frenkel defects and Schottky defects could be expected to alter the optical, thermal, electronic, and mechanical properties of ionic materials.

14.32 *Ferromagnetic shape-memory alloys* change their shape in response to a magnetic field. The shape change is not due to change in structure as in conventional shape-memory alloys, but a change in the arrangement of the martensite structure when a magnetic field is applied, as shown in Figure 14.28. Compositions showing ferromagnetic shape-memory properties include Fe-Pd, Fe-Pt, Co-Ni-Al,

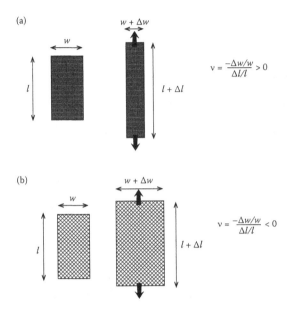

FIGURE 14.25
When a material is extended, if the Poisson ratio, ν, is (a) positive, it thins laterally, but if the Poisson ratio is (b) negative, it expands laterally.

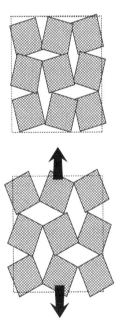

FIGURE 14.26
A material composed of flexible cells can be auxetic, that is, exhibit a negative Poisson ratio, ν (lateral expansion when extended).

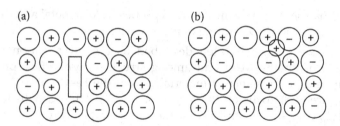

FIGURE 14.27
(a) The missing cation–anion pair, shown here as a box, is referred to as a Schottky defect; (b) when an ion moves from its lattice site to an interstitial position, it forms a Frenkel defect.

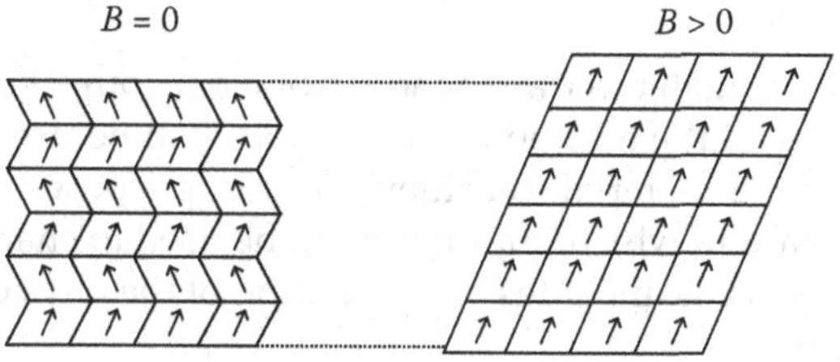

$B = 0$ $B > 0$

FIGURE 14.28
When the magnetic field, B, is imposed on a ferromagnetic shape-memory alloy, the cells align in the magnetic field, leading to a strain (extension shown here).

Co-Ni-Ga, and Ni-Mn-Ga. The latter can show strains up to 10% and has a high Curie temperature (103 °C). What properties would be required to make these alloys useful for applications as actuators?

14.33 In a traditional ceramic, such as a clay, the firing process removes water from between the inorganic layers in the structure. How does the firing process change the Young's modulus of the ceramic, compared with the so-called *green body* (unfired ceramic)?

14.34 Microelectromechanical systems (MEMS) are machines with components ranging in size from 0.02 to 1 mm. MEMS became practical once they could be fabricated with technologies normally used to make semiconductor-based electronic devices. Si-based MEMS devices are almost perfectly elastic, showing no hysteresis when flexed. Why is this lack of hysteresis so important?

14.35 3-D printing is a process whereby computer-controlled ejection of a material can be used to create a three-dimensional object. Because the materials add together to create the object (rather than being removed by carving, lathing, etc.), this process is also referred to as *additive manufacturing*. Additive manufacturing techniques can lead to less waste than subtractive techniques. Many different types of materials can be used in 3-D printing. Consider the role of microstructure in the mechanical properties of a metal and discuss challenges for 3-D printing of metal parts.

14.36 Materials science is often described as the investigation of the relationships between and among structure, properties, processing, and performance of materials. From the Fe–C phase diagram (Figure 9.38), focus on the region of 1% C at 1000 °C (which gives steel), compared with 3% C at the same temperature (which gives cast iron), and discuss the differences in each of those two regions with regard to: (i) structure, (ii) properties, (iii) processing, and (iv) performance.

14.37 Give examples of hysteresis for three different types of physical processes. Include descriptions of the processes and labelled diagrams for each.

14.38 Many useful materials are made from one pure single component. However, the addition of impurities can be used to alter the properties of a material to achieve certain purposes. Explain four different ways in which impurities can be used to change the properties of materials. Include an explanation of the physical principles involved.

14.39 Conjugated polymers offer a unique combination of electrical, optical, and mechanical properties.

 a. To make them processable by techniques such as extrusion and rolling, sidechains are required. Why?

 b. However, sidechains can make the conjugated units such as aromatic rings rotate out of the plane of the long polymer chain direction. How can this influence the electrical properties?

14.40 A glass material can be produced under compression by quick cooling (also known as quenching).

 a. Explain whether a higher thermal expansion coefficient would increase or decrease the compression.

 b. Explain whether a higher thermal conductivity would increase or decrease the compression.

 c. Would this glass material in compression have a higher or lower Griffith length than if it was not under compression? Explain your reasoning.

14.41 Cutting tools, including scissors, knife blades, and lathe tools, need to be made of materials that will remain sharp under all possible conditions. Some of the best materials used for lathe tools include silicon carbide (SiC) and silicon nitride (Si_3N_4). To be useful, they need to be hard so that they cut well. In addition, cutting tools need to conduct heat well so that frictional heat created in their use on the lathe will be efficiently dissipated, and ideally they should have low thermal expansion coefficients so that their dimensions do not change much as they heat up during use.

    a. Is a hard material likely to have a higher or lower thermal conductivity than a soft material? Explain your reasoning.

    b. Is a hard material likely to have a higher or lower thermal expansion coefficient than a soft material? Explain your reasoning.

14.42 The *grain size* (i.e., the average size of microcrystalline domains) in a material can influence its properties.

    a. Explain how the grain size can change the mechanical properties of material. Consider a particular mechanical property for a material with small grains, compared with the same material with larger grains.

    b. How would the size of the grains change on heating? Explain.

    c. How can the grain size influence the transparency of the material?

14.43 The precursor to a *thermoset plastic* is a liquid that becomes hardened by thermally induced formation of chemical bonds, making crosslinks between the large molecules. This process is not reversible as the crosslinks are permanent chemical bonds. Thermosets are usually placed in a mold while in their liquid form, and the resultant crosslinked product conforms to the shape of the mold. A common thermoset is epoxy, which is crosslinked by the addition of a curing agent that typically contains reactive $-NH_2$ groups. In contrast, *thermoplastics* have polymer chains that are not crosslinked. They are rigid at low temperature but become more flexible, adapting new shapes, above their glass transition temperature, T_g. This transition is reversible. Sketch Young's modulus of a typical thermoset as a function of time on cooling from above its cure point. On the same graph, show Young's modulus of a typical thermoplastic on cooling from above T_g.

14.44 In determination of the thermal expansion coefficient of sintered ceramics, it is common to see the effects of crack healing in the first heating of a sample, but not in subsequent heating or cooling. Sketch the strain of such a sample with (normal) positive thermal expansion as a function of temperature from room temperature to a high temperature at which crack healing is taking place, and then for subsequent cooling and then a second heating, all on the same graph.

14.45 *Nacre* is a biocomposite, the iridescent material found inside shells, and sometimes called mother-of-pearl. It has a "brick-and-mortar" structure, as shown in Figure 14.29. The "bricks" are calcium carbonate in the aragonite form, about 10 µm long and 250 nm thick. These plates form about 95% of the mass of the structure and are surrounded by an organic matrix composed of proteins and polysaccharides.

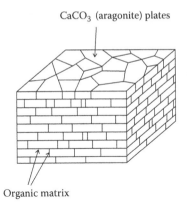

CaCO$_3$ (aragonite) plates

Organic matrix

FIGURE 14.29
Nacre is a natural composite with a brick-and-mortar structure in which the inorganic component is surrounded by a matrix of organic material.

    a. Explain why the structure of nacre leads to iridescence.

    b. Nacre is about 3000 times tougher than pure aragonite. Explain
        how the structure can be tough, and especially how it could pre-
        vent crack propagation.

Further Reading

General References

ASM International, 2002. *Atlas of Stress-Strain Curves*, 2nd ed. ASM International,
    Materials Park, OH.
Elastic strain engineering. Special issue. J. Li, Z. Shan, and E. Ma, Eds. *MRS Bulletin*,
    February 2014, 108–162.
Interatomic potentials for atomistic simulations. Special issue. A. F. Voter, Ed. *MRS
    Bulletin*, February 1996, 17–49.
Mechanical properties in small dimensions. Special issue. R. P. Vinci and S. P. Baker,
    Eds. *MRS Bulletin*, January 2002, 12–53.
Mechanical behavior of nanostructured materials. Special issue. H. Kung and T.
    Foecke, Eds. *MRS Bulletin*, February 1999, 14–58.
Materials for sports. Special issue. F. H. Froes, S. Haake, S. Fagg, K. Tabeshfar, and X.
    Velay, Eds. *MRS Bulletin*, March 1998, 32–58.
Superhard coating materials. Special issue. Y.-W. Chung and W. D. Sproul, Eds. *MRS
    Bulletin*, March 2003, 164–202.
S. Ashley, 2003. Alloy by design. *Scientific American*, July, 24.
D. R. Askeland and W. J. Wright, 2015. *The Science and Engineering of Materials*, 7th ed.
    CL Engineering.

P. Ball, 1994. *Designing the Molecular World*. Princeton University Press, Princeton, NJ.

P. Ball, 1997. *Made to Measure; New Materials for the 21st Century*. Princeton University Press, Princeton, NJ.

D. J. Barber and R. Loudon, 1989. *An Introduction to the Properties of Condensed Matter*. Cambridge University Press, Cambridge.

J. D. Birchall and A. Kelly, 1983. New inorganic materials. *Scientific American*, May, 104.

W. D. Callister, Jr. and D. G. Rethwisch, 2013. *Materials Science and Engineering: An Introduction*, 9th ed. John Wiley & Sons, Hoboken, NJ.

D. J. Campbell and M. K. Querns, 2002. Illustrating negative Poisson's ratios with paper cutouts. *Journal of Chemical Education*, 79, 76.

R. Cotterill, 2008. *The Material World*. Cambridge University Press, Cambridge.

J. P. Den Hartog, 1952. *Advanced Strength of Materials*. Dover, New York.

A. B. Ellis, M. J. Geselbracht, B. J. Johnson, G. C. Lisensky, and W. R. Robinson, 1993. *Teaching General Chemistry: A Materials Science Companion*. American Chemical Society, Washington, DC.

J. Gilman, 2003. *Electronic Basis of the Strength of Materials*. Cambridge University Press, Cambridge.

J. E. Gordon, 1984. *The New Science of Strong Materials*, 3rd ed. Princeton University Press, Princeton, NJ.

A. Grant, 2017. Printing glass in 3D. *Physics Today*, June, 24.

P. Grünberg, 2001. Layered magnetic structures: History, highlights, applications. *Physics Today*, May, 31.

P. Hess, 2002. Surface acoustic waves in materials science. *Physics Today*, March, 42.

R. A. Higgins, 1994. *Properties of Engineering Materials*, 2nd ed. Industrial Press, New York.

D. Hull and D. J. Bacon, 2011. *Introduction to Dislocations*, 5th ed. Butterworth-Heinemann, Oxford.

R. E. Hummel, 2004. *Understanding Materials Science*. 2nd ed. Springer, New York.

D. H. R. Jones and M. F. Ashby, 2012. *Engineering Materials 2: An Introduction to Microstructures, Processing and Design*, 4th ed. Butterworth-Heinemann, Oxford.

B. H. Kear, 1986. Advanced materials. *Scientific American*, October, 158.

J. Krim, 1996. Friction at the atomic scale. *Scientific American*, October, 74.

D. W. Pashley, Ed., 2001. *Imperial College Inaugural Lectures in Materials Science and Engineering*. Imperial College Press, London.

W. E. Pickett and J. S. Moodera, 2001. Half metallic magnets. *Physics Today*, May, 39.

M. A. Porter, P. G. Kevrekidis, and C. Daraio, 2015. Granular crystals: Nonlinear dynamics meets materials engineering. *Physics Today*, November, 44.

C. N. R. Rao and K. J. Rao, 1992. Chapter 8: Ferroics. In *Solid State Chemistry: Compounds*, A. K. Cheetham and P. Day, Eds. Clarendon Press, Oxford.

G. L. Schneberger, 1983. Metal, plastic and inorganic bonding: Practice and trends. *Journal of Materials Education*, 5, 363.

J. F. Shackelford, 2004. *Introduction to Materials Science for Engineers*, 6th ed. Prentice Hall, Upper Saddle River, NJ.

W. F. Smith and J. Hashemi, 2009. *Foundations of Materials Science and Engineering*, 5th ed. McGraw-Hill, New York.

M. A. Steinberg, 1986. Materials for aerospace. *Scientific American*, October, 66.

P. A. Thrower, 1995. *Materials in Today's World*, 2nd ed. McGraw-Hill, New York.

A. H. Tullo, 2017. 3-D printing: A tool for production. *Chemical and Engineering News*, October 16, 30.

J. D. Verhoeven, 2001. The mystery of Damascus blades. *Scientific American*, January, 74.

M. A. White and P. MacMillan, 2003. The cymbal as an instructional device for materials education. *Journal of Materials Education*, 25, 13.

Adhesion

Materials science of adhesives: How to bond things together. Special issue. C. Creton and E. Papon, Eds. *MRS Bulletin*, June 2003, 419–454.

Polymer first is sticky, then it isn't, 1999. *Chemical and Engineering News*, August 23, 42.

The ties that bind abalone shells, 1999. *Chemical and Engineering News*, June 28, 23.

A. R. C. Baljon and M. O. Robbins, 1996. Energy dissipation during rupture of adhesive bonds. *Science*, 271, 482.

L. A. Bloomfield, 1999. Instant glue. *Scientific American*, June, 104.

H. R. Brown, 1996. Adhesion of polymers. *MRS Bulletin*, 21(1), 24.

R. Dagani, 2001. Sticking things to carbon nanotubes. *Chemical and Engineering News*, May 7.

N. A. de Bruyne, 1962. The action of adhesives. *Scientific American*, April, 114.

C. Gay and L. Leibler, 1999. On stickiness. *Physics Today*, November, 48.

G. F. Krueger, 1983. Design methodology for adhesives based on safety and durability. *Journal of Materials Education*, 5, 411.

J. R. Minkel Jr., 2006. Bubble adhesion. *Scientific American*, October, 32.

W. J. Moore, 1967. *Seven Solid States*. W.A. Benjamin, New York.

P. Patel, 2018. Kirigami creates strong, removable adhesive. *Chemical and Engineering News*, February 28, 8.

J. T. Rice, 1983. The bonding process. *Journal of Materials Education*, 5, 95.

R. V. Subramanian, 1983. The adhesive system. *Journal of Materials Education*, 5, 31.

Ceramics

Glass-ceramics. Special issue. M. J. Davis and E. D. Zanotto, Eds. *MRS Bulletin*, March 2017, 195–240.

H. K. Bowen, 1986. Advanced ceramics. *Scientific American*, October 1986, 168.

R. E. Newnham, 1984. Structure-property relations in electronic ceramics. *Journal of Materials Education*, 6, 807.

K. M. Prewo, J. J. Brennan, and G. K. Layden, 1986. Fibre reinforced glasses and glass-ceramics for high performance applications. *American Ceramic Society Bulletin*, 65, 305.

Composites

T.-W. Chou, R. L. McCullough, and R. B. Pipes, 1986. Composites. *Scientific American*, October, 192.

J. Hay and S. Shaw, 2001. Into the labyrinth. *Chemistry in Britain*, November, 34.

M. B. Jakubinek, C. Samarasekera, and M. A. White, 2006. Elephant ivory: A low thermal conductivity, high strength nanocomposite. *Journal of Materials Research*, 21, 287.

T. Kelly and B. Clyne, 1999. Composite materials—Reflections on the first half century. *Physics Today*, November, 37.

C. O. Oriakhi, 2000. Polymer nanoncomposition approach to advanced materials. *Journal of Chemical Education*, 77, 1138.

P. Scott, 2002. Heads on tails. *Scientific American*, April, 24.

A. Tullo, 1999. New DVDs provide opportunity for polymers. *Chemical and Engineering News*, December 20, 14.

Elasticity and Strength

S. Ashley, 2003. A stretch for strong copper. *Scientific American*, January, 31.

S. Borman, 2018. Process makes wood stronger than steel. *Chemical and Engineering News*, February 12, 4.

R. A. Guyer and P. A. Johnson, 1999. Nonlinear mesoscopic behavior: Evidence for a new class of materials. *Physics Today*, April, 30.

M. Jacoby, 2018. Chemically strengthened glass's next act: Car windshields. *Chemical and Engineering News*, January 15, 16.

M. Reibold, P. Paufler, A. A. Levin, W. Kochmann, N. Pätzke, and D. C. Meyer, 2006. Carbon nanotubes in an ancient Damascus sabre. *Nature*, 444, 286.

M. Reisch, 1999. What's that stuff? [Spandex]. *Chemical and Engineering News*, February 15, 70.

A. M. Rouhi, 1999. Biomaterials for women. *Chemical and Engineering News*, June 28, 24.

M.-F. Yu, B. S. Files, S. Arepalli, and R. S. Ruoff, 2000. Tensile loading of ropes of single wall carbon nanotubes and their mechanical properties. *Physical Review Letters*, 84, 5552.

Fracture

Atomistic theory and simulation of fracture. Special issue. R. L. B. Selinger and D. Farkas, Eds. *MRS Bulletin*, May 2000, 11–50.

S. Ashley, 2004. Cryogenic cutting. *Scientific American*, March, 28.

M. Bain and S. Solomon, 1991. Cracking the secret of eggshells. *New Scientist*, March 30, 27.

M. J. Buehler and H. Gao, 2006. Dynamical fracture instabilities due to local hyperelasticity at crack tips. *Nature*, 439, 307.

M. E. Eberhart, 1999. Why things break. *Scientific American*, October, 66.

J. J. Gilman, 1960. Fracture in solids. *Scientific American*, February, 95.

D.G. Holloway, 1968. The fracture of glass. *Physics Education*, 3, 317.

M. Marder and J. Fineberg, 1996. How things break. *Physics Today*, September, 24.

T. A. Michalske and B. C. Bunker, 1987. The fracturing of glass. *Scientific American*, December, 122.

J. W. Provan, 1988. An introduction to fracture mechanics. *Journal of Materials Education*, 10, 325.

C. N. Reid, 1988. An introduction to crack resistance (R-Curves). *Journal of Materials Education*, 10, 483.

R. O. Ritchie, 2014. In pursuit of damage tolerance in engineering and biological materials. *MRS Bulletin*, October, 880.

I. Sage, 2001. Seeing the light. *Chemistry in Britain*, February, 24.

E. Wilson, 1997. New ceramic bends instead of breaking. *Chemical and Engineering News*, September 8, 12.

Hardness and Toughness

F. P. Bundy, 1974. Superhard materials. *Scientific American*, August, 62.

R. Dagani, 1995. Superhard-surfaced polymers made by high-energy ion irradiation. *Chemical and Engineering News*, January 9, 24.

H. Schubert and G. Petzow, 1988. What makes ceramics tougher? *Journal of Materials Education*, 10, 601.

E. Stoye, 2014. Hardest diamonds ever made from carbon onions. *Chemistry World*, June 11.

D. M. Teter, 1998. Computational alchemy: The search for new superhard materials. *MRS Bulletin*, January, 22.

Smart Materials, Including Shape-Memory Materials

Responsive materials. Special issue. *MRS Bulletin*, C. A. Alexander and I. Gill, Eds. September 2010, 659–683.

Science and technology of shape memory alloys: New developments. Special issue. K. Otsuka and T. Kakeshita, Eds. *MRS Bulletin*, February 2002, 91–136.

R. Abeyaratne, C. Chu, and R. D. James, 1996. Kinetics of materials with wiggly energies: theory and application of twinning microstructures in a Cu-Al-Ni shape memory alloy. *Philosophical Magazine A*, 73, 457.

S. Ashley, 2001. Shape-shifters. *Scientific American*, May, 20.

S. Ashley, 2003. Artificial muscles. *Scientific American*, October, 53.

S. Ashley, 2006. Muscling up color. *Scientific American*, November, 27.

M. Brennan, 2001. Suite of shape-memory polymers. *Chemical and Engineering News*, February 5, 5.

K. S. Brown, 1999. Smart stuff. *Scientific American Presents: Your Bionic Future*, 72.

E. Cooper, 2016. Camel hair shows shape memory. *Chemistry World*, August 26.

R. Dagani, 1997. Intelligent gels. *Chemical and Engineering News*, June 9, 26.

M. Fischetti, 2003. Cool shirt. *Scientific American*, October, 92.

L. Fisher, 2017. Shape memory polymers get a grip. *Chemistry World*, March 6.

M. Freemantle, 2003. Stressed polymers change color. *Chemical and Engineering News*, December 22, 10.

K. R. C. Gisser, M. J. Geselbracht, A. Cappellari, L. Hunsberger, A. B. Ellis, J. Perepezko, and G. C. Lisensky, 1994. Nickel-titanium memory metal. A "smart" material exhibiting a solid-state phase change and superelasticity. *Journal of Chemical Education*, 71, 334.

T. C. Halsey and J. E. Martin, 1993. Electrorheological fluids. *Scientific American*, October, 58.

A. J. Hudspeth and V. S. Markin, 1994. The ear's gears: Mechanoelectrical transduction by hair cells. *Physics Today*, February, 22.

T. W. Lewis and G. G. Wallace, 1997. Communicative polymers: The basis for development of intelligent materials. *Journal of Chemical Education*, 74, 703.

C. Modes and M. Warner, 2016. Shape-programable materials. *Physics Today*, January, 32.

R. E. Newnham, 1997. Molecular mechanisms in smart materials. *MRS Bulletin,* May, 20.

K. Otsuka and C. M. Wayman, 1999. *Shape Memory Metals.* Cambridge University Press, Cambridge.

J. Ouellette, 1996. How smart are smart materials? *The Industrial Physicist,* December, 10.

C. A. Rogers, 1995. Intelligent materials. *Scientific American,* September, 154.

L. McDonald Schetky, 1979. Shape-memory alloys. *Scientific American,* November, 98.

M. Schwartz, Ed., 2008. *Smart Materials.* CRC Press, Boca Raton, FL.

R. Schweinfest, A. T. Paxton, and M. W. Finnis, 2004. Bismuth embrittlement of copper is an atomic size effect. *Nature,* 432, 1008.

A. G. Smart, 2016. This phase-transforming metal never gets old. *Physics Today,* August, 18.

A. V. Srinivasan and D. M. McFarland, 2001. *Smart Structures: Analysis and Design.* Cambridge University Press, Cambridge.

S. Stinson, 2001. Plastic mends its own cracks. *Chemical and Engineering News,* February 19, 13.

D. W. Urry, 1995. Elastic biopolymer machines. *Scientific American,* January 1995, 64

T. A. Witten, 1990. Structured fluids. *Physics Today,* July 1990.

Metals

A. H. Cottrell, 1967. The nature of metals. *Scientific American,* September, 90.

R. D. Doherty, 1984. Stability of the grain structure in metals. *Journal of Materials Education,* 6, 841.

J. Schroers, 2013. Bulk metallic glasses. *Physics Today,* February, 32.

Polymers

E. Baer, 1986. Advanced polymers. *Scientific American,* October, 192.

F. L. Buchholz, 1994. A swell idea. *Chemistry in Britain,* August, 652.

G. B. Kauffman, S. W. Mason, and R. B. Seymour, 1990. Happy and unhappy balls: Neoprene and polynorbornene. *Journal of Chemical Education,* 67, 199.

C. O'Driscoll, 1996. Spinning a stronger yarn. *Chemistry in Britain,* December, 27.

W. Peng and B. Reidl, 1995. Thermosetting resins. *Journal of Chemical Education,* 72, 587.

R. B. Seymour and G. B. Kauffman, 1990. Piezoelectric polymers. *Journal of Chemical Education,* 67, 763.

Spider Silk

S. Chang, 2017. Spider dragline silk's surprising twist. *Physics Today,* August, 23.

A. King, 2017. Spinning out spider silk research. *Chemistry World,* May 12.

A. H. Simmons, C. A. Michal, and L. W. Jelinski, 1996. Molecular orientation and two-component nature of the crystalline fraction of spider dragline silk. *Science,* 271, 84.

S. Trohalaki, 2012. Spider webs rely on nonlinear material behavior and architecture. *MRS Bulletin,* March, 182.

P. Vollrath, 1992. Spider webs and silks. *Scientific American*, March, 70.

F. Vollrath, D. Porter, and C. Holland, 2013. The science of silks. *MRS Bulletin*, January, 73.

Viscoelastic Behavior

R. J. Hunter, 1993. *Introduction to Modern Colloid Science*. Oxford University Press, New York.

T. W. Huseby, 1983. Viscoelastic behavior of polymer solids. *Journal of Materials Education*, 5, 491.

A. Widener, 2015. Magic sand and kinetic sand. *Chemical and Engineering News*, March 23, 41.

Websites

For links to relevant websites, see PhysicalPropertiesOfMaterials.com.

Appendix 1: Fundamental Physical Constants

Quantity	Symbol	Value with Units
Speed of light in vacuum	c	$299792458\,\text{m s}^{-1}$
Permeability of vacuum	μ_0	$4\pi\times10^{-7}\,\text{N A}^{-2}$
Permittivity of vacuum	ε_0	$8.854187817\times10^{-12}\,\text{F m}^{-1}$
Newtonian constant of gravitation	G	$6.67259(85)\times10^{-11}\,\text{m}^3\,\text{kg}^{-1}\,\text{s}^{-2}$
Planck's constant	h	$6.626070040(81)\times10^{-34}\,\text{J s}$
Avogadro constant	N_A	$6.022140857(74)\times10^{23}\,\text{mol}^{-1}$
Molar gas constant	R	$8.314472(15)\,\text{J K}^{-1}\,\text{mol}^{-1}$
Boltzmann constant, R/N_A	k	$1.38064852(79)\times10^{-23}\,\text{J K}^{-1}$

Appendix 2: Energy Unit Conversions

	Energy/J	$\tilde{v}$/m^{-1}	v/s^{-1}	T/K	E/eV
1 J =	1	$1/(hc)$	$1/h$	$1/k$	$1/e$
		5.03411	1.50919	7.242973×10^{22}	6.241506
		$\times 10^{24}$	$\times 10^{33}$		$\times 10^{18}$
1 m^{-1} =	(hc)	1	c	$(hc)/k$	$(hc)/e$
	1.98645		299,792,458	0.0143877	1.2398242
	$\times 10^{-25}$				$\times 10^{-6}$
1 s^{-1} =	h	$1/c$	1	h/k	h/e
	6.62607	3.335640952		4.799245	4.1356692
	$\times 10^{-34}$	$\times 10^{-9}$		$\times 10^{-11}$	$\times 10^{-15}$
1 K =	k	$k/(hc)$	k/h	1	k/e
	1.380649	69.50346	2.083661		8.617386
	$\times 10^{-23}$		$\times 10^{10}$		$\times 10^{-5}$
1 eV[a] =	e	$e/(hc)$	e/h	e/k	1
	1.602177	806,555	$2.417990 \times$	11,604.53	
	$\times 10^{-19}$		10^{14}		

[a] 1 eV corresponds to 806,555 m^{-1}.

Appendix 3: The Greek Alphabet

Name	Lower Case	Upper Case
Alpha	α	A
Beta	β	B
Gamma	γ	Γ
Delta	δ	Δ
Epsilon	ε	E
Zeta	ζ	Z
Eta	η	H
Theta	θ	Θ
Iota	ι	I
Kappa	κ	K
Lambda	λ	Λ
Mu	μ	M
Nu	ν	N
Xi	ξ	Ξ
Omicron	o	O
Pi	π	Π
Rho	ρ	P
Sigma	σ	Σ
Tau	τ	T
Upsilon	υ	Υ
Phi	φ	Φ
Chi	χ	X
Psi	ψ	Ψ
Omega	ω	Ω

Appendix 4: Sources of Lecture Demonstration Materials

Arbor Scientific
   arborsci.com/
Edmund Scientific
   scientificsonline.com/
Educational Innovations
   teachersource.com
Flinn Scientific Ltd.
   flinnsci.com
Institute for Chemical Education
   ice.chem.wisc.edu
NADA Scientific Ltd.
   nadascientific.com
Sayal Electronics
   sayal.com

The shops at many museums and science centers also carry many interesting items that can be used to illustrate physical properties of materials.

Glossary

3-D printing: process whereby computer-controlled ejection of a material can be used to create a three-dimensional object; also called additive manufacturing

absorbed species: species in the interior of a material

acceptor impurity: an impurity in a semiconductor that is deficient in electrons relative to the major material (i.e., creates holes)

acceptor level: additional energy level in the energy gap, present in a p-doped semiconductor and not present in the pure semiconductor; see Figure 12.14

additive manufacturing: see 3-D printing

adhesion: process by which two different materials attach to each other

adhesive: a material used to bind two materials together

adhesive strength: measure of interactions between an adhesive and substrate

adsorbate: see adsorbed species

adsorbed species: species adhered to surface of a material; also known as adsorbate

aerogel: porous ultra low-density material derived from a gel, in which the liquid component of the gel has been replaced with a gas

AFM: see atomic force microscope

allochromic: property in which a material has no inherent color, but can be colored due to impurities

allotrope: one of a number of structures of an element

alloy: metallic combination of metals, or metal with another element, as single phase (solid solution) or multiple phases

amorphous: shapeless, without form; for a material, this generally means noncrystalline solid

amorphous metal: see metallic glass

amphiphilic: part hydrophilic, part hydrophobic

anharmonic: interactions not represented well by Hooke's law; see Equation 7.8

anisotropic: different in different directions

anneal: to heat a material to a high temperature, without melting, followed by controlled cooling, thereby changing its properties

antiferroelectric: material showing microscopic ferroelectric regions that are exactly balanced by regions with the opposite polarity, giving no net polarization; see Figure 12.28

antiferromagnet: material showing microscopic ferromagnetic regions that are exactly balanced by regions with the opposite magnetic polarity, giving no net magnetization

aragonite: polymorph of calcium carbonate

arrest point: see halt point

atomic force microscope (AFM): instrument to investigate topology of a surface via the force of attraction between the surface atoms and a tip in close proximity to the surface

atomic transition: electronic transition that is intrinsic to a given element; can give rise to color by emission

auxetic: having a negative Poisson ratio

auxochrome: electron-withdrawing or electron-donating functional groups used to change the electronic energy levels of organic compounds and thereby change their color

avalanche effect: amplification of current in a diode

azeotrope: constant boiling mixture in a binary system, i.e., boils at a fixed temperature; see Figures 9.25 and 9.26

band gap: forbidden energy region between the valence band and the conduction band in an insulator or semiconductor

band theory: theory describing energy levels in a bulk solid with conducting electrons, for which the energy levels are so plentiful and therefore so close together that they form bands, not discrete energy levels

BCS theory: after Bardeen, Cooper, and Schrieffer; theory that attributes the lack of electrical resistivity in the superconducting state to interactions between pairs of electrons (Cooper pairs), working cooperatively and with phonon assistance, to give zero resistivity; see Figure 12.32

binary: having two components

bioluminescence: fluorescence initiated by chemical reactions in a living organism that give chemical products in an excited state and thereby fluoresce

biomaterials: materials designed to, by themselves, or as part of a complex system, control interactions with components of living systems; can be natural materials or lab-produced

biomimetic material: synthetic material imitating natural material

birefringence: optical property of double refraction in an anisotropic material arising from different refractive indices in different directions in the material; see Figures 5.5 and 5.6

black-body radiation: a broad emission of electromagnetic radiation from an idealized material that absorbs light of all wavelengths (and hence is black) for which the intensity and peak wavelength of emitted light depend on the temperature; see Figure 2.7 and Equation 2.3

BNNT: see boron nitride nanotubes

boron nitride nanotubes (BNNT): polymorph of boron nitride with cylindrical structures of nm-scale diameters and considerably longer lengths, similar to CNT structures

Bose–Einstein condensation: state of matter in which bosons (particles with zero or integral spin) are cooled to very low temperatures and a large fraction occupy the ground state

Bragg's law: equation describing condition of constructive interference of x-ray (or neutron) beam reflecting from different planes of atoms in a crystal structure such that constructive interference results; see Equation 7.21

brass: alloy that is primarily copper, with added zinc

Bravais lattices: 14 distinct types of crystal structures, each different depending on relations between unit cell dimensions and angles, that can be used to describe crystal structures; see Table 1.3 and Figure 1.4

break point: temperature in a cooling or warming curve at which the number of degrees of freedom changes from a non-zero value to another non-zero value

Brinell hardness number: hardness index based on the area of penetration of a very hard ball into the surface of a material under a standardized load

brittle: property of a material in which the reversible stress–strain curve ends abruptly in failure without first undergoing deformation; see Figure 14.9

bronze: alloy that is primarily copper, with added tin and sometimes other elements

calamitic liquid crystal: material that has its liquid crystalline properties by virtue of the rod-shaped molecules from which it is composed; see Figure 4.22

calcite: polymorph of calcium carbonate

capillary effect: rise (or depression) of a liquid in a capillary; see Figures 10.8 and 10.9

carbon nanotube (CNT): nanoscale tubes composed exclusively of carbon; see Figure 9.17

cast iron: a heterogeneous iron-carbon alloy with greater than 2% carbon

casting: forming by pouring material into a mold

cathodoluminescence: emission of light as a result of cathode rays (electron beam) energizing a phosphor to produce luminescence

CCD: see charge-coupled device

CD: see circular dichroism

cement: a material made with calcined lime and clay and used with other materials to make concrete

ceramic: a solid material composed of a mixture of metallic, or semi-metallic, and nonmetallic elements, giving a material that is hard, durable, and resistant

cermet: composite of a ceramic and a metal, generally with properties quite different from either the ceramic or metal from which it is composed

charge-coupled device (CCD): device in which light is transformed into electrical signals that are then converted to digital data

chemical potential: rate of change of Gibbs energy per mole of that species in a multicomponent system; see Equation 9.13

chemical vapor deposition (CVD): chemical process to produce a thin film of a material on a substrate, from precursor chemicals in the vapor phase

chemiluminescence: fluorescence initiated by chemical reactions that give products in an excited state

chemisorption: chemical adsorption arising from strong (covalent) interactions between adsorbate and surface

chiral molecule: a molecule that is nonsuperimposable on its mirror image

chromatic aberration: color patterns (usually unwanted) such as color fringes that arise because matter has different refractive indices for different wavelengths of light, therefore preventing all the colors from being focused at a single point

chromophore: literally "color bearer" used to impart color; see dye and also pigment

circular dichroism (CD): difference in light absorption coefficient for left- and right-circularly polarized light, giving rise to different colors depending on the handedness of the circularly polarized light; see Figure 5.8

circularly polarized light: light in which the direction of polarization rotates in the plane perpendicular to the direction of light propagation, either clockwise or anticlockwise; see Figure 5.7

Clapeyron equation: equation relating the p–T phase coexistence line of any two phases in a pure material to thermodynamic properties; see Equation 9.7

clathrate: multicomponent crystalline material in which one component forms cages in which the other component(s) reside and cannot exit; see Figure 11.6

Clausius–Clapeyron equation: relationship between vapor pressure and temperature for a substance; see Equation 9.25

cleave: process of splitting a crystal along certain crystallographic planes

cmc: see critical micellar concentration

CNT: see carbon nanotube

coefficient of thermal expansion (CTE, α): rate of change of volume of a material (relative to initial volume) with respect to change in temperature; see Equation 6.71; units are K^{-1}; see also linear thermal expansion coefficient

coefficient of thermal conductivity: see thermal conductivity

coercive field (H_c): after magnetization of a ferromagnetic material, this is the reverse magnetic field required to completely remove the induction; see Figure 13.7

cohesion: interactions by which a given material holds together

cohesive strength: measure of interactions within a material

cold welding: joining of two materials without heating

colligative property: physical property that depends only on the number of species in solution, not their composition; includes freezing point depression, boiling point elevation, vapor pressure lowering, and osmotic pressure

colloid: system of particles, typically 10 to 1000 nm in size

colossal magnetoresistance: exceptionally large magnetoresistance, typically of orders of magnitude

color centre: defect in crystalline lattice due to absence of ion and consequent presence of extra electron or hole for which electronic energy levels allow absorption of visible light and thereby impart color; also called F-center (F stands for *Farbe*, which is German for color)

columnar liquid crystal: liquid crystalline phase in which the disc-shaped molecules are arranged in columns; see Figure 4.22

complementary colors: the colors opposite each on the color wheel (diagram in which all colors of visible light are written in a circle); see Figure 2.3

composite: material made from two or more different materials with different properties, and for which the overall properties of the composite are different from the properties of either of the materials from which it is composed

compressive strength: maximum stress that can be experienced by a material during a test in which it is being compressed; units are Pa

concrete: a material made from cement with water and aggregate such as sand and rocks

conduction band: range of energy bands that are unfilled in a nonmetal at $T = 0$ K; electrons can be promoted to these bands by heat or light

conduction electrons: electrons with sufficient energy (i.e., in the conduction band) to conduct electrical current and also heat

conductivity (σ): parameter to quantify ability of a medium to conduct electrical current; units are $\Omega^{-1} m^{-1} \equiv S\ m^{-1}$

congruent melting: process by which a compound formed by more than one component melts directly to the liquid; see Figure 9.39

conjugated: type of chemical bond in which formal double bonds alternate with formal single bonds, such that π-bonds are shared (delocalized) over many atoms

constructive interference: process by which waves exactly in phase produce a wave of the same wavelength and summed amplitudes of the incoming waves; see Figure 4.10

contact potential: electrical potential arising from dissimilar metals in contact; see Figure 12.3

Cooper pairs: pair of electrons of opposite spin, moving in a coordinated, phonon-assisted manner, giving a BCS superconductor zero resistance

Cotton effect: presence of both circular dichroism and optical rotatory dispersion

crack propagation: growth of a crack (fracture) in a material

creep: move slowly under continuous stress

critical angle: angle of incidence of light on an interface, beyond which there is no refraction, resulting in internal reflection; see Figure 4.6

critical magnetic field: magnetic field above which a superconductor loses its property of superconductivity

critical micellar concentration (cmc): minimum concentration of surfactant required for micelle formation

critical opalescence: onset of cloudiness in an otherwise transparent mixture near its critical point

critical point: point in pressure and temperature beyond which (i.e., at higher T and p) liquid and vapor phases are no longer distinguishable, and phase is a supercritical fluid; see Figure 7.2

critical pressure: pressure beyond which the liquid and vapor phases are no longer distinguishable, and phase is a supercritical fluid; see Figure 7.2

critical temperature: temperature beyond which the liquid and vapor phases are no longer distinguishable, and phase is a supercritical fluid; see Figure 7.2

crosslinking: formation of chemical bonds between polymer molecules, thereby changing the physical properties of the overall polymer

crystal field: the electric field resulting from the arrangement of ions in a crystal

crystallite: very small crystal

CTE: see coefficient of thermal expansion

cubic: a crystal class in which all the unit cell dimensions are identical and all the angles are 90°; see Table 1.3 and Figure 1.4

Curie law: temperature-dependence of magnetic susceptibility of a Pauli paramagnet; see Equation 13.8

Curie temperature (T_C): temperature above which the spins in a ferromagnet have too much thermal energy to behave cooperatively, and the material becomes paramagnetic

Curie-Weiss law: temperature-dependence of magnetic susceptibility of a ferromagnetic material in the regime where it is a Pauli paramagnet; see Equation 13.9

CVD: see chemical vapor deposition

Debye characteristic temperature (θ_D): thermal energy (kT) corresponding to the highest frequency of an atomic vibration in the Debye model of the heat capacity of a solid; units are K; see Equation 6.30

Debye heat capacity model: temperature-dependent heat capacity of a solid arising from treating each atom as vibrating harmonically on its lattice site, with a range of frequencies; see Equation 6.29

Debye-T^3 law: T^3 temperature dependence of the heat capacity of an insulating solid as $T \rightarrow 0$ K

degrees of freedom: the number of independent factors required to specify a system at equilibrium; can apply to types of motion of atoms or collections of atoms (translational, rotational, vibrational), or to physical parameters (T, p, etc.) required to specify phases of system (see phase rule)

destructive interference: process by which waves exactly out of phase will exactly cancel each other; see Figure 4.11

devitrification: loss of vitreous (glassy) state by crystallization

diamagnetic: property of material in which all spins are paired; material is expelled from magnetic field

dichroism: two different colors of material when viewed in different directions

dielectric: insulating material that reduces the Coulombic force between two charges; a material that when placed between the plates of an electrical capacitor increases its capacitance; see Equation 12.13

dielectric constant: ratio of the total electric field within the material to the applied electric field of the light; see Equations 5.5 and 12.13

differential scanning calorimetry (DSC): analytical technique to determine how much energy is associated with heating (or cooling) a sample, relative to a reference material

differential thermal analysis (DTA): analytical technique to determine how quickly a sample heats (or cools), relative to a reference material

diffraction grating: optical component with regular slits or grooves that disperse light by interference effects, such that different colors have their maximum intensity in different directions

diffuse reflection: process by which light scatters in many directions on reflection, usually from a rough surface; also called diffuse scattering; see Figure 3.5

diffusion: property describing general mass transport

dilatometer: instrument that measures change in dimension of a sample as a function of temperature

diode: n,p-semiconductor device that exhibits the electrical property of rectification; see Figure 12.16

discotic liquid crystal: material that has its liquid crystalline properties by virtue of the disc-shaped molecules from which it is composed; see Figure 4.22

dislocation: packing irregularity in a crystal structure

dispersion: heterogeneous mixture of one solid phase in another (different) solid phase; also called suspension

dispersion of light: spreading of different wavelengths of light so they travel in different directions as the light passes through a material, arising from different refractive indices for different wavelengths; see Figure 4.5

domain: uniform region, separated from other different uniform regions by domain walls; see Figure 13.5

domain wall: interface between domains in a material; see Figure 13.5

donor impurity: an impurity in a semiconductor that carries extra electrons

donor level: additional energy level in the energy gap, present in an n-doped semiconductor and not present in the pure semiconductor; see Figure 12.13

dopant: impurity in a material, usually added purposefully

doped semiconductor: semiconductor with impurities

double refraction: see birefringence

DSC: see differential scanning calorimetry

DTA: see differential thermal analysis

ductility: strain of a material at its failure point, usually expressed as %; see Figure 14.1

Dulong–Petit law: the molar heat capacity for a nonmetallic solid at room temperature and constant volume is ~ $3R$, independent of composition

dye: liquid used to impart color

edge dislocation: layer of extra atoms within a crystal structure interrupting regular periodicity; see Figure 14.14

effusion: property describing motion of mass out of a small orifice

Einstein characteristic temperature (θ_E): thermal energy (kT) corresponding to the frequency of an atomic vibration in the Einstein model of the heat capacity of a solid; units are K; see Equation 6.24

Einstein heat capacity model: temperature-dependent heat capacity of a solid arising from treating each atom as vibrating harmonically on its lattice site at a fixed frequency, independent of its neighbors; see Equation 6.25

elastic limit: stress at which elongation is no longer reversible on removal of stress; units are Pa; see Figure 14.1

elastic modulus: see Young's modulus

electrical conductivity: see conductivity

electrical resistivity: see resistivity

electrochromic: appearing with a different color or degree of transparency depending on the applied electric field

electroluminescence: emission of light as a result of electrical stimulus

electromagnetic spectrum: the range of all possible frequencies (and energies and wavelengths) of electromagnetic radiation, from high-energy gamma rays, to low-energy radio waves, including visible light; see Figure 2.1

electronic transition: transition between electronic energy levels, e.g., by absorption or emission of light

electrooptic effect: dependence of the refractive index of a nonlinear optical material on an applied voltage

embodied energy: energy required to make a material from its raw materials

emissivity: ratio of thermal radiation emitted from the surface of a material to the radiation emitted from a perfect black-body radiator, which thereby quantifies effectiveness in emitting energy as thermal radiation; is dimensionless and values range from 0 (perfect reflector) to 1 (perfect emitter)

emulsion: heterogeneous material composed of one liquid dispersed in another liquid

enantiomers: a pair of chiral molecules that are mirror images of each other

enthalpy of fusion: heat required to melt a material; $\Delta_{fus}H$

enthalpy of vaporization: heat required to vaporize a material; $\Delta_{vap}H$

entropy of fusion: entropy change associated with melting of a material; $\Delta_{fus}S$

entropy of vaporization: entropy change on vaporization of a material; $\Delta_{vap}S$

equipartition theory: for temperatures high enough that translational, rotational and vibration degrees of freedom are all fully excited, each type of energy contributes equally, $\frac{1}{2} kT$, to the internal energy per degree of freedom

eutectic: mixture of substances in a multicomponent system that melts and solidifies at a fixed temperature that is lower than the melting points of the separate components; see Figure 9.33

eutectic composition: composition corresponding to eutectic point

eutectic halt: temperature at which there are zero degrees of freedom on a heating or cooling curve through the region of a eutectic

eutectic temperature: temperature corresponding to eutectic point

exchange interactions: forces that tend to keep magnetic spins aligned in a parallel fashion (↑↑)

extensive property: a property that depends on the size of the system (i.e., extends with the size of the system)

extrinsic semiconductor: see doped semiconductor

F-centre: see color centre

Fermi energy (E_F): the energy for which probability of occupation of the energy level is exactly $\frac{1}{2}$

Fermi-Dirac distribution: temperature-dependent probability distribution function for electronic states in a metal; see Equation 3.1

ferrimagnetism: property of material showing microscopic ferromagnetic regions that are somewhat but not completely balanced by regions with the opposite magnetic polarity, giving nonzero magnetization; see Figure 13.13

ferroelectric: pyroelectric material for which the spontaneous polarization can be reversed by the application of a small electric field; see Figure 12.28

ferromagnetic: paramagnetic material in which the spins work cooperatively, having a high susceptibility to magnetization, with the strength depending on the applied magnetic field, and magnetization persisting after the applied field is removed

fiber optic: flexible, transparent fiber that can transmit light efficiently by total internal reflection; also called optical fiber

Fiberglass®: fibers of acid-resistant glass (generally soda lime glass with added Al_2O_3) reinforced with an organic plastic

field-effect liquid crystal display: liquid crystal display in which the optical activity of the liquid crystal depends on the electric field, giving rise to light absorption or reflection, depending on the electric field; see Figure 5.4

field-effect transistor: a transistor in which most current is carried along a path for which the resistance can be controlled by a transverse electric field

first-order phase transition: phase transition for which the first derivatives of G (namely V and S) are discontinuous; see Figure 6.15

flocculation: coalescence of dispersed particles such as a colloidal dispersion

fluorescence: rapid emission of light from a cool body after absorption of light of higher energy and decay to an intermediate energy level; see Figure 2.14

flux density, magnetic (B): magnitude of the magnetic induction

foam: heterogeneous material composed of a solid dispersed in a gas

free electron gas model: simple model in which electrons in a metal are considered as a gas in a sea of nuclei

freezing point depression: reduction in freezing point of a material due to the addition of a second component to the pure material

Frenkel defect: defect arising from an ion in an interstitial location and its absence from its regular lattice site; see Figure 14.27

frequency doubling: see second harmonic generation

Fresnel lens: flat lens with concentric stepped prisms that act as lenses to focus light by refraction; see Figure 4.8

Fullerene: molecule of carbon with polyhedral structure, such as C_{60} (see Figure 9.15) or C_{70}

giant magnetoresistance (GMR): very large magnetoresistance effect

Gibbs phase rule: see phase rule

glass: non-crystalline amorphous solid with compositions ranging from inorganic to organic to metal

glass ceramic: material with an amorphous phase and one or more crystalline phases; can be processed like glass but has the special properties of ceramics, especially in being able to withstand high temperatures

glass transition temperature (T_g): temperature below which an amorphous supercooled liquid becomes a rigid glass; a kinetic temperature (depends on thermal treatment), not an equilibrium, thermodynamic temperature

GMR: see giant magnetoresistance

Gorilla® glass: thin, light, damage-resistant alkali-aluminosilicate glass

grain boundary: interface between homogeneous regions of the material

grain size: particle size of individual grains or crystallites in a material

graphene: one-atom thick layer of hexagonally bonded carbon atoms; see Figure 10.2

Griffith length (l_G): length beyond which a crack will grow spontaneously

Hall effect: production of a potential drop across an electrical conductor when a magnetic field is applied in a direction perpendicular to that of the flow of current; see Figure 13.19 and Equation 13.15

halt point: constant temperature over which phase(s) transform, with zero degrees of freedom

hard magnet: magnet with a broad *B-H* hysteresis loop (see Figure 13.9) because large magnetic fields are required to achieve saturation induction, due to difficulty in moving domain walls

hardness: resistance of a material to penetration of its surface

harmonic approximation: interactions are closely represented by Hooke's law; see Equation 7.4

heat capacity: amount of energy required to increase the temperature by 1 K, at constant pressure (C_p) or constant volume (C_V); can be expressed per unit mass (specific heat), per unit volume, or per mole (molar heat capacity, C_m)

Herfindahl–Hirschman index (HHI): a way to quantify the market availability of any item; used in materials science to quantify availability of the elements

heterogeneous: not uniform

hexagonal: a crystal class in which two of the unit cell dimensions are identical and two of the angles are 90° and the third is 120°; see Table 1.3 and Figure 1.4

HHI: see Herfindahl–Hirschman index

hole: absence of an electron, therefore with effective positive charge, that can be treated as a pseudoparticle in terms of transport and other properties

HOMO: highest unoccupied molecular orbital

homogeneous: uniform throughout

Hume-Rothery rules: empirical rules describing the conditions under which an element can dissolve into a crystalline metallic solid to form a solid solution

hydrophilic: attracted to water

hydrophobic: repelled by water

hydrothermal crystal growth: growth of crystals in a supercritical fluid

hyperpolarizability: nonlinear optical property; see Equation 5.9

ideal gas: a theoretical concept, giving rise to the equation of state $pV = nRT$, in which the gas molecules do not interact with each other, nor do the molecules themselves contribute to the volume of the system

ideal solution: solution for which the vapor pressure is a linear function of the mole fraction; see Equation 9.21

idiochromatic: property in which material color is invariant to how finely divided the material is

immiscible: do not mix

impurity precipitate: small quantity of crystalline impurity as heterogeneity in another crystalline solid; see Figure 14.15

incandescence: emission of light from a material due to its high temperature; see black-body radiation

inclusion compound: multicomponent material in which one component forms a host in which other component(s) reside; see Figure 11.6

incongruent melting: process by which a compound formed by more than one component melts to give a liquid and also a solid of a different composition from the compound composition, followed by a higher-temperature transformation to one liquid phase; see Figure 9.40

index of refraction: see refractive index

inhomogeneous: see heterogeneous

insulator: a material with very low electrical conductivity

intensive property: a property that is independent of the size of the system

intercalate: species absorbed into a material

interface: a surface where two materials meet

interference: interactions between waves based on their different phases; constructive interference leads to an increase in amplitude; destructive interference leads to a decrease in amplitude; see Figures 4.10 and 4.11

interference colors: colors resulting from a thin film of a material in which light reflected from the front face interferes with light reflected from the back face, such that some wavelengths interfere constructively (giving color) and others interfere destructively (and are absorbed); see Figure 4.15

interstitial impurity: presence of an impurity atom in the interstices in a crystal; see Figure 14.12

intrinsic resistivity: see resistivity

intrinsic semiconductor: pure semiconductor

Invar: an alloy of Ni (35%) and Fe (65%) that has invariant dimensions when the temperature is changed, i.e., zero thermal expansion

iridescence: optical property in which colors depend on viewing direction

isothermal bulk modulus (K): reciprocal of isothermal compressibility; units are Pa; see Equation 14.16

isothermal compressibility (β_T): rate of shrinkage of volume of a material (relative to initial volume) with respect to pressure at constant temperature; units are Pa^{-1}; see Equation 6.72

isotropic: the same in all directions

Joule–Thomson coefficient (μ_{JT}): the rate of change of temperature of a gas per change in pressure, at constant enthalpy; units are K Pa^{-1}; see Equation 6.20; it is zero for an ideal gas and thereby provides a quantitative measure of nonideality of a gas

kinetic theory of gases: descriptive and quantitative theory of gases as particles in constant motion with a distribution of speeds

Krafft temperature (T_K): minimum temperature of surfactant solution required for micelle formation

Langmuir–Blodgett film: highly organized film of monolayers of amphiphilic molecules on the surface of water or transferred onto a substrate; see Figures 10.11 and 10.14

laser: a device that emits bright light by stimulated emission; "lase" from Light Amplification by Stimulated Emission; see Figure 2.15

lattice parameter: the dimension (length) of the unit cell (*a, b* or *c*) of a crystalline material

lattice vacancy: absence of an atom at a regular lattice site in a crystal; see Figure 14.12

lattice wave: see phonon

LCA: see life cycle analysis

lead crystal: glass to which lead has been added to increase the refractive index; *not crystal*!

LED: see light-emitting diode

left-handed materials: materials with negative index of refraction

lever principle: principle by which relative amounts of different phases can be determined from a phase diagram, either in terms of moles or mass; see Equation 9.24 and Figure 9.30

lever rule: see lever principle

life cycle analysis (LCA): assessment of all the environmental impacts associated with a product or material, from the extraction of the raw materials, to the production and processing of the product, through its useful life, and including its ultimate demise

light-emitting diode (LED): n,p-semiconductor device (i.e., diode) that emits light with high efficiency as the electrons drop in energy from the conduction band to the valence band in the presence of an electric field; see Figure 3.10

light scattering: change in direction of incoming light, to continue on paths in many directions

line defect: defect in a crystal due to irregular placement of whole rows of atoms; see Figure 2.10

linear coefficient of thermal expansion: see linear thermal expansion

linear defect: groups of atoms in irregular places in a crystal lattice; see Figures 14.13 and 14.14

linear polarizability: see polarizability

linear thermal expansion (α): rate of change of one dimension of a material (relative to initial dimension) with respect to temperature; units are K^{-1}; see Equation 7.20; see also thermal expansion coefficient

linearly polarized light: light waves in which the electric field only oscillates in one plane; see Figure 5.1

liquid crystal: mesophase of matter that exhibits some properties of liquid (e.g., flow) and some of solid (e.g., iridescence); see Figure 4.22

liquid crystal display: display device that relies on the presence of liquid crystals and their special properties; see also field-effect liquid crystal display

low-E glass: see low-emissivity glass

low-emissivity glass: a type of glass that has low emissivity and therefore reflects most of the radiation incident on it

lower consolute temperature: temperature below which an immiscible mixture becomes miscible; see Figure 9.22

luminescence: emission of light from a cool body subsequent to absorption of light or another form of energy; includes fluorescence, phosphorescence, bioluminescence and chemiluminescence

LUMO: lowest occupied molecular orbital

lyotropic liquid crystal: liquid crystal that has liquid crystalline properties as a result of addition of solvent

magnetic dipole interactions: forces that tend to align spins antiparallel ($\uparrow\downarrow$)

magnetic field strength $(\bar{H})$: vector that measures the force acting on a unit pole placed at a fixed point in a magnetic field in vacuum; measures the magnitude and direction of the magnetic field; units are A m^{-1}

magnetic susceptibility (χ): dimensionless ratio of the magnetization of a material to the magnetic field strength; useful to define categories of magnetic materials; see Equation 13.6

magnetization: magnetic polarization

magnetocaloric effect: reversible change in the temperature of a thermally isolated magnetic material in a magnetic field, caused by variation in magnetic field; see Figure 13.18

magnetoresistance: relative change in electrical conductivity of a material when it is placed in a magnetic field; see Equation 13.14

Matthiessen's rule: empirical conclusion that the total resistivity of a crystalline metal is the sum of the resistivity due to thermal motions of the metal atoms in the lattice and the resistivity due to the presence of imperfections in the crystal

mean free path: average distance travelled (e.g., by molecule in gas or by phonon in solid) before colliding with another

Meissner effect: expulsion of a magnetic field by a superconductor

memory metals: see shape-memory materials

MEMS: see microelectromechanical device

mesophase: a phase that is intermediate between other phases, usually at an intermediate temperature; e.g., a liquid crystal is a mesophase as it forms a crystal at lower temperatures and an isotropic liquid at higher temperatures

metal: a material with very high, but not infinite, electrical conductivity

metallic glass: amorphous material that is metallic, generally with very high fracture strength; also known as amorphous metal

metallurgy: investigation of metals and their properties

metamaterial: engineered material that derives its properties from its manipulated structure, rather than the properties of its components

metastability: a position of local stability, but not global stability; see Figure 6.9

micelle: three-dimensional structure of an aggregate of amphiphilic molecules; see Figure 11.2

microelectromechanical device (MEMS): machines with components ranging in size from 0.02 to 1 mm

microlithography: a method to print or pattern matter on a substrate, with dimensions of ~10^{-6} m

microstructure: structure of a material at sub-mm length scale

Mie scattering: scattering of light from particles that are about 0.1 to 50 times the wavelength of light, resulting in color

miscible: to mix completely in all proportions

moiré pattern: large-scale interference pattern when two repeating patterns are placed on top of one another but not aligned; see Figure 4.12

molecular engineering: methods for the design and synthesis of novel molecules or materials with desirable physical properties or functionalities

molecular materials: materials in which discrete molecules can be identified within the structure

monochromatic: single color

monochromator: optical component that produces only single color (single wavelength) of light

monoclinic: a crystal class in which each unit cell dimension is unique and two of the angles are 90°; see Table 1.3 and Figure 1.4

monolayer: single-molecule-thick layer

MOSFET: metal-oxide-semiconductor field-effect transistor

MWCNT: multiwalled carbon nanotube; see Figure 9.17

N-Process: see Normal process

nacre: iridescent composite material found inside shells; also called mother-of-pearl

nanomaterial: material with at least one of its dimensions less than 100 nm, usually with different (or very different) properties from the bulk material

Néel temperature (T_N): temperature below which a material can be an antiferromagnet

negative thermal expansion: see thermomiotic

nematic liquid crystal: liquid crystalline phase in which the molecules are aligned along their long axis, as fibers in a thread; see Figure 4.22

Neumann–Kopp law: generalized experimental finding that in many cases the heat capacity of a solid can be well approximated by the sum of the heat capacities of its constituent elements, weighted by their molar contribution to the total composition

neutron scattering: scattering of a neutron beam from a sample either elastically (neutron diffraction) or inelastically (inelastic neutron scattering)

Nitinol: nickel-titanium alloy, named for **N**ickel **T**itanium **N**aval **O**rdnance **L**aboratory, with shape-memory properties

nonideal gas: a gas for which $pV = nRT$ does not hold

nonlinear effect: nonlinear proportionality of a response to an input

nonlinear optical effect: optical effect in which the induced polarization of the material is not a linear function of the electric field of the light; see Figure 5.12

non-Newtonian behavior: behavior in which viscosity depends on rate of stress

Normal process (N-process): process by which two phonons interact to give a resultant phonon with motion in the general motion of the incident phonons, carrying heat; see Figure 8.6

n-type semiconductor: a semiconductor that has been doped with impurities that carry extra electrons compared with the base material

OLED: organic light-emitting diode

optical activity: ability of a material to rotate the electric field oscillation direction of linearly polarized light

optical fiber: see fiber optic

optical rotatory dispersion (ORD): difference in refractive index for left- and right-circularly polarized light; see Figure 5.8

ORD: see optical rotatory dispersion

orientationally disordered solid: crystalline molecular solid in which all of the molecules sit on lattice sites, but they are rotationally dynamically disordered; see Figures 9.9 and 9.16 for examples

orthorhombic: a crystal class in which each of the unit cell dimensions is unique and all the angles are 90°; see Table 1.3 and Figure 1.4

osmotic pressure: minimum pressure that would need to be applied to a solution to prevent influx of solvent across a semipermeable membrane

p,n-junction: see diode

packing fraction: fraction of space that is filled; see Equation 1.4

paraelectric: material that is not polarized but can be polarized at lower temperature; see Figure 12.28

paramagnetism: property of material in which some spins are unpaired; material is attracted to magnetic field

Pauli paramagnet: paramagnetic material in which the magnetic centers act independently, not cooperatively

PCM: see phase change material

Peltier effect: an effect in which heat is emitted or absorbed when electrical current passes across a junction of dissimilar materials, converting electrical energy into a temperature gradient

peritectic: composition and temperature of liquid that is formed at the temperature of melting an incongruently melting compound; see Figure 9.41

peritectic halt: temperature at which there are zero degrees of freedom on a heating or cooling curve through the region of incongruent melting

permanent magnet: hard magnet that can maintain magnetization easily

phase: homogeneous region in matter, separated from other homogeneous regions by a phase boundary

phase change material (PCM): material that stores energy via a phase change (usually, but not necessarily, melting), for later recovery and use

phase rule: relationship between number of degrees of freedom related to phase equilibria, number of components, and number of phases; see Equation 9.20

phonon: quantum of crystal wave energy in a lattice travelling at the speed of sound; quantized lattice wave; see Figure 8.3

phosphor: material that shows luminescence, including phosphorescence or fluorescence

phosphorescence: slow emission of light from a cool body after absorption of light of higher energy and decay to an intermediate energy level; slower than fluorescence due to change in spin between excited state and ground state

photochromism: property of material that changes color in the presence of light

photonic material: material designed to have structure manipulate light, e.g. to have periodic variation in refractive index

photorefractive effect: dependence of the index of refraction of a material on the local electric field

photovoltaic (PV) material: material in which voltage and electrical current are created on exposure to light

physisorption: physical adsorption; weak interaction between adsorbate and substrate

piezoelectric: material exhibiting the piezoelectric effect

piezoelectric effect: effect by which electric field causes a noncentrosymmetric ionic crystal to change dimensions, and vice versa (e.g., compression causes an electric field); see Figure 14.19

piezomagnetism: induction of a magnetic moment by stress

piezooptic effect: stress on a crystal changes its degree of birefringence

piezoresistivity: variation in electrical resistivity of a material produced by applied mechanical stress

pigment: solid, often suspended in a solvent, used to impart color

plane polarized light: see polarized light

plastic crystal: see orientationally disordered solid

plastic deformation: permanent strain when a material has been extended beyond its yield strength; also called plastic strain

plastic strain: see plastic deformation

plasticizer: a substance (typically a solvent) added to a polymer to make it more flexible and less brittle

pleochroic: appearing to be a different color depending on the viewing direction

Pockels effect: see electrooptic effect

point defect: atoms missing or substituted or in irregular places in a crystal lattice; see Figure 14.12

Poisson ratio: measure of lateral contraction of a material relative to its extension under tension; dimensionless; see Equation 14.5 and Figure 14.25

polarizability: ability of matter to exhibit instantaneous (induced) dipole

polarized light: see linearly polarized light

polarizer: material that only passes one polarization of light; see Figure 5.1

polymer: material made of many, many chemically linked molecular units called monomers

polymorph: one of a number of structures of a material

polytype: one of a number of forms of a layered crystalline substance that differ only in the periodicity of the layers; special form of polymorph

Prince Rupert drop: glass drop formed by quench cooling in water, resulting in a very strong drop with the outer surface in compression; the stress can be relieved explosively by snipping off the tail

processing: the way in which materials are prepared, which can directly influence their structure (e.g., microstructure) and properties

proper interstitial: presence of an atom (not impurity) in the interstices in a crystal; also called self interstitial; see Figure 14.12

p-type semiconductor: a semiconductor that has been doped with impurities that are deficient in electrons compared with the base material

PV: see photovoltaic material

Pyrex®: soda lime glass with added B_2O_3 to increase thermal shock fracture resistance

pyroelectric: crystal that can have a nonzero electric polarization even in the absence of an applied electric field; see Figure 12.28

quantum confinement: restriction of electron and phonon mean free path by virtue of a material having structure at the nm length scale; greatly changes properties of the material compared to a macro-scale material of the same composition

quantum dot: semiconductor particle of the order of a few nm in dimension, with properties different from macroscopic particles of the same composition

quasicrystal: crystalline material with quasiperiodic structure; see Figure 9.18

quenching: cooling very rapidly

Rayleigh scattering: scattering of light from small particles that are less than 10% of the wavelength of light, resulting in color; see Equation 4.5

refraction: bending of light as it changes from one medium to another, due to different speeds in different media

refractive index: ratio of speed of light in vacuum compared to speed of light in medium; see Equation 4.1

refractometer: instrument that measures refractive index

refrangibility: see refraction

relative permittivity: see dielectric constant

remanent induction (B_r): magnetic induction in a ferromagnetic material that remains when the magnetizing field has been reduced to zero; see Figure 13.7

resistivity (ρ): reciprocal of electrical conductivity; units are Ω m

rhombohedral: a crystal class in which all of the unit cell dimensions are identical and none of the angles is 90°; see Table 1.3 and Figure 1.4

rule of mixtures: quantitative description of the property of a material in terms of the properties of the materials from which it is composed, weighted by their contributions to the composition; see Equation 11.5

SAM: see self-assembled monolayer

saturation induction (B_s): maximum induction in a ferromagnetic material; see Figure 13.7

scanning tunneling microscope (STM): instrument to investigate topology of a surface via the tunneling current between the surface atoms and a tip in close proximity to the surface

Schottky defect: missing cation–anion pair in an ionic lattice; see Figure 14.27

screw dislocation: defect arising in a crystal lattice in which atoms spiral around an axis; see Figure 14.13

second harmonic generation (SHG): production of light of twice the frequency of the incident light; a nonlinear optical effect; also known as frequency doubling

second-order phase transition: phase transition for which the first derivatives of G (namely V and S) are continuous, but the second derivatives of G are discontinuous; see Figure 6.16

Seebeck coefficient: see thermoelectric power

self interstitial: see proper interstitial

self-assembled monolayer (SAM): single-molecule thick layer of bi-functional organic molecules chemisorbed to a substrate; see Figure 10.16

semiconductor: a material with higher electrical conductivity than an insulator but lower than a metal

sensible heat storage: energy storage via the heat capacity of a material, excluding any phase change

SHG: see second harmonic generation

shape-memory material: material that can return from its temporary shape to its permanent shape by application of a stimulus such as heat

siemens (unit): reciprocal Ohms; $1 \text{ S} = 1 \text{ }\Omega^{-1}$

sinter: to heat (and possibly also compress) a porous mass to make it coalesce and increase in density, without melting it

slip plane: plane between domains in a crystal in which the domains differ by the presence of an edge dislocation; see Figure 14.14

smart material: material designed to change its properties in response to changes in its environment, such as stress, electric field, magnetic field, pH or temperature

smectic liquid crystal: liquid crystalline phase in which the molecules are arranged in layers; see Figure 4.22

soda lime glass: common glass

soft magnet: magnet with a narrow *B-H* hysteresis loop (see Figure 13.9) because only weak magnetic fields are required to achieve saturation induction due to easy motion of the domain walls

sol: heterogeneous material composed of a solid dispersed in a liquid

solar cell: device in which sunlight is converted to electrical power (photovoltaic cell) or thermal energy (thermal cell)

sol-gel processing: method used to prepare materials; process involves preparation of a sol (colloidal dispersion) of a precursor that forms a network gel that can then be dried to produce the product

solid dispersion: heterogeneous material composed of one solid dispersed in another solid

solid solution: homogeneous solid mixture of components

specific heat: heat capacity per unit mass

specular reflection: reflection for which the angle that the reflected light makes with the surface is the same as the angle at which the incident light hit the surface; see Figure 3.5

speed of sound (*v*): speed at which sound propagates in a medium

spintronics: shortened form of "spin transport electronics"; use of the fundamental property of spin for information processing

sputter coating: physical vapor deposition method to produce a thin film of material on a substrate, by ejection of materials (usually metals) from a target by bombarding with high-energy gas molecules or ions; target materials then deposit on substrate

stainless steel: steel with a significant amount of chromium which imparts corrosion resistance

steel: alloy that is primarily iron with carbon and other elements

stiffness: resistance to strain; see Young's modulus

STM: see scanning tunneling microscope

strain (ε): change in length of a material under strain, relative to its original length; see Equation 14.2

strength, tensile: maximum stress that can be experienced by a material during a test in which it is being extended; units are Pa

strength, compressive: maximum stress that can be experienced by a material during a test in which it is being compressed; units are Pa

stress (σ): force on a material per unit area; units are Pa; see Equation 14.1

stress birefringence: birefringence induced by stress in a material that is isotropic when no stress is present; birefringence arises from stress-induced asymmetry; see Figure 14.6

substitutional impurity: presence of an impurity atom at a regular lattice site in a crystal; see Figure 14.12

substrate: surface on which deposition of other species takes place, by physisorption or chemisorption

sum frequency generation: production of light for which the frequency is the sum of the frequencies of two beams of incident light; a nonlinear optical effect

superconductor: phase of a material in which electrical resistivity is zero

supercritical fluid: fluid beyond the critical point (neither gas nor liquid)

superfluid: liquid with zero viscosity

supramolecular material: material in which the properties derive from the assembly of the molecules, not the individual molecules

surface active agent: see surfactant

surface energy: energy of a surface of a material, relative to energy of bulk; units are J m^{-2}

surface tension (γ): energy required to increase the surface area; units are J m^{-2}; see Equation 10.1

surfactant: surface active agent, influences surface tension of a liquid

suspension: see dispersion

SWCNT: single-walled carbon nanotube; see Figure 9.17

tempering: controlled heating, usually used to change mechanical properties of a material

tensile strength: maximum stress that can be experienced by a material during a test in which it is being extended; units are Pa

ternary: containing three components

tetragonal: a crystal class in which two of the unit cell dimensions are identical and all the angles are 90°; see Table 1.3 and Figure 1.4

T_g: see glass transition temperature

TGA: see thermogravimetric analysis

thermal conductance (K): an extensive property related to thermal conductivity; see Equation 8.21

thermal conductivity (κ): parameter to quantify ability of a medium to conduct heat; units are W m^{-1} K^{-1}

thermal diffusivity (a): property that quantifies speed of heat flow through a material; see Equation 8.20

thermal expansion coefficient: see coefficient of thermal expansion

thermal gravimetric analysis (TGA): analytical technique to determine mass of a sample as a function of temperature

thermal resistance: reciprocal of thermal conductance

thermal shock fracture resistance (R_s): parameter to quantify fracture resistance of a material when undergoing large temperature changes; see Equation 7.22

thermochromism: ability of a material to change color with temperature

thermoelectric: relating thermal and electrical properties

thermoelectric figure of merit (ZT): dimensionless parameter to quantify efficiency of thermoelectric material, relating its Seebeck coefficient, electrical conductivity, and thermal conductivity; see Equation 12.18

thermoelectric material: material that can be used to convert a temperature gradient to electrical power or vice versa

thermoelectric power (S): parameter relating temperature gradient of a material to its voltage gradient, also known as Seebeck coefficient; see Equation 12.14

thermoluminescence: emission of light initiated by heating a material

thermomiotic: exhibiting negative thermal expansion, i.e., shrinking on heating

thermoplastic: plastic (polymer) that can be repeatedly softened when heated and hardened when cooled

thermoset: plastic (polymer) that becomes irreversibly hard on heating due to thermally induced crosslinking, and does not soften on cooling

thermotropic liquid crystal: liquid crystal that has liquid crystalline properties only in a certain temperature range

topological insulator: material in which the interior is an insulator and the exterior is a conductor

toughness: measure of the energy required to break a material; related to the area under the stress–strain curve up to failure

transistor: device with n,p,n or p,n,p semiconductor configuration, capable of amplification or switching

transmission of light: the passage of light through a material without change in direction or intensity

transport properties: properties describing motion of mass (effusion if out of a small orifice, diffusion if general mass transport), momentum (viscosity), charge (electrical conductivity), or heat (thermal conductivity)

triboelectric effect: build up of static electricity arising from friction between dissimilar materials

triboluminescence: emission of light as a result of mechanical stimulation (e.g., crushing)

triclinic: a crystal class in which each of the unit cell dimensions is unique and none of the angles is 90°; see Table 1.3 and Figure 1.4

trigonal: see rhombohedral

Trouton's rule: experimental finding that most materials have a similar entropy change on vaporization, about 90 J K^{-1} mol^{-1}

twisted nematic liquid crystal: see cholesteric liquid crystal

Tyndall scattering: scattering of light from small particles that are more than 50 times the wavelength of light, resulting in color

U-process: see Umklapp process

Umklapp process (U-process): process by which two phonons interact to give a resultant phonon with a component of motion in a direction opposite to the general motion of the incident phonons, giving thermal resistance; see Figure 8.6

unary: containing one component, i.e., pure

unit cell: smallest building block for a crystalline structure that will fill space by repetitive translation, giving a macroscopic structure

upper consolute temperature: temperature above which an immiscible mixture becomes miscible; see Figure 9.21

valence band: range of energy bands that are filled in a nonmetal at $T = 0$ K

valence electrons: low-energy (core) electrons in the valence band and thereby having insufficient energy to contribute to thermal or electrical conduction

viscosity: property describing magnitude of internal friction

visible light: light that we can see with our eyes, in the approximate wavelength range 400 to 700 nm

wavenumber ($\tilde{\nu}$): reciprocal of wavelength; see Equation 6.23

Wiedemann–Franz law: relationship between electrical and thermal conductivity of a metal; see Equation 12.7

work function (Φ): minimum energy required to remove an electron into vacuum far from the atom or material; see Figure 12.3

work hardening: process whereby a ductile material undergoes cycles of increasing and decreasing stress until it becomes brittle due to dislocation entanglement and pinning

work of fracture (W): energy (work) required to form a fresh surface of a material by fracture

x-ray diffraction: elastic scattering of an x-ray beam from a crystalline material, giving regions in space in which interference from different planes of atoms is constructive and other areas in which interference is destructive; can be interpreted via Bragg's law (Equation 7.21) to discern structure of crystals

yield strength: value of stress when the strain is 1.002 times the extrapolated elastic stress; see Figure 14.1

Young's modulus (E): proportionality between stress on a material and its strain, at low stress; a measure of stiffness; also known as elastic modulus; units are Pa; see Equation 14.4

Zener diode: diode with little current flow at low reverse bias as usual for a diode, but more flow at large reverse bias due to tunneling; see Figure 12.21

zone refinement: procedure by which solid can be purified by regional melting and resolidification; see Figure 9.51

ZT: see thermoelectric figure of merit

Index

Page numbers followed by f, t, n and g indicate figures, tables, footnotes and glossary, respectively.

A

Absorbed species, 276, 455g
Acceptor impurity, 55, 63, 455g
Acceptor level, 57f, 335, 337f, 455g
Acetone, 238
Acoustic phonons, 164, *see also* Phonons
Active-matrix, 108–109
Additive manufacturing, 437, 455g
Adhesion, 420–422, 421f, 455g
Adhesive strength, 420, 455g
Adsorbate, 275, 278, 289f, 455g
Adsorbed species, 275, 455g
Adsorption, 275, 276f, 278
Aerogel, 210, 312, 455g
Aerosol, 300t
AFM, *see* Atomic force microscope
Alexandrite, 29
Allochromatic mineral, 29
Allochromic, 29, 455g
Allotrope, 220, 268, 325, 455g
Alloy, 3, 185, 263, 380, 402, 411, 414, 415, 425, 427–429, 430, 455g
Alnico, 380, 381t
Aluminum, heat treatment of, 415
Amethyst color, 31
Ammolite, 96
Amorphous, 9, 145, 206, 230, 241, 455g
Amorphous metals, 147, 402
Ampère, André Marie, 322, 322n
Amphiphilic molecules, 285, 285n, 301, 455g
Anharmonic potential, 115, 178, 455g
Anisotropic materials, 107, 109, 405n, 455g
Annealing, 411, 414f, 455g
Antiferroelectric materials, 347, 348, 348f, 455g
Antiferromagnetic materials, 347, 348t
Antifogging, 284–285

Aquamarine, 29
Aragonite, 226, 439, 440, 456g
Arrest point, 234, 243, 456g
Arrhenius' theory, 236
Artificial bones, 406
Atomic force microscope (AFM), 278, 424, 455g, 456g
Atomic transition, 22–24, 456g
Audio speakers, 143
Aurora borealis, 23
Austenite phase, 425–427
Auxetic materials, 435, 436f, 456g
Auxochrome, 32, 456g
Avalanche effect, 340, 456g
Avogadro, Amedeo, 71n
Avogadro's number, 199, 303
Azeotrope, 237, 237f–239f, 456g

B

Backlit liquid crystal display, 108
Band
  gap, 49f, 53, 54t, 55, 57f, 60, 118, 325, 327, 331, 332, 335, 336f, 341, 356f, 456g
  structure, 49–50, 342, 358f
  theory, 321–328, 456g (*see also* Electrical properties)
    insulators, 328
    intrinsic semiconductors, 330–335
    metals, 329–330
    semiconductors, 330–342
    superconductors, 347–352
Bardeen, John, 351n
Bartholin, Erasmus, 109n
Baseball, elasticity, 407
BCS theory, 351–352, 456g
Bednorz, J. Georg, 350n
Beer, August, 27n
Beer–Bouger–Lambert law, 27

Beryl, 28
Big Bang, 26
Bimetallic strip in thermostat, 187f
Binary system, 235, 236, 240, 456g,
    see also Thermal stability
Binnig, Gerd, 278n
Bioluminescence, 34, 456g
Biomaterials, 310, 456g
Biomimetic gels, 305
Biomimetic material, 4, 456g
Birefringence, 109–110, 456g
Black body
    definition of, 24
    radiation, 24–26, 456g
Black phosphorus, 200t
Blue and green azurite, 29
Blue sapphire, 33
Bohr magneton, 385
Boiling point diagram, 239f, 240f
Bolometer, 359
Boltzmann, Ludwig, 48f, 48n
    constant, 39, 329
    distribution, 146, 331
Boron nitride nanotube (BNNT), 335, 456g
Bose–Einstein condensation, 226, 226n,
    456g
Bouger, Pierre, 27n
Boyle, Willard S., 63n
Bragg diffraction condition for x-rays, 183f
Bragg, Sir William Henry, 182n
Bragg's law, 182, 187, 457g
Bragg, William Lawrence, 182n
Brass, 9t, 210, 355, 428, 430, 457g
Bravais lattices, 11, 11f, 12, 457g
Break point, 243, 410, 457g
Brinell hardness number (BHN),
    399, 457g
Brinell, Johan August, 399n
Brittle materials, 408, 416, 417, 429, 457g
Bronze age, 3, 427, 457g
Bubbles, 31, 277, 292f, 403
Bulk modulus, 403
Bulletproof vests, 410–411

C

C_{60}, 143, 224, 227, 228f, 229
Calamitic liquid crystal, 89, 89f, 457g,
    see also Liquid crystals

Calcite, 110f, 226, 457g
Calorie, 144, 144n
Calorimetry
    adiabatic, 162
    differential scanning, 161, 461g
Capacitance dilatometer, 184
Capillarity and surface tension, 279–285
Capillary depression, 283
Capillary effect, 281, 457g, *see also*
    Surface tension and capillarity
Carbon dioxide, 219
Carbon fibers, 418
Carbonless copy paper, 34
Carbon monoxide sensor, 360
Carbon nanotube (CNT), 229, 229f,
    332–333, 418, 457g, 458g
    electrical properties of, 332–334
Carlson, Chester, 59
Casting, 206, 247, 429, 457g
Cast iron, 247, 259, 438, 457g
Catalysis, 287
Cathodoluminescence, 56, 457g
Cationic surfactants, 310
CCD, *see* Charge-coupled device
CD, *see* Circular dichroism
Cement, 207, 457g
Cementite, 242f, 247
Ceramics, 207, 209, 301, 457g
    glass, 419–420
    processing of, 301
Cermets, 419, 420, 457g
Chameleon, 86
Charge-coupled device (CCD), 63, 63n,
    457g
Charge delocalization, 32–35
Chemical potential, 230–232, 457g
Chemical vapor deposition (CVD), 288,
    289, 458g, 460g
Chemiluminescence, 33, 34, 458g
Chemisorption, 275, 289, 458g
Ching-Wu (Paul) Chu, 350
Chirality, 105
Chiral molecule, 91, 105, 120, 333f, 359,
    458g
Chiral vector, 332–333, 334f
Chocolate-caramel binary phase
    diagram, 256f
Cholesteric liquid crystal, 91, 91f, 105,
    106f, 107, 107f

Cholesteryl benzoate, 88
Chromatic aberration, 76, 100, 458g
Chromium dioxide, 388
Chromophore, 32–33, 86, 118, 264, 458g
Cinnabar, 53
Circular dichroism (CD), 110–112, 457g, 458g
Circularly polarized light, 110–112, 111f, 458g
Citrine quartz, 29
Clapeyron, Benoit, 218n
Clapeyron equation, 216–219, 458g
Clathrate, 306, 306f, 307, 458g
Clathrate hydrates, 307, 311
Clausius–Clapeyron equation, 260, 458g
Clausius, Rudolph Julius Emmanuel, 260, 260n
Cleave, 424, 458g
Colored liquid crystal displays, 92
Closed-shell electron configuration, 20
CMC, *see* Critical micellar concentration
CMOS, *see* complementary metal oxide semiconductor
CNTs, *see* carbon nanotubes
Coefficient of thermal conductivity, 196, 458g
Coefficient of thermal expansion (CTE), 157, 173, 458g, 460g
Coercive field (H_c), 378, 379f, 380, 381t, 458g
Coherent films, 287
Cohesion, 458g
Cohesive strength, 420, 458g
Cold welding, 188, 458g
Collector, 343
Colligative property, 304, 458g–459g
Colloid, 299–301, 459g
Colloidal dispersion compositions, 300t
Colloidal sponges, 300
Color(s), 35
    of amethyst, 31
    of doped semiconductors, 54–59
    in materials, 19
    of pure semiconductors, 52–54
    vibrational transitions and, 26–27
"Color bearing," 32
Color blindness, 34
    of Dalton, 34

Color centers, 29, 459g
    definition of, 31
    F-centers, 29–32
Color-changing
    due to polishing, 50
    of Hercules beetle, 86
    species, 86–87
Colored liquid crystal displays, 92
Color-matching of chameleon, 86–87
Colors and band gaps, in pure semiconductors, 54t
Colossal magnetoresistance, 387, 459g
Columnar liquid crystal, 90, 459g
Commercial thawing tray, 205
Complementary colors, 21, 22, 33, 459g
Complementary metal oxide semiconductor (CMOS), 63
Composites, 309–310, 459g
Compound formation, 248–252
Compressibility factor, 173
    isothermal, 403
    variation with temperature and pressure, 174f
Compressive strength, 406, 459g
Concrete, 432–433, 459g
Conducting polymers, 341
Conduction band, 52–55, 57, 57f, 59, 61, 325, 326f, 328, 330, 331, 335, 336f, 339f, 340f, 459g
    definition of, 52, 52f
Conduction electrons, 47, 324, 329, 331, 359, 374, 459g
Conductivity (σ), 322, 322t, 328, 459g
    anisotropy, 385
    electrical, 327
    graphite, 327
    insulators, 328
    metals, 324–325
    polymers, 341
    selected materials, 345t
    semiconductors, 325–327
    superconductors, 374
    temperature dependence, 328–335
Congruently melting compound, 248–249
Congruent melting, 250, 459g
Conjugation, 32, 40, 97, 341, 438, 459g
Consolute temperature, 235
Constant boiling mixture, *see* Azeotrope

Constructive interference, 80, 82, 82f, 87,
        88, 91, 96, 97, 182, 183f, 459g
    definition of, 79
    of two identical waves, 79f
Contact mode, 278
Contact potential, 324, 325f, 339, 459g
Cooling curve, 243, 244f, 246f
Cooper, Leon, 351n
Cooper pairs, 351, 358, 374, 459g
Copolymerization, 283
Copper
    nickel phase diagram, 241
    resistivity a function of temperature,
        325
    thermal conductivity, 190
    tin phase diagram, 428f
Core electrons, 20
Corningware, 419
Corundum, 261
Cotton effect, 112, 459g
Crack propagation, 415–420, 459g
Creep, 225, 409, 459g
Cristobalite, 261
Critical angle, 76, 77f, 94, 460g
Critical magnetic field, 374, 460g
Critical micellar concentration (CMC),
        301, 458g, 460g
Critical opalescence, 266, 460g
Critical point, 174, 215, 216f, 219, 225, 257,
        266, 460g
Critical pressure, 174, 174f, 460g
Critical temperature, 174, 175t, 215, 216,
        350, 374, 382, 460g
    definition of, 345
Critical volume, definition of, 174
Crosslinking, 402, 406, 419, 439, 460g
Crutzen, Paul, 138n
Cryogen, 225
Crystal
    classes, 10, 11, 11t
    defects in, 29
    field colors, 27–29, 460g
    field strength, effect of, 28f
    growth, 258
Crystallite, 65, 192, 248, 406, 419, 460g
CTE, *see* Coefficient of thermal
        expansion
Cubic lattice, 223f, 228f, 460g
Cupping process, 429

Curie, Jacques-Paul, 382n, 422
Curie law, 382, 460g
Curie, Pierre, 382n
Curie temperature (T_C), 382, 383, 383t,
        460g
Curie–Weiss law, 382, 460g
CVD, *see* Chemical vapor deposition
Cymbals, 427–430

D

Dalton, John, 34
Dark current, 359
Davy, Sir Humphrey, 300n
Debye characteristic temperature (θ_D),
        141, 148, 164, 204, 460g
Debye equation for thermal
        conductivity, 199
Debye, Peter Joseph Wilhelm, 140n
    equation for thermal conductivity, 199
    heat capacity theory, 144
    model, 140–143
Debye temperature, 140, 143
    speed of sound, 140, 143
Debye-T^3 law, 143, 460g
Defects in crystals, 29, 29f, 411–414
De Fermat, Pierre, 72n
Degrees of freedom, 127, 128, 231–234,
        243–247, 460g
Demagnetization, 380, 381f
Density of states, definition of, 49f
Dental fillings, 210
Depression, capillary, 282, 283
Destructive interference, 79, 461g
Devitrification, 411, 461g
Diamagnetic behavior, 374–375, 461g
Diamagnetic material, 378
Diamond
    band gap, 53, 54t
    Debye temperature, 142t
    dielectric constant, 345t
    dispersion of light, 76, 77f
    doped, 55
    fracture, 434
    phase stability, 226, 227f
    refractive index of, 76
    resistivity, 323
    speed of sound, 143
    structure, 227f

tensile strength, 400t
thermal conductivity, 200t
thermal expansion, 183t
thermoluminescence, 40
Young's modulus, 400t
Diamond structure, periodic trends in, 54t
Dichroism, 110–112, 461g
Dielectric constant, 113, 344–347, 461g
Differential scanning calorimetry (DSC), 160, 161, 461g, 462g
Differential thermal analysis (DTA), 156, 160, 461g, 462g
Diffraction grating, 87–88, 461g
  colors, 88
  liquid crystal, 90
Diffusely scattered light, 50
Diffusion, 196, 332, 461g
  reflection, 50, 51f, 461g
Dilatometer, 461g
4,4'-dimethoxyazoxybenzene (DMAB), 266
Diode, 59, 330, 338, 338f, 339, 461g
Dirać, Paul A. M., 48n
Discharge lights, high-intensity, 37
Discotic liquid crystal, 89, 461g
Dislocation, 409, 411–416, 413f, 418, 419, 461g
Dispersed phase, 299
Dispersing medium, 299
Dispersion of light, 76, 461g
Displaced field, 113
Distillation, 239f
Domains, 376, 461g
Donor impurity, 55, 461g
Donor level, 335, 336f, 462g
Dopants, 29, 462g
Doped semiconductor, 55, 335, 336, 340, 462g, *see also* Semiconductors
  colors of, 54–59
Double refraction, 109, 110f, 462g
  effect of, 110f
Dragline silk, 406
Drop weight method, 282t
"Dry copying," 59
DSC, *see* Differential scanning calorimetry
DTA, *see* Differential thermal analysis
Ductility, 398, 462g

Dulong–Petit law, 138–139, 462g
Dulong, Pierre Louis, 138n
Duralumin, 415
Dye *versus* pigments, 32, 462g

E

Eclipse, 97
Edge dislocations, 411, 462g
Effusion, 196, 462g
  definition of, 192
Ehrenfest classification of phase transitions, 149, 153
Ehrenfest, Paul, 153n
Einstein, Albert, 139n
  characteristic temperature (θ_E), 140, 462g
  heat capacity model, 139–140, 462g
Elasticity, 87n, 402–407
  limit, 407–411
Elastic limit, 402, 462g
Elastic modulus, 314, 462g
Electrical anisotropy, 328
Electrical conductivity, 196, 322t, 462g
  temperature dependence of, 328–335
Electrical resistivity, 322t, 324, 330f, 462g
Electrochemistry, 236n
Electrochromic device, 92, 462g
Electroluminescence, 57, 462g
Electromagnetic spectrum, 20f, 462g
Electromechanical properties, 422–430
Electronic band structure, 57f, 333
Electronic coefficient of heat capacity, 329
Electronic energy, 28, 339
Electronic excitation, 29–30
Electronic transitions, 22–24, 462g
Electron paramagnetic resonance (EPR), 385
Electron spin resonance (ESR), 385
Electrooptic effect, 117, 462g
Electrophoresis, 300
Embodied energies of common materials, 7, 9t
Embodied energy, 7, 9t, 462g
Emeralds, 28
Emissivity, 288, 462g
Emittance, 26f
Emitter, 24, 343
Emulsion, 61, 463g

Enantiomers, 120, 463g
Energy bands
    in an insulator, 55f
    in a metal, 55f
    in a semiconductor, 55f, 326f
Energy efficiency, 379
Energy, equipartition of, 127–130
Energy flux, 196–198
Enthalpy
    changes on vaporization, 132t
    definition of, 130
    of fusion, 169, 463g
    of vaporization, 131, 463g
Entropy
    changes on vaporization, 132t
    of fusion, 223, 463g
    of vaporization, 463g
Equation
    Clapeyron, 216–219
    Clausius–Clapeyron, 260, 458g
    Debye heat capacity, 141, 142
    Debye thermal conductivity, 195, 199
    Einstein heat capacity, 140f
    fundamental, 151
    of state for an ideal gas, 133, 173
    van der Waals, 176
    virial, 176, 177
Equipartition of energy, 127–130
    heat capacity of a monatomic gas, 129
    heat capacity of a nonlinear triatomic
        gas, 129–130
Equipartition theory, 129, 130, 138, 143,
        196, 463g
Ethanol–methanol–water phase
        diagram, 253, 253f
Eutectic composition, 245, 463g
Eutectic halt, 250, 463g
Eutectic solders, melting points of, 245t
Eutectic temperature, 245, 463g
Eutectoid, 247
Exchange interactions, 375, 376f, 383, 463g
Exfoliated graphite, 275, 276
Extensive property, 157, 463g
Extra charge centers, 31
Extrinsic electrical conductivity, 338
Extrinsic semiconductor, 335, 336f, 337f,
        463g
Extrinsic (doped) semiconductors,
        see also Semiconductors

electrical devices using, 336–344
properties of, 335, 336
transistors, 342–344
Eye shadow, 96

F

Failure, 36, 398, 399f, 409, 417, 418, 420
Faraday, Michael, 300, 300n
F-centers, 29–32, 463g
    definition of, 31
Feather color, 99
Fermat's principle of least time, 72
Fermi–Dirac distribution function, 48,
        63, 463g
    for electrons in a metal, 48f
    probability distribution for electrons,
        48f
Fermi energy (E_F), 48, 49, 52f, 55f, 143,
        324f, 331, 375, 463g
Fermi, Enrico, 48n
Ferrites, 242f, 247, 384
Ferroelectricity, 346, 347, 347, 348t, 463g
    crystal, 346, 347
    materials, 346f, 347f, 348t, 390
Ferroelectric liquid crystals (FLCs), 359
Ferromagnetism, 383, 384, 463g
    materials, 376f, 377f, 378, 380f, 391f
    shape-memory alloys, 435
Ferroxdur, 381t
Fiberglass®, 147, 464g
Fiber optics, 93–94, 463g
Fibers, added strength in, 418
Field-effect liquid crystal display (LCD),
        106–109, 107f, 464g
Field-effect transistor (MOSFET), 343,
        343f, 464g
Films, 285–289
    colors in, 41, 62, 81
Fireworks, 23
First law of thermodynamics, 134, 136
First-order phase transition, 154, 218,
        261, 464g
Flash bulb, 56
Flocculation, 300, 464g
Fluorescence, 28, 33, 55, 464g
    lights, 36, 37, 41, 108, 227–228, 228f
    schematic representation of, 33f
Fluorite, 31

Flux density, magnetic (*B*), 372, 373, 374f, 464g
Flux lines, 371
Foam, 300t, 305, 464g
Fogging, 284
Fog lights, 99
Forward-bias voltages, 354f
Fourier analysis, 117f
Fourier, Jean Baptiste Joseph, 197n
Fourier's first law of heat flux, 197
Four-wave mixing, 116
Fracture mechanics, 424
Fraunhofer, Joseph, 23n
Free electron gas model, 47, 464g
Freezing point depression, 242, 243f, 464g
Frenkel defect, 435, 437f, 464g
Freons, 138
Frequency doubling, 116, 464g
Fresnel, Augustin Jean, 77n
Fresnel lens, 77, 78f, 464g
Fringe patterns, 87f
Fry, Art, 421
Fullerene, 227, 228, 464g
Fuller, R. Buckminster, 228
Fundamental equations, 151

G

Gabor, Dennis, 84n
Garnet, 29
Gas
    constant, 129
    discharge tubes, 36
    hydrates, 311
    thermal conductivity of, 195–200
    thermal expansion of, 173–177
    thermodynamic stability of, 215
Gate, 343
Geodes, 258f
Germanium, 325, 326f, 330f
Giant magnetoresistance (GMR), 386, 387f, 464g
Gibbs energy, 153f, 215, 230, 231, 251, 265, 277, 277f, 279, 280
Gibbsite layers, 308f
Gibbs, Josiah Willard, 131n
Gibbs phase rule, 230, 464g, *see also* Phase rule

Glass, 9t, 72, 75, 147, 464g
    ceramics, 192, 419, 464g
    definition of, 145
    heat capacity of, 145–149
    low-emissivity, 288, 468g
    transition temperature, 148f, 162, 189, 403, 431, 464g
Glass transition temperature (T_g), 148f, 149, 403, 409, 475g
Glue, 420
GMR, *see* Giant magnetoresistance
Gold film, 51
    light transmission by, 51
    reflectance spectra of, 51f
Gold-lead phase diagram, 264
Gold sols, 299
Gorilla® glass, 410, 464g
Grain boundary, 414, 414f, 464g
Grain size, 411, 414, 439, 464g
Graphene, 229, 276, 464g
Graphite, 186
    band structure of, 358f
    energy bands, 358f
    exfoliated, 275–276
    intercalates, 275, 276f, 306f, 308, 312, 323f
    phase stability, 227f
    structure, 227f
    thermal conductivity, 200t
    thermal expansion, 183t, 186
Graphite intercalates, 275
Griffith, Alan A., 417n
Griffith length (l_G), 417, 465g

H

Hair care products, 310–311
Hall effect, 391, 465g
Halogen gas, 36
Halogen lamp, 36–37
Halt point, 234, 243, 465g
Hard magnets, 379, 380, 380f, 465g
    properties of, 381t
Hardness, 209, 211, 399, 399n, 414, 414f, 420, 422, 444, 465g
Harmonic approximation, 113f, 177, 178, 465g
Harmonic potential, 113f, 115f, 177, 178f, 179f, 184

Heat capacity, 142f, 143, 147, 148f, 465g
  acoustic phonons, 164
  at boiling point, 132, 133f
  constant pressure, 133
  constant volume, 133
  diatomic gas, 130
  Dulong–Petit law, 138
  Einstein model, 139–140
  electronic, 143–144
  of glasses, 145–149
  lattice, 139, 140
  liquids, 144–145
  metals, 143–144
  monatomic gas, 129
  nonlinear gas, 129–130
  nonlinear triatomic gas, 129–130
  optic phonons, 164
  phase transition, 133f, 154f, 155f
  of solids, 138–144
  solids, 138–149
  water, 144–145
Heat content of real gases, 130–138,
      145–149
Heater, 264f
Heat flux, 360, *see also* Fourier's first law
      of heat flux
Heat storage materials, 158–160
Heat treatment of aluminum, 415
Helium, 225–226, 225f, 347
  neon laser, 35
Helmholtz energy, 150
Hercules beetle, 86–87
Herfindahl–Hirschman index (HHI),
      13, 465g
Hertz, Heinrich Rudolf, 139n
Heterogeneous, 247, 299, 309, 465g
Heterogeneous dispersion, 242
Hexagonal, 109, 110f, 220, 222f, 225f, 465g
High-energy excitation, 26
Highest occupied molecular orbital
      (HOMO), 32, 341, 465g
  LUMO transition, 32
High-intensity discharge lights, 37
Holes, 30–31, 465g
  definition of, 30
Holograms, 84
Homogeneous, 231, 241, 299, 465g
Hooke, Robert, 177n
Hooke's law, 177–178, 202, 399f

Hougen–Watson plot, 175
Hume-Rothery rules, 412, 465g
Humidity in baseball, 407
Huygens, Christiaan, 109n
Hydrogen bonds, 407
Hydrogen, metallic, 358
Hydrophilic, 81, 285, 301, 465g
Hydrophobic, 81, 285, 301, 302, 465g
Hydrothermal crystal growth, 258,
      465g
Hydroxyapatite, 406
Hyperpolarizability, 115, 116, 465g
Hypervalent impurity definition of,
      30f
Hysteresis, 346, 346f, 378, 379, 379f, 380,
      380f, 408f, 409

I

Ice, color, 26–27
Ice, hexagonal structure of, 220, 222f
Iceland spar, 109, 110f
Ideal binary solution, 238
Ideal gas, 130, 133–135, 137, 158, 173, 176,
      177, 215, 465g
Ideal solution, 236, 237f, 240, 240f, 465g
Ideochromatic mineral, 29
Idiochromatic, 29, 465g
Immiscible liquids, 255, 465g
Impurities, 411
  concentration, 264f
Impurity precipitate, 466
Incandescence, 24, 33f, 466
Incandescent light bulbs, 25, 36, 40, 209
Inclusion compounds, 306–311, 323f,
      466
  applications of, 311
Incongruently melting compound,
      249–250, 249f
Incongruent melting, 466
Index of refraction, 72, 466, *see also*
      Refraction index
  definition of, 72
Indium tin oxide, 7, 107, 108, 121
Induced dipole, 113, 115, 116f, 345
Induction, magnetic, 371–381, 378f,
      379f–381f, 379t, 381t
Inhomogeneous dispersion, 242, 466
Inkjet printer, 424

Insulating solids, thermal conductivities of, 200–204
Insulators, 321–328, 466
Integrated circuits, 343
Intensive property, 157, 466
Intercalate, 466
Interface, 466
Interfacial phenomena, *see also* Surface and interfacial phenomena
    liquid films on surfaces, 285–289, 285f–287f
    surface energetics, 277–278, 277f, 278t
    surface investigations, 278–279, 279f
    surface tension and capillarity, 279–285, 280f, 281t, 282f
Interference, 79–84, 466
    colors in soap films, 81–82
Interference colors, 466
Intermolecular potential, 178, 179f, 189f
Internal energy, 128, 134, 135, 143–145, 416
Interstitial impurity, 412f, 466
Interstitial octahedral sites, 247
Intrinsic electrical conductivity, 336, 337, 356f
Intrinsic resistivity, 321, 466
Intrinsic semiconductors, 330–335, 333f, 334f, 466
Invar, 185, 466
Invariant point, 233, 247
Iridescence, 84, 440, 466
Iron phase diagram, 247–248, 248f
Irradiation, 419
Isobaric condensation, 216
Isothermal bulk modulus (K), 403, 466
Isothermal compressibility (β_T), 157, 403, 466
Isotherms, 215, 216f
Isotropic liquid phase, 91, 466
Ivory, 313, 314

J

Jade, 29, 38
Joule, James Prescott, 134n
    experiment, 134–135
    heating, 344
    Thomson experiment, 135–138

Joule–Thomson coefficient (μ_{JT}), 163, 164, 174, 466
Joule–Thomson experiment, 135–138
Joule–Thomson inversion temperature, 137, 138

K

Kamerlingh Onnes, Heike, 176n, 347n
Kao, Charles, 93n
Kaolinite, 308f
Kelvin, Lord, 135n
Kevlar, molecular structure, 410, 411f
Kinetic theory of gases, 48n, 129, 133, 151n, 195, 199, 466g
Krafft temperature (T_K), 301, 302f, 467g
Kroto, Sir Harold W., 227
Kwolek, Stephanie, 410

L

Lambda transition, 155
Lambert, Johann Heinrich, 27, 27n
Landé splitting factor, 385
Langmuir-Blodgett films, 286–288, 286f, 287f, 467g
Langmuir, Irving, 287, 287n, 288, 299
Lasers, 23, 34–35, 35f, 115, 117, 467g
    printer, 60–61
Latex paints, 305
Lattice imperfections, 29, 31, 325, 411, 414
Lattice parameter, 53, 181, 202, 467g
Lattice vacancy, 412f, 467g
Lattice waves, 200, 467g, *see also* Phonons
    depiction in a two-dimensional solid, 201f
Law of atomic heats, 138
Lead crystal, 147, 227n, 467g
Lecture demonstration materials, sources of, 453
Lecture demonstrations, 453
LEDs, *see* Light-emitting diodes
Lee, David M., 226
Left-handed materials, 78, 467g
Lever principle, 239–240, 240f, 467g
Lever rule, 269, 467g
Lexan, 403

Life cycle analysis (LCA), 36, 467g
Ligand field, definition of, 28
Light-emitting diodes (LEDs), 37, 57, 58, 467g
　colors of, 56t
　definition of, 57
　new directions in, 58–59
　passive matrix, 108
　reflective, 108
Light scattering, 85–86, 285, 303f, 467g
Lindemann, Frederick Alexander, 202n
Lindemann's law of melting, 202
Linear coefficient of thermal expansion, 182, 183f, 467g
Linear defect, 411, 467g
Linear electrooptic (LEO) effects, 117
Linearly polarized light, 105, 119, 404, 467g
Linearly (or plane) polarized light, 105
Linear optical effects, 112, 114f
Linear polarizability, 113, 115, 467g
Linear thermal expansion (α), 189, 190, 191, 467g
Line defects, 29, 467g
Liquefaction, 215
Liquid
　binary phase diagrams, 234–236
　crystal, 88–94, 467g
　　active matrix, 108–109
　　backlit, 108
　　calamitic, 89, 89f, 90f
　　cholesteric, 91, 91f, 105–106, 106f, 107f
　　colors of, 85, 105
　　columnar, 90f
　　device, 107f
　　discotic, 89, 90f
　　display, 92, 468g
　　lyotropic, 89
　　nematic, 89–92, 90f
　　pitch, 91, 91f
　　smectic, 89, 90f, 91, 109
　　structures, 90f
　　thermotropic, 89
　films on surface, 285–290, 285f–287f
　heat capacity of, 144–145
　helium, 225
　solid binary phase diagrams, 241–248
　vapor binary phase diagrams, 236–239

Liquid-liquid binary phase diagrams, 234–236, 235f, 236f
Liquid Crystal Displays, 81, 92, 107, 468g
Liquid-solid binary phase diagrams, 241–248, 241f–244f, 245t, 246f, 248f
Liquid-vapor binary phase diagrams, 236–239, 237f–239f
Localized electrons, 20
London, Fritz, 226
Lorenz constant, 329, 330
Low-boiling gases, 138
Low-E glass, 294, 468g
Low emissivity coatings, 288
Low-emissivity glass, 287–288, 468g
Lower consolute temperature, 235, 235f, 236f, 468g
Lowest unoccupied molecular orbital (LUMO), 32, 341, 468g
Luminescence, 33, 33n, 468g
Lungs, surface tension in, 306
Lyotropic liquid crystal, 89, 468g

M

Magnetic behavior, origin of, 371–377, 372f, 374f, 376f, 377f
Magnetic devices, 387–389
Magnetic dipole interactions, 375, 468g
Magnetic field strength (H), 349f, 353, 372, 378, 378f, 379f, 468g, *see also* Induction, magnetic
Magnetic induction, as function of field strength, 378–381
Magnetic levitation, 349, 349f
Magnetic materials, 371n
Magnetic properties
　induction (field strength), 378–381, 378f, 379f, 380f
　magnetic devices, 387–389
　magnetization, 378–381, 378f–381f, 379t
　origin of magnetic behavior, 371–377, 372f, 374f, 376f, 377f
Magnetic resonance, 385–386
Magnetic storage devices, 387
Magnetic susceptibility (χ), 373, 382f, 468g
Magnetic tape, 290, 371n, 388

Magnetization, 373, 468g
   temperature dependence of, 382–389,
      382f, 383t, 384f, 385f
Magnetocaloric effect, 391, 468g
Magnetoresistance, 386–387, 387f, 468g
Malachite, 29
Martensite phase, 248, 425, 426f
Mass-averaged molecular mass, 303
Master holographic image, 84
Materials
   contraction while heating, 186
   with no thermal expansion, 185
   science
      daily living, impact on, 6
      future developments in, 6–7
      future materials and
         sustainability issues, 6–9
      recent trends, 4–5
      structures of, 9–11
   smart, 425
   structure of, 9–11
   sustainability, 6–9
   thermal conductivities of, 207–208
Matthiessen's rule, 325, 468g
Maxwell, James Clerk, 151, 151n, 152
Mean free path, 195, 197, 199, 200, 204,
      207, 329, 468g
   of phonons, 200, 202, 203f
Mean phonon speed, 202
Mechanical properties, 397–402, 399f,
      400t, 401f
   adhesion, 420–422, 421f
   beyond elastic limit, 403f, 405f,
      407–411, 407f
   crack propagation, 415–420, 416f
   cymbals, 427–430, 428f
   defects and dislocations, 411–415,
      412f, 413f, 414f, 415f
   elasticity, 402–407, 403f, 405f, 407f
   electromechanical properties,
      422–425, 423f
   exercises on, 431–440
   memory metals, 425–427, 426f, 427f
Meissner effect, 349, 349f, 350, 468g
Melting
   congruent, 249, 249f
   incongruent, 248, 249f
Melting point diagram, 241f, 246f, 249f
Melting points of eutectic solders, 245t

Memory metals, 425–427, 426f, 427f,
      468g
Mercury, 283f, 348f
Mesophase materials, 89, 468g
Metallic glass, 402, 468g
Metallic luster, 47–51
Metallurgy, 4, 414, 430, 468g
Metal oxide semiconductor field-effect
      transistor (MOSFET), 343,
      344f, 469g
Metals, 321–328, 324f, 325f, 330, 330f,
      468g
   heat capacities of, 143–144
   thermal conductivity of, 204–208
Metamaterials, 78, 469g
Metastability, 145, 469g
Methane, 221, 222, 223f, 224, 307f
Micelles, 301–304, 302f, 469g
Microcapsules, 34
Microelectromechanical device
      (MEMS), 437, 468g, 469g
Microgels, 300
Microlithography, 287, 469g
Microstructure, 411, 469g
Mie scattering, 51, 469g
Miracle Thaw, 205
Mirages, 74, 74f
Miscibility gap, 235
   partial, 245
Miscible liquids, 235, 469g
Modeling fracture, 424
Moiré pattern, 80, 80f, 469g
Moissanite, 211
Molecular
   engineering, 316, 469g
   magnets, 377
   materials, 4, 309, 469g
   orbitals, 32–37
   self-assembly, 308, 309f
   sieves, 307
   speeds, distributions of, 196f
Mole fraction
   of component, 232, 236, 237, 252
   definition of, 232
Molina, Mario, 138
Monatomic gas, 23
   heat capacity of, 129
Monatomic species, 128
Monochromatic light, 72, 469g

Monochromator, 88, 469g
Monoclinic lattice, 11f, 11t, 469g
Monolayer, 285, 286, 289, 469g
Müller, K. Alexander, 350
Mullite, 261
Multicomponent system, 235, 239, 411
Multiple internal scattering, 50
Multiwalled carbon nanotubes
        (MWCNT), 229, 229f, 469g
Mumetal, 379
Musical instruments
    cymbals, 427–430
    woodwinds, 185n

N

Nacre, 226, 439, 440f, 469g
Nanomaterials, 289–290, 469g, *see also*
        Carbon nanotubes
Natural abundance of elements, 8t
Néel, Louis Eugène, 383n
Néel temperature (T_N), 383, 469g
Negative refractive index, 78
Negative thermal expansion (NTE), 186,
        469g
Nematic liquid crystal, 89, 91, 469g
Neon emission, 23
Neon lights, 37
Neoprene ball, 407f
Neumann–Kopp law, 167, 469g
Neutron scattering, 143, 184, 470g
Newton, Isaac, Sir, 75n
Nickel oxide, 383
Nippon Telephone & Telegraph (NTT),
        291
Nitinol, 425, 426f, 427f, 470g
NLO effect, *see* Nonlinear optical effect
NMR spectroscopy, *see* Nuclear
        magnetic resonance
        spectroscopy
Nobel prizes, 47n, 48n, 63n, 84, 85n, 93n,
        138n, 139n, 140n, 176n, 182n,
        185, 226n, 227n, 230n, 276n,
        278n, 287n, 341, 347n, 350n,
        351n, 375n, 382n, 383n, 386
Non-centrosymmetric structure,
        422–423
Noncontact mode, 278
Noncrystalline structures, 9

Nonideal binary solution, 237
Nonideal gas, 133, 135, 137, 173–174, 177,
        470g
Nonlinear effect, 112, 112n, 470g
Nonlinear optical (NLO) effect, 112–117,
        116f, 470g
Nonlinear triatomic gas, heat capacity
        of, 129–130
Non-Newtonian behavior, 421, 470g
Nonstick coating, 283
Normal process (N-process), 203f, 469g,
        470g
Northern lights, 23
n-type semiconductor, 54–55, 57, 470g,
        see also Semiconductors
    extrinsic, 335
Nuclear magnetic resonance (NMR)
        spectroscopy, 385, 386f
Number-averaged molecular mass, 304

O

Ohm, Georg Simon, 321n
Ohm's law, 321, 321n
One-dimensional conductors, 328
Opacity, 118
Opal, 88–89
Optical activity, 105–109, 470g
    and related effects, 105–109
Optical fiber, 93–94, 470g
Optical rectification, 116
Optical rotatory dispersion (ORD),
        110–112, 470g
Orbital magnetism, 373, 375
Ordinary paramagnets, 375
Organic light-emitting diodes (OLEDs),
        58, 470g
Organic materials, color of, 32
Orientational disorder, 222–224, 223f,
        224f, 228, 470g
Orthorhombic lattice, 11f, 11t, 220, 470g
Osheroff, Douglas D., 226n
Osmometer, 304, 304f
Osmotic pressure, 304, 304f, 470g

P

Packing fraction, 12, 301f, 470g
Paraelectric crystal, 346, 470g

Paraelectric material, 346, 347
Paramagnetism, 375, 376, 377t, 470g
Partial miscibility, 245
Passive-matrix, 108–109
Pauli exclusion principle, 47, 375n
Pauli paramagnets, 377n, 382, 470g
Pauli, Wolfgang, 47n, 375n
Pearlite, 242f, 247
Peierls, Sir Rudolf Ernst, 203, 203n
Peltier, Jean Charles Athanase, 344, 344n, 470g
Penrose, Sir Roger, 230
Peritectic halt, 250, 251, 470g, 471g
Peritectic point, 250
Peritectic temperature, 250
Permalloy, 379t
Permanent magnets, 380, 390, 471g
Permeability of vacuum, 372
Petit, Alexis Thérèse, 138n
Phase, 79, 90f, 471g
    equilibria in pure materials, 216–219
    of matter, 299–311
    proportions of, 239–241
    rule, 230–234, 233f, 234f, 471g
    stability, 149–155
    stability and transitions, 149–155
    transitions, 149–155
Phase change material (PCM), 159, 168, 470g, 471g
Phase diagrams, 219–230
    aluminum-copper, 330f
    carbon dioxide, 219, 219f
    copper-nickel, 241, 241f
    copper–tin, 428f
    helium, 348f, 350
    liquid-liquid binary, 234–236, 235f, 236f
    liquid-solid binary, 241–248, 241f, 242f, 243t–246t, 248f
    liquid-vapor binary, 236–239, 237f–239f
    methane, 221–224, 307f
    phenol-water-acetone, 254
    of pure materials, 219–230, 219f–225f, 227f–230f
    sulfur, 220, 220f
    ternary (three component), 252–256, 253f–255f
    water, 220, 221f
Phonon-phonon collisions, 203, 204
Phonons, 164, 200–208, 201f, 203f, 204f, 255, 323, 358, 471g
Phosphor, 37, 55, 471g
Phosphorescence, 34, 55, 56f, 471g
Phosphors, 37, 55
    thermal conductivity, 200t
Photochromism, 65, 471g
Photocopying process, 59–61
Photonic crystals, 93
Photonic material, 93, 471g
Photorefractive effect, 117, 471g
Photovoltaic (PV), 472g
    material, 471g
    solar cells, 59
Physical constants, 10, 447
Physisorption, 275, 471g
Piezoelectric effect, 422–425, 471g, *see also* Electromechanical properties
Piezomagnetism, 433, 471g
Piezooptic effect, 433, 471g
Piezoresistivity, 433, 471g
Pigments, 32, 97, 471g
    and dye, difference between, 32
Planck, Max Karl Ludwig, 24n
Plane polarized light, 471g
Plastic crystal, 223, 471g
Plastic deformation, 398, 399f, 471g
Plasticizers, 167, 471g
Plastic strain, 398, 471g
Pleochroic material, 92, 472g
Pleochromic dyes, 92, 93f
Plexiglass, 405f, 410
p,n-junctions, 57, 342, 342f, 470g
    electric potential to, 28
Pockels, Agnes, 286, 286n
Pockels effect, 117, 472g
Point defect, 29, 411, 412f, 472g
    in crystal, 412f
Poisson ratio, 398, 400t, 435, 436f, 472g
Polarizability, 85, 108, 113, 115, 116, 303, 345, 346f, 347f, 472g
Polarized light, 105, 106f, 107f, 108–112, 404, 472g
    seeing stress with, 405
Polishing changes color, 50–51
Poly(N-isopropylacrylamide), 300

Poly(*p*-phenyleneterephthalamide), 410, 411f
Polymer(ization), 287, 420, 472g
  adhesive, 420
  bonding energy diagram, 49f
  bullet-proof, 397
  classifications, 418–419
  colloidal microgels, 300
  conducting, 341
  creep, 225, 409
  crosslinked, 276, 406, 410, 419, 420
  depolymerization, 7
  elasticity, 402
  fracture, 408
  Fresnel lens, 77, 78f
  gels, 300, 305
  of Langmuir–Blodgett films, 287
  in LEDs, 58
  optical fibers, 93, 94
  plasticizers and glass transition
      temperature, 167
  stretching, 403–404
  super-hard, 419
  tensile strength, 399f
  thermoplastics, 418, 419
  thermoset, 418, 419
  viscoelastic behavior, 421
Polymorphism, 211, 219–221, 226, 264, 472g
Polynorbornene, 406, 407f
Polytypism, 211, 472g
Popping corn, 206
Potash feldspar, 269
Pressing process, 430
Pressure-temperature phase diagram, 153f, 219–221, 219f, 221f, 227f
Prince Rupert drops, 410, 472g
Principle of equipartition, 128
Processing, 4f, 301, 472g
Proper interstitial, 412f, 472g
p-type semiconductor, 55, 55f, 57, 59, 334f, 336, 338, 339f, 472g
  extrinsic, 335
Pure materials
  phase diagrams of, 219–230
  phase equilibria of, 216–219
Pure semiconductors, colors of, 51–54, 54t
Push-rod dilatometer, 184
Pyrex®, 147, 186, 472g

Pyroelectric crystal, 345, 346, 347f, 355, 472g

Q

Quantity calculus, 10
Quantum confinement, 327, 333, 472g
Quantum dots, 327, 472g
Quantum mechanics, 7, 48n, 139, 153, 203n, 222, 373
Quartz halogen, 36–37, 261, 424
Quasicrystals, 8–9, 230, 230f, 472g
Quenching, 402, 429, 472g

R

Racemic mixture, 265
Rainbow, 25, 75, 84, 85, 98
Raoult, François Marie, 236n
Raoult's law, 236, 236n
Rare earth magnets, 390, 381t
Rayleigh, Lord, 85n, 286
Rayleigh scattering, 85, 86, 472g
Rectification, 116, 336, 337f, 338, 339
Reduced pressure, definition of, 174–175
Reduced temperature, definition of, 174–175
Reduced volume, definition of, 174–175
Reflection, 21–22
Refraction, 71–78, 472g
Refractive index, 72–78, 472g
  definition of, 72
  gradients, 185
  measurement, 73–74
  negative, 78
  for some common materials at 25°C, 73t
Refractometer, 75, 473g
Refrangibility, 75, 473g
Refrigerants, 138
Reinitzer, Friedrich, 88n
Relative permeability, 344, 373, 473g
Remanent induction (B_r), 378, 379, 473g
Resistivity, 321, 322, 330f
Retinal, 33
Rhinoceros horn, structure of, 6
Rhombohedral lattice, 11f, 11t, 473g
Richardson, Robert C., 226n
Rigid glass, 148
Rohrer, Heinrich, 278n

Rolling, 390f, 429
Rowland, F. Sherwood, 138n
Rubber, 166, 300, 398, 400t, 401f, 420–422
Ruby glass, 51, 120, 191, 294
Rule of mixtures, 309, 473g
Ruska, Ernst, 278n

S

Salting out, 255
SAM, *see* Self-assembled monolayer
Saturation induction (B_s), 378, 380, 473g
Scanning tunneling microscope (STM), 278, 279f, 473g, 474g
Scattering of light, 85–87
Schottky defect, 435, 473g
Schrieffer, John Robert, 351n
Screw dislocation, 411, 413f, 473g
Second-harmonic generation (SHG), 116, 119, 473g
Second law of thermodynamics, 150, 404
Second-order phase transition, 150, 154, 473g
Seebeck coefficient, 353, 360, 473g
Seebeck, Thomas Johann, 353n
Self-assembled monolayer (SAM), 289, 473g
Self-assembly, 308, 309f
Self interstitial, 411, 473g
Semiconductors, 52–53, 57, 59, 118, 321–328, 322t, 325, 327, 335, 354, 388, 473g
    band gaps of pure, 54t
    colors of doped, 54–59
    colors of pure, 52–53
    effect of pressure, 233
    electrical conductivity, 196
    energy bands, 52f
Sensible heat storage, 154, 158, 473g
Sensors, 63
SFG, *see* Sum frequency generation
Shape-memory alloys, 425–427
Shape-memory material, 425–427, 473g
Shear stress, 421f
SHG, *see* Second-harmonic generation
Shirakawa, Hideki, 341
Shockley, William, 351n
Siemens unit, 322n, 473g
Silicon, 265, 332

Silicon-based semiconductors, 332
Silicon dioxide, 258
Silly Putty, 421
Silver–lanthanum isobaric binary phase diagram, 264
Silver, Spence, 421
"Single-color producer," 88
Single-crystalline silicon materials, 59
Single-walled carbon nanotube (SWCNT), 229, 229f, 475g
Sinter, 207, 411, 473g
Sky, color, 85
Slip plane, 413f, 414f, 473g
Smart glass, 92
Smart material, 6, 425, 474g
Smectic liquid crystal, 88, 89, 474g
Snell's law, 77f
Snell van Royen, Willebrod, 76
Soap films, 81–84
    interference colors in, 81
Soda lime glass, 147, 474g
Sodium acetate trihydrate, 159, 160
Soft magnets, 373–374, 379–380, 474g
    properties of, 379t
Sol, 301, 474g
Solar cells, 59, 474g
Solder, 245, 246f
Sol-gel processing, 301, 474g
Solid
    heat capacity of, 138–144
    thermal expansion of, 177–187
Solid dispersion, 242f, 474g
Solid solution, 241, 245, 474g
Sound interference, 80
Sound speed, 191, 403
Space-filling model, 223f, *see also* Methane
Space vehicle tiles, 206–207
Sparkling diamonds, 76–77
Specific heat capacity, 158n, 169, 474g
Specular reflection, 50, 474g
    definition of, 49
Speed of sound (v), 143, 166, 403, 474g
Spider silk, 6, 406, 409
    strength of, 406
Spin magnetism, 373
Spintronics, 334, 387, 474g
Sputter coating, 288, 474g
Stainless steel, 4, 210, 474g

Stars
　temperature, 41
　twinkling, 97
State function, 131, 151, 152
Static dipole, 115
Steel, 210, 247, 474g
Stiffness, 398, 474g
STM, *see* Scanning tunneling
　　microscope
Strain (ε), 397–398, 474g
Strength, compressive, 431, 474g
Strength of crystalline fibers, 418
Strength, tensile, 433, 474g
Stress (σ), 397, 474g
　controlled, 410
　polarization, 405f
Stress birefringence, 405, 474g
Stress–strain relations, 399, 408, 408f
Substitutional impurity, 412f, 475g
Substrate, 287, 420, 475g
Sulfur, 219
Sum frequency generation (SFG), 117, 475g
Superalloy, 400t, 433
Superconductivity, 347–355
Superconductors, 348–350, 475g
　as thermal switches, 374
Supercooled liquid, 145
Supercritical fluids, 219, 257–258, 475g
Superexchange, 383, 384f
Superfluid, 225, 226, 475g
Super-hard polymers, 404, 419
Supramolecular materials, 306, 475g
Surface
　energetics, 277–278
　investigations, 278–279
　liquid films on, 285–290
　tension and capillarity, 279–285
Surface active agent, 285n, 305, 475g
Surface and interfacial phenomena
　liquid films on surfaces, 285–289
　nanomaterials, 289–290
　surface energetics, 277–278
　surface investigations, 278–279
　surface tension and capillarity,
　　279–285
Surface energy, 277–278, 408, 475g
Surface investigations, 278–279
Surface pressure, 286
Surface tension (γ), 279–285, 475g

Surface tension and capillarity, 279–285
Surfactants, 285n, 304–306, 475g
Suspension, 242, 300t, 475g
SWCNT, *see* Single-walled carbon
　　nanotube

T

Tate's law, 281
Taylor series expansion, 115
TCNQ, *see* Tetracyanoquinodimethane
Teflon, 284, 420
Television, phosphors, 55, 65
Temperature
　eutectic, 245
　peritectic, 250
Temperature dependence of electrical
　　conductivity, 328–335
Tempering, 429, 475g
Tempering process, 430
Tensile strength, 398, 399, 475g
Ternary phase diagrams, 252–258, 475g
Ternary system, 253
Tesla, Nikola, 373n
Tetracyanoquinodimethane (TCNQ),
　　327–328
Tetragonal, 11t, 11f, 109, 475g
Tetrahedral bonding, 326n
Tetrahedral interstitial sites, 247
Tetrahydrofuran (THF), 311
Tetrathiafulvalene (TTF), 327–328
T_g, *see* Glass transition temperature
TGA, *see* Thermogravimetric analysis
Theory of equipartition of energy, 127
Thermal analysis, 160–162
Thermal conductance, 211–212, 475g
Thermal conductivity, 195–208, 475g
　amorphous materials, 409
　anharmonicity, 203
　ceramics, 209
　Debye equation, 206
　electronic, 228
　of gases, 195–200
　　pressure independence of, 199
　inclusion compounds, 306–310
　of insulating solids, 200–204
　of materials, 207–208
　metals, 197, 204–208
　　temperature dependence of, 197f

Miracle Thaw, 205
popcorn, 206
selected materials, 322t
of a single crystal, 204f
solids, 177–186
superconductor, 374
Thermal diffusivity (*a*), 211, 475g
definition of, 211
Thermal energy, 219
storage materials, 158–160
Thermal expansion, 173–187, 207
anharmonicity, 203
coefficient, 157, 475g
Corningware, 419
gases, 173–177, 195–200
negative values, 216
selected materials, 322t
solids, 177–187
Thermal gravimetric analysis (TGA), 160, 475g
Thermal resistance, 203, 475g
Thermal shock fracture resistance (R_s), 147, 185, 475g
Thermal stability binary systems, 255–256
compound formation, 248–250
lever principle, 239–241
liquid–liquid binary phase diagrams, 234–236
liquid–solid binary phase diagrams, 241–248
liquid–vapor binary phase diagrams, 236–239
phase equilibria in, 216–219
phase rule, 230–234
pure gases, 215–216
pure materials phase diagrams of, 219–230
ternary (three-component) phase diagrams, 252–256
Thermal switches, 374
Thermistor, 354
Thermochromism, 90, 475g
Thermocouple, 353
Thermodynamics
manipulations, 155–162
of pizza, 149
stability of gas, 215
Thermoelectric, 344, 353–360, 475g

Thermoelectric figure of merit (*ZT*), 360, 476g
Thermoelectric material, 138, 361, 476g
Thermoelectric power (*S*), 353, 476g
Thermogravimetric analysis (TGA), 160, 475g
Thermoluminescence, 40, 476g
Thermometry, 352–355
Thermomiotic, 186, 192, 476g
Thermoplastics, 410, 418–419
Thermopower, 354, 360
Thermoset, 418, 419, 476g
Thermoset polymers, 419
Thermostat, 187
bimetallic strip inside, 187f
Thermotropic liquid crystal, 89, 90, 476g
THF, *see* Tetrahydrofuran
Thomson, William, 135n
Three-component phase diagrams, 252–258
3-D printing, 437, 455g
Three-wave mixing process, 116
Tie lines, 240, 245
Tin–lead phase diagram, 245, 246t, 246f
Tin–lead solder, 96, 246f
Titanic, 417
Titanium oxides, 419
Topaz, 42
Topological insulator, 342, 476g
Toys, color-changing, 93
Transducers, 424
Transformer, 184, 380, 384
Transistor, 341–344, 476g
Transition metals, compounds containing, 388
Transmission, 21–22, 51, 94
Transmission of light, 93n, 476g
Transparency, 118
Transport property, 196, 476g
Triangular coordinate graph, 248
Triboelectric effect, 360, 476g
Triboluminescence, 56, 476g
Triclinic lattice, 11t, 476g
Tridymite, 261, 269
Trigonal, 11, 476g
Triple point, definition of, 229
Trouton, Frederick Thomas, 132n
Trouton's rule, 132, 476g
TTF, *see* Tetrathiafulvalene

Tungsten, 36
Twisted nematic liquid crystal, 91, 357,
    476g
Tyndall, John, 85n
Tyndall light scattering, 85, 86
Tyndall scattering, 86, 476g

U

Umklapp process (U-process), 203,
    476g–477g
Unary, 241, 252, 477g
Unit cell, 10, 11, 164, 182, 477g
    definition of, 9
Unit pole, 371, 372
Units
    conversions, 449
    and unit presentation, 10
Upper consolute temperature, 235, 236f,
    477g
U-process, 203, 476g
Unattached electrons, 30
UV light, 27, 32, 36

V

Valence band, 52, 53, 55, 57, 61, 325, 330,
    477g
    definition of, 52
Valence electrons, 20, 27, 143, 144, 335,
    376f, 477g
van der Waals equation, 175–176, 215,
    216f
van der Waals, Johannes Diderik,
    176n
Vermilion, 53
Vibrational transitions
    and color, 26–27
    as source of color, 26–27
Virial coefficients, 177
Virial equation, 176
Virtual transitions, 113
Viscosity, 219, 257, 477g
    definition of, 196
Visible light, 32, 36, 53, 302, 477g
Volta, Count Alessandro, 322n

Von Helmholtz, Hermann Ludwig
    Ferdinand, 150n
Von Siemens, Ernst Werner, 322n

W

Water, 26–27, 31, 36, 73t, 88, 130
Wavenumber, 21f, 139, 163, 477g
Weiss constant, 384, 385f
Weiss, Pierre, 382n
Wiedemann–Franz law, 329–330, 477g
Wint-O-Green Lifesavers®, 56, 422
Work function (Φ), 324, 325f, 339, 477g
Work-hardened material, 414
Work hardening, 408f, 414, 430, 477g
Work of fracture (W), 416, 477g
"Write-black" process, in laser printer,
    60, 61

X

Xerography, 59–61
X-ray, 17, 31, 182, 184, 288
    diffraction studies, 9, 477g
    holograms, 84
    versus neutron scattering, 184

Y

Yield strength, 398, 399f, 408f, 430, 477g
Young's modulus (E), 398, 400, 400t, 401f,
    402–403, 409, 410, 417, 418, 425,
    477g
    relationship to interatomic potential,
        431
    relationship to speed of sound, 403
    selected materials, 345t, 383t
Young, Thomas, 87n

Z

Zener diode, 339, 340f, 477g
Zeolites, 306f, 307, 307f, 313
Zirconium tungstate, 186
Zone refinement, 265, 477g
ZT, 360, 361, 477g

Printed in the United States
by Baker & Taylor Publisher Services